LONDON MATHEMATICAL SOCIETY LECTURE NOTE SERIES

Managing Editor: Professor J.W.S. Cassels, Department of Pure Mathematics and Mathematical Statistics, University of Cambridge, 16 Mill Lane, Cambridge CB2 1SB, England

The titles below are available from booksellers, or, in case of difficulty, from Cambridge University Press.

London Mathematical Society Lecture Note Series. 255

Symmetries and Integrability of Difference Equations,

Edited by

Peter A. Clarkson
University of Kent at Canterbury

Frank W. Nijhoff
University of Leeds

 CAMBRIDGE
UNIVERSITY PRESS

CAMBRIDGE UNIVERSITY PRESS
Cambridge, New York, Melbourne, Madrid, Cape Town, Singapore, São Paulo

Cambridge University Press
The Edinburgh Building, Cambridge CB2 2RU, UK

Published in the United States of America by Cambridge University Press, New York

www.cambridge.org
Information on this title: www.cambridge.org/9780521596992

First published 1999

A catalogue record for this publication is available from the British Library

ISBN-13 978-0-521-59699-2 paperback
ISBN-10 0-521-59699-8 paperback

Transferred to digital printing 2006

Contents

6 Symmetries of Difference Equations

7 Numerical Methods and Miscellaneous

8 Cellular Automata

9 q-Special Functions and q-Difference Equations

10   Quantum Aspects and Yang-Baxter Equations

Preface

The second international conference on "Symmetries and Integrability of Difference Equations" (SIDE II) was held at the University of Kent at Canterbury, July 1-5, 1996. It was the successor of a first meeting on the same topic held in Estérel (Québec, Canada), under the auspices of the Centre de Recherches Mathématiques (CRM) of the University of Montréal in 1994, cf. the Proceedings of that meeting, [1]. Like in Estérel, this SIDEII meeting aimed at bringing together researchers working in the general field of discrete systems and difference equations with emphasis on symmetries and integrability.

The subject area of the meeting is relatively young: in the last decade, and particularly during the last few years, a great deal of progress has been made on the mathematical aspects of discrete integrable systems, including integrable dynamical mappings, ordinary and partial difference equations, lattice solitons, discrete versions of the Painlevé equations, symmetry approaches and singularity analysis, and applications to numerical analysis, computer science and Physics. The two SIDE meetings have brought together many leading experts in the various aspects of this field, coming from quite different backgrounds. As such the interdisciplinary nature of the meeting is reflected in the present volume, marking contributions in a large variety of fields.

It is important to give some explanation as to why we believe that combining the efforts in the fields of discrete systems is significant, especially in this day and age. From a general perspective, research into discrete systems has greatly lagged behind some other fields, notably the far more developed theories of ordinary and partial differential equations. One should not forget, of course, the ambitious programme of the Birkhoff school at the beginning of the century, to develop the analytic aspects of linear difference equations (cf. [8]). The great tradition of the early 1900's, with the works of Painlevé and his school [19, 20, 21, 22], dealing with the classification of ordinary differential equations, was in line with the spirit of developing analogous theories for difference equations. Furthermore, a vast amount of knowledge, dating back to the classic works of the 19th century and earlier, has been accumulated, mostly relating to finite-element methods and the classic techniques of numerical analysis (cf. [9]). It is somehow tragic that this tradition doesn't seem to have been continued in recent years, with the notable exceptions of the fields of q-special functions and orthogonal polynomials (cf. [23, 24, 25, 26]).

From the point of view of algebra and geometry, symmetry approaches have been at the heart of the development of effective methods for integrating differential equations (cf. [12, 13, 14]). However, no comprehensive theory of symmetry methods exists to date for difference equations, although new impetus has been given to such a quest in recent years (cf. [15]). It is only now that symmetry reduction techniques, which have nowadays become so very much a standard tool in the analysis of partial differential equations, are being developed for difference and differential-difference equations.

One may ask what is the reason for this difference in development between the comparable theories for differential equations on the one hand and difference equations on the other hand. One explanation is probably the lack of motivation. Whereas in the "classical era", mathematicians such as Birkhoff naturally considered difference equations alongside the "allied" theories of (continuous) differential equations, after the second world war mathematical theories were dominated by continuum models. Thus there was no longer as much interest in studying the admittedly more difficult theory of difference equations. Also in physics, there has been an overall domination of continuum models to describe the processes of nature. It is understandable from one perspective: the continuum is the state of matter as it appears to us in our daily observation of nature, and in order to capture the smoothness of all movement and the coherence of all material we need the mathematical tools of differential calculus. This point of view reached its cumulation in theoretical high energy physics, especially in the fifties and sixties, with the significant developments of relativistic quantum field theory, a continuum theory *in extremis*, which was expected to provide the basis for the description for the fundamental forces in nature. However, it is known now that this theory has severe limitations exactly because of the difficulties of uniting on the one hand quantum mechanics, which is inherently an algebraic and discrete theory, with the notions of continuum functions on the other. One step has been the development of string theory, and more recently the theory of membranes, where in a partial fashion the pointwise structure of quantum fields has been generalised by allowing extended objects as our fundamental building blocks of nature. But even this may not be enough and we should contemplate taking the bold step of acknowledging the assumption that at the quantum mechanical level the continuum description of space-time actually is no longer valid and that an inherently discrete description should take precedence. This point of view, revolutionary as it might seem, has actually been proposed on many occasions (cf. [2, 3]), and there have been recent attempts to put these ideas into practice (cf. [7]). From a general perspective, one needs a much more developed theory of difference equations to be able to classify ordinary and partial difference equations, to develop the analytical tools to study their solutions and to get control over the intrinsically nonlocal features of their behaviour. For this one needs examples that can be rigorously treated. Thus we are led to study exactly integrable discrete systems in all their appearances.

The SIDE meetings, the second of which the present volume gives an account, try to bring together and unite the various developments in the different areas of research where discrete systems play a role. In such a way, we have seen in recent years on the discrete level the unification of the theory of ordinary differential equations, notably the Painlevé equations, with the theory of orthogonal polynomials, [24, 25], the unification of numerical analysis [26] and convergence algorithms with the theory of integrable discrete lattices, the unification of difference geometry [16], with graph theory and combinatorics, unification of quantum field theory with the theory of q-special functions and the unification of soliton equations and cellular automata viz. neural networks. This is the aim of the present volume: to demonstrate the linking elements between the various disciplines dealing with discrete systems. The contributions are written in such a way that they give a brief overview of the state of the art whilst reporting some original research in the subject area. In this way, we hope that the volume can on the one hand assist to familiarise young researcher with this relatively new field of research, and on the other hand serve as a benchmark for the present-day understanding in the various fields. Hopefully, the volume forms an inspiration for further research and so help to establish the links between the various communities working on discrete systems.

The contributions to the present volume cover roughly the following topics:

1. **Special Functions and Difference Equations.**
   Discrete Painlevé equations, difference- and q-difference orthogonal polynomials, separation of variables.

2. **Algebraic aspects.**
   Quantum algebras and representation theory, associated special functions.

3. **Computational and Numerical Aspects.**
   Formal theory of orthogonal polynomials, soliton cellular automata, symplectic and volume-preserving integrators, integrable versus non-integrable discretisations in computation.

4. Symmetry aspects.

Symmetries of difference equations, similarity reductions, integration techniques through symmetries.

5. Analytic aspects.

Analysis of difference equations, isomonodromic deformation theory for discrete systems, asymptotics of orthogonal polynomials.

6. Geometry.

Discrete curves and surfaces, connections with discrete soliton systems, visualisation techniques.

7. Applications.

Neural networks, coding theory and cryptology, data compression, mathematical biology and economics, integrable algorithms.

All these subjects are highly interconnected, and it is the objective of the present volume to demonstrate clearly to the readers how much all these issues are intertwined.

We would like to thank Dr Elizabeth Mansfield for her considerable assistance in the organisation of the meeting and doing most of the "running around", which was especially appreciated as one of us couldonly get around in a wheelchair at the time of the meeting. We also thank Pilar Gordoa, Koryn Grant, Andrew Hicks, Michael Ody, Andrew Pickering, Thomas Priestley, Barry Vowden and Helen Webster for their help with the meeting.

The meeting was supported by grants from the London Mathematical Society and the University of Kent at Canterbury Research Fund, which are both gratefully acknowledged.

Peter A Clarkson (University of Kent at Canterbury)
Frank W Nijhoff (University of Leeds)

References

[1] D. Levi, L. Vinet and P. Winternitz (Editors), *"Symmetries and Integrability of Difference Equations"*, Proceedings of the first SIDE meeting, Estérel, Québec, Canada, May 1995, CRM Proceedings and Lecture Notes Series, **9** American Mathematical Society, Providence, RI (1996).

[2] A. Einstein, *Physics and Reality*. Printed in: *Essays in Physics*, Philosophical Library, New York, (1950).

[3] R.P. Feynman, *Simulating Physics with Computers*, Int. J. Theor. Phys. **21** (1982) 467.

[4] Y. Nambu, *Field Theory and Galois' Fields*, Int. J. Theor. Phys. **21** (1982) 625–636.

[5] T.D. Lee, *Can time be a discrete variable?* Phys. Lett. **122B** (1983) 217–220.

[6] G. 't Hooft, *Quantization of Discrete Deterministic Theories by Hilbert Space Extension*, Nucl. Phys. **B342** (1990) 471–485.

[7] G. 't Hooft, K. Isler and S. Kalitzin, *Quantum Field theoretic behaviour of a deterministic cellular automaton*, Nucl. Phys. **B386** (1992) 495–519.

[8] G.D. Birkhoff, *General theory of linear difference equations*, Trans. Amer. Math. Soc. **12** (1911) 243–284; *The generalized Riemann Problem for linear differential equations and the allied problems for linear difference and q-difference equations*, Proc. Am. Acad. Arts Sci. **49** (1913) 521–568.

[9] N.E. Nörlund, *Vorlesungen über Differenzenrechnung*, Kopenhagen (1923).

[10] E.L. Ince, *"Ordinary Differential Equations"*, Dover, New York (1956).

[11] R.L. Devaney, *"An Introduction to Chaotic Dynamical Systems"*, Addison-Wesley, New York (1989).

[12] S. Lie, *"Theorie der Transformationengruppen"*, B.G. Teubner Verlag, Leipzig (1888, 1890, 1893).

[13] P.J. Olver, *Applications of Lie Groups to Differential Equations*, Second Edition, *Graduate Texts Math.*, vol. **107**, Springer-Verlag, New York (1993).

[14] G.W. Bluman and S. Kumei, *"Symmetries and Differential Equations"*, *Appl. Math. Sci.*, vol. **81**, Springer-Verlag, New York (1989).

[15] S. Maeda, *The similarity method for difference equations* IMA J. Appl. Math. **38** (1987) 129–134.

[16] R. Sauer, *"Differenzengeometrie"*, Springer-Verlag, Berlin (1970).

[17] L. Bianchi, *"Lezioni di geometria differenziale"*, Zanechelli, Bologna (1929).

[18] G. Darboux, *"Leçons sur la théorie générale des surfaces et les applications géometriques du calcul infinitesimal"*, vol. 1–4, Gauthier-Villars, Paris (1887–1896).

[19] P. Painlevé, *Memoire sur les équations différentielles dont l'intégrale générale est uniforme*, Bull. Soc. Math. France **28** (1900) 201–261; *Sur les équations différentielles du second ordre et d'ordre supérieur dont l'intégrale générale est uniforme*, Acta Math. **25** (1902) 1–85.

[20] B. Gambier, *Sur les équations différentielles du second ordre et du premier degré dont l'intégrale générale est à points critiques fixés*, Acta Math. **33** (1909) 1–55.

[21] J. Chazy, *Sur les équations différentielles du troisième ordre et d'ordre supérieur dont l'intégrale générale a ses points critiques fixes*, Acta Math. **34** (1911) 317–385.

[22] M.R. Garnier, *Sur des équations différentielles du troisième ordre dont l'intégrale est uniforme et sur une classe d'équations nouvelles d'ordere supérieur dont l'intégrale a ses points critiques fixes*, Ann. Sci. de l'ENS vol. **XXIX**, # 3, (1912), 1–126.

[23] G. Gasper and M. Rahman, *"Basic Hypergeometric Series"*, Cambridge University Press, Cambridge (1990).

[24] G. Szegö, *"Orthogonal Polynomials"*, Amer. Math. Soc. Colloq. Publ., vol. **23**, Providence, RI (1982).

[25] G. Freud, *"Orthogonal Polynomials"*, Pergamon Press, New York (1971).

[26] C. Brezinski, *"Padé-Type Approximation and General Orthogonal Polynomials"*, (Birkhäuser, Boston, 1980).

Participants

Name	Affiliation
Ablowitz, Mark	University of Colorado, Boulder, USA
Atakishiyev, Natig	Universidad Nacional Autonoma de Mexico, Mexico
Bobenko, Alexander	Technische Universität Berlin, Germany
Bullough, Robin	UMIST, Manchester, UK
Capel, Hans	University of Amsterdam, Netherlands
Cieslinski, Jan	Warsaw University Division in Bialystok, Poland
Clarkson, Peter	University of Kent, Canterbury, UK
Common, Alan	University of Kent, Canterbury, UK
Doliwa, Adam	Warsaw University, Poland
Dorodnitsyn, Vladimir	Keldysh Institute of Applied Mathematics, Moscow, Russia
Estevez, Pilar	Universidad de Salamanca, Spain
Fokas, Athanassios	Imperial College, London, UK
Fordy, Allan	University of Leeds, UK
Gandarias, Maria	Universidad de Cadiz, Spain
Gibbons, John	Imperial College, London, UK
Gilson, Claire	University of Glasgow, UK
Gordoa, Pilar	University of Kent, Canterbury, UK
Grammaticos, Basil	Université Paris VII, France
Grant, Koryn	University of Kent, Canterbury, UK
Grünbaum Alberto	University of California, Berkeley, USA
Harnad, John	Concordia University, Montreal, Canada
Hesameddini, Esmaeil	University of Kent, Canterbury, UK
Hicks, Andrew	University of Kent, Canterbury, UK
Hietarinta, Jarmo	University of Turku, Finland
Hu, Xing-Biao	Chinese Academy of Sciences, Bejing, China
Hydon, Peter	University of Surrey, Guildford, UK
Its, Alexander	Indiana University-Purdue University at Indianapolis, USA
Kajiwara, Kenji	Doshisha University, Kyoto, Japan
Kashaev, Rinat	Steklov Mathematical Institute, St. Petersburg, Russia
Kruskal, Martin	Rutgers University, New Brunswick, USA
Kuznetsov, Vadim	University of Leeds, UK
Lambert, Franklin	Vrije Universiteit Brussel, Belgium
Levi, Decio	Università di Roma III, Italy
Luther, Gregory	University of Notre Dame, USA

Lutzewisch, Shanna	Universität Paderborn, Germany
Ma, Wen-Xiu	Universität Paderborn, Germany
Magnus, Alphonse	Université Catholique de Louvain, Belgium
Mansfield, Elizabeth	University of Kent, Canterbury, UK
Mikhailov, Alexandre	University of Leeds, UK
Mironov, Andrei	Lebedev Physics Institute, Moscow, Russia
Mugan, Ugurhan	Bilkent University, Turkey
Musette, Micheline	Vrije Iniversiteit Brussel, Belgium
Nijhoff, Frank	University of Leeds, UK
Ody, Michael	University of Kent, Canterbury, UK
Oevel, Walter	Universität Paderborn, Germany
Papageorgieu, Vassilis	Technical University of Crete, Chania, Greece
Pickering, Andrew	University of Kent, Canterbury, UK
Priestley, Thomas	University of Kent, Canterbury, UK
Ragnisco, Orlando	Università di Roma III, Italy
Satsuma, Jumkichi	University of Tokyo, Japan
Schief, Wolfgang	University of New South Wales, Sydney, Australia
Smart, Nigel	University of Kent, Canterbury, UK
Sorace, Emanuele	Università di Firenze, Italy
Springael, John	Vrije Universiteit Brussel, Belgium
Suris, Yuri	Universität Bremen, Germany
Takhtajan, Leon	State University of New York at Stony Brook, USA
Tamizhmani, Kilkothur	Pondicherry University, India
Umeno, Ken	Institute of Physical and Chemical Research, Saitama, Japan
Veselov, Alexander	Loughborough University, UK
Viallet, Claude	Université Paris VI, France
Volkov, Alexandre	Steklov Mathematical Institute, St. Petersburg, Russia
Vowden, Barry	University of Kent, Canterbury, UK
Vu, Khai	Monash University, Melbourne, Australia
Webster, Helen	University of Kent, Canterbury, UK
Willox, Ralph	Vrije Universiteit Brussel, Belgium
Winternitz, Pavel	Université de Montreal, Canada

Chapter 1

Partial Difference Equations

Discrete linearisable Gambier equations

A K Common[†], E Hesameddini[†] and M Musette[‡]
† Institute of Mathematics and Statistics,
University of Kent,
Canterbury, CT2 7NF, UK.
‡ Dienst Theoretische Natuurkunde,
Vrije Universiteit Brussel,
Pleinlaan 2, B-1050 Brussels, Belgium

Abstract

We propose two candidates for discrete analogues to the nonlinear Ermakov equation. The first discretisation of degree two possesses the two features which characterise its linearisation in the continuum while the second form of degree one is directly linearisable into a third order equation.

We extend our procedure of discretisation, based on an association with a two dimensional conformal mapping, to a nonlinear equation of the Gambier classification linearisable into a fourth order equation.

1 Introduction

In 1906, Gambier [8] reported that the second order nonlinear differential equation

$$zz_{xx} - \frac{p_x}{p}zz_x - \frac{1}{2}z_x^2 + 2qz^2 - cp^2 = 0, \tag{1}$$

where $c \neq 0$ is a constant and p, q two arbitrary functions of x, is linearisable into a third order equation. Making the transformation $z = py$, one obtains for y the following equation

$$yy_{xx} - \frac{1}{2}y_x^2 + 2fy^2 - c = 0, \tag{2}$$

where

$$f(x) = q(x) + \frac{1}{2}\left(\frac{p_{xx}}{p} - \frac{3}{2}\left(\frac{p_x}{p}\right)^2\right), \tag{3}$$

which by derivation, is simply related to the third order linear equation

$$y_{xxx} + 4fy_x + 2f_xy = 0 .$$ (4)

The expression of the general solution of (2) in terms of two solutions (ψ_1, ψ_2) of the second order linear equation

$$\psi_{xx} + f(x)\psi = 0$$ (5)

was previously given by Ermakov [6, 4] in 1880 as

$$y(x) = \alpha\psi_1^2(x) + 2\beta\psi_1(x)\psi_2(x) + \gamma\psi_2^2(x)$$ (6)

with $\alpha\gamma - \beta^2 = -cW^{-2}/2$, the constant W representing the Wronskian of the two solutions of (5).

At this time, Appell [2] also found the relation (6) in searching for the link between the solution of the third order linear equation (4) and the square of the general solution of (5).

Moreover, the transformation

$$y(x) = \varphi(x)Y(X(x)), \quad X_x \neq 0$$ (7)

converts the equation (2) to another equation with constant coefficients

$$YY_{XX} - \frac{1}{2}Y_X^2 = \frac{c}{K^2}, \quad K \text{ arbitrary}$$ (8)

under the conditions

$$\varphi(x) = \frac{K}{X_x(x)}, \quad \{X; x\} = 2f(x)$$ (9)

where $\{X; x\}$ is the Schwarzian derivative of X with respect to x. Therefore, one can represent the solution of (2) either by the formula (6) or by

$$y(x) = \sqrt{-2c}\,\frac{X}{X_x}$$ (10)

with X solution of the third order equation

$$\frac{X_{xxx}}{X_x} - \frac{3}{2}\left(\frac{X_{xx}}{X_x}\right)^2 = 2f(x).$$ (11)

We concentrate our attention on the equation (2) and notice that one can set, without implying any restriction, the function $p(x)$ equal to a constant.

In §2, we establish a very useful association between this equation and a particular form of a 2-dimensional conformal Riccati system which again reveals the link between the nonlinear equation (2) and the linear equations (4) and (5).

In the first part of §3, we consider the mappings leading to the difference analogues of the conformal Riccati systems which are linearisable. A discrete version of (2) is built from a particular conformal mapping which coincides, in the continuum limit, with its analogous 2-dimensional Riccati system. The resulting equation of degree two is linearisable into a discrete analogue of (4) and its general solution is related to a discrete analogue of (5) by the quadratic expression (6).

In the second part of §3, we obtain a second form of discrete Ermakov equation by using its link with a discrete Schwarzian explicitly invariant for the homographic transformation. Contrary to the expression given in the first part of the present section, this second form corresponds to a scheme of discretisation of first degree. It is also linearisable into a discrete third order equation with continuous limit identical to (4) but a relationship in a compact form with the solutions of a second order linear difference equation has not yet been achieved.

In §4, we extend our procedure of discretisation, based on a two dimensional conformal mapping, to a nonlinear equation of the Gambier classification linearisable into a fourth order equation.

2 Conformal Riccati system

Two main classes of linearisable coupled Riccati systems are ([1]) :

(1) Projective Riccati equations:

They have the matrix form

$$\boldsymbol{\omega}_x = \mathbf{a} + B\boldsymbol{\omega} + \boldsymbol{\omega}(\mathbf{c}, \boldsymbol{\omega}) \tag{12}$$

where the elements of $N \times N$ matrix B and N dimensional vectors $\mathbf{a}, \mathbf{c}$ are functions of the independent variable x.

(2) Conformal Riccati equations:

$$\boldsymbol{\omega}_x = \boldsymbol{\beta} + E\boldsymbol{\omega} + a\boldsymbol{\omega} + \boldsymbol{\omega}(\boldsymbol{\gamma}, \boldsymbol{\omega}) - \tfrac{1}{2}\boldsymbol{\gamma}(\boldsymbol{\omega}, \boldsymbol{\omega}) \tag{13}$$

with $N \times N$ matrix E satisfying

$$E\tilde{I} + \tilde{I}E^T = 0 \tag{14}$$

where $\tilde{I}$ is such that

$$(\delta, \alpha) \equiv \delta^T \tilde{I} \alpha = \delta_1 \alpha_1 \ + \ \delta_2 \alpha_2 + \cdots$$
$$+ \ \cdots + \delta_p \alpha_p - \delta_{p+1} \alpha_{p+1} - \cdots - \delta_N \alpha_N \tag{15}$$

and $1 \le p \le N$.

One can show [11] that specific transformations on the dependent and independent variables of the set of equations (13) lead to a standard form which in case $N = 2$, $\tilde{I} \equiv \begin{pmatrix} 1 & 0 \\ 0 & \varepsilon^2 \end{pmatrix}$ with $\varepsilon^2 = \pm 1$, is

$$u_x = A_1 + D_2 uv \tag{16}$$
$$v_x = A_2(x) - \frac{\varepsilon^2}{2} D_2 \left(u^2 - \varepsilon^2 v^2 \right) \tag{17}$$

where A_1, D_2 are constant and $A_2(x)$ a function of x. Eliminating v from the two coupled first order differential equations (16-17), we obtain for $u(x)$

$$u u_{xx} - \frac{3}{2} u_x^2 + 2 A_1 u_x - \frac{1}{2} A_1^2 - A_2 D_2 u^2 + \frac{\varepsilon^2}{2} D_2^2 u^4 = 0, \tag{18}$$

or in the variable $y = u^{-1}$

$$y y_{xx} - \frac{1}{2} y_x^2 + 2 A_1 y^2 y_x + \frac{1}{2} A_1^2 y^4 + A_2 D_2 y^2 - \frac{\varepsilon^2}{2} D_2^2 = 0. \tag{19}$$

We see that this is equivalent to (2) when

$$A_1 \equiv 0, \quad A_2(x) = \frac{2f(x)}{D_2}, \quad c = \frac{\varepsilon^2 D_2^2}{2}. \tag{20}$$

Let us note that in the variables $w = -(u + v)/2$, $y = u^{-1}$ and for $\varepsilon^2 = -1$ the system (16-17) becomes

$$w_x + D_2 w^2 + \frac{A_1 + A_2(x)}{2} = 0 \tag{21}$$
$$y_x + A_1 y^2 - D_2(2yw + 1) = 0 \tag{22}$$

which corresponds to the system given by Gambier [9] in association with the equation of class XXVII in the particular case $n = 2$.

If $A_1 \equiv 0$, instead of having two Riccati equations in "cascade", the second equation of the set (21-22) is linear and possesses the particular solution $y = \psi_1 \psi_2$ where (ψ_1, ψ_2) are two solutions of the linear equation

$$\psi_{xx} + \frac{D_2 A_2(x)}{2} \psi = 0. \tag{23}$$

with $w = (\text{Log}\psi_1)_{,x}/D_2$ and the Wronskian $W(\psi_1, \psi_2) = D_2$. Hence performing a linear transformation on (ψ_1, ψ_2), one recovers the solution (6) with $\alpha\gamma - \beta^2 = -1/4$.

On the other hand, the standard way to linearise the conformal Riccati system (13) is to first of all convert it to a projective Riccati system of one higher dimension. For the two dimensional case (16-17) we make the definition

$$w \equiv u^2 + \varepsilon^2 v^2 \tag{24}$$

and obtain

$$
\begin{align}
u_x &= A_1 + D_2 uv \tag{25} \\
v_x &= A_2 - \frac{1}{2}\varepsilon^2 D_2 w + D_2 v^2 \tag{26} \\
w_x &= 2\varepsilon^2 A_2 v + D_2 wv + 2A_1 u \tag{27}
\end{align}
$$

which is a 3-dimensional projective Riccati system. This is equivalent to the linear system

$$
\begin{pmatrix} \psi_{1,x} \\ \psi_{2,x} \\ \psi_{3,x} \\ \psi_{4,x} \end{pmatrix} = \begin{pmatrix} 0 & 0 & 0 & A_1 \\ 0 & 0 & -\frac{1}{2}\varepsilon^2 D_2 & A_2 \\ 2A_1 & 2\varepsilon^2 A_2 & 0 & 0 \\ 0 & -D_2 & 0 & 0 \end{pmatrix} \begin{pmatrix} \psi_1 \\ \psi_2 \\ \psi_3 \\ \psi_4 \end{pmatrix} \tag{28}
$$

when $u = \psi_1/\psi_4$, $v = \psi_2/\psi_4$, $w = \psi_3/\psi_4$.

In the particular case $A_1 \equiv 0$, the linear system degenerates in three coupled equations ($\psi_1 \equiv K = $ constant) and the variable ψ_4 with the identification $D_2 A_2(x) = 2f(x)$ satisfies the third order linear equation (4).

3   Discrete forms of the Ermakov equation

There are of course infinitely many ways of constructing nonlinear difference equations which tend to the equation (2) in the continuum. We will discuss two approaches based on the connection with the conformal Riccati system and Schwarz derivative respectively. Firstly, we will obtain a discrete form whose general solution is connected to a linear second order difference equation in the same way as in the continuum.

The conformal Riccati equations in (13) are the infinitesimal counterpart [1] of the discrete conformal transform of the vector ω in $\mathbb{R}^N$ given by

$$\omega \rightarrow e^\rho \Lambda \frac{[\omega + \gamma\omega^2]}{[1 + 2(\omega, \gamma) + \omega^2\gamma^2]} + \alpha \tag{29}$$

where Λ is a general Lorentz transformation and $\boldsymbol{\gamma}, \boldsymbol{\alpha}$ are in $\mathbb{R}^N$ and ρ is a scalar. The most obvious and natural way to discretise (13) is then to consider (29) as a mapping from $\boldsymbol{\omega}(n)$ to $\boldsymbol{\omega}(n+1)$. To obtain the discrete analogy of the equation (2) we take $N = 2$ as in §2 and the standard form of (29) with $\Lambda = I$, $\rho = 0$, $\boldsymbol{\gamma} = \binom{0}{\gamma_2}$, $\boldsymbol{\alpha} = \binom{\alpha_1}{\alpha_2}$. The mapping (29) may then be written in the component form:

$$u(n+1) - \alpha_1(n) = \frac{u(n)}{\{1 + 2\varepsilon^2\gamma_2(n)v(n) + [u^2(n) + \varepsilon^2 v^2(n)]\varepsilon^2\gamma_2^2(n)\}}, \quad (30)$$

$$v(n+1) - \alpha_2(n) = \frac{\{v(n) + \gamma_2(n)[u^2(n) + \varepsilon^2 v^2(n)]\}}{\{1 + 2\varepsilon^2\gamma_2(n)v(n) + [u^2(n) + \varepsilon^2 v^2(n)]\varepsilon^2\gamma_2^2(n)\}}, \quad (31)$$

and one recovers the standard form (16–17) by setting

$$x = nh, \quad u(n) \to u(x), \quad v(n) \to v(x)$$

$$\gamma_2(n) = -\frac{h\varepsilon^2 D_2}{2}, \quad \alpha_2(n) = hA_2(x), \quad \alpha_1(n) = hA_1, \quad (32)$$

and taking the continuum limit $h \to 0$.

To linearise the set (30-31) we make the definition

$$w(n) = u^2(n) + \varepsilon^2 v^2(n) \quad (33)$$

$$u(n) = \frac{\xi_1(n)}{\xi_4(n)}, \quad v(n) = \frac{\xi_2(n)}{\xi_4(n)}, \quad w(n) = \frac{\xi_3(n)}{\xi_4(n)} \quad (34)$$

and obtain the four linear recurrence relations:

$$\begin{pmatrix} \xi_1(n+1) \\ \xi_2(n+1) \\ \xi_3(n+1) \\ \xi_4(n+1) \end{pmatrix} = \mathcal{A}(n) \begin{pmatrix} \xi_1(n) \\ \xi_2(n) \\ \xi_3(n) \\ \xi_4(n) \end{pmatrix} \quad (35)$$

with

$$\mathcal{A} = \begin{pmatrix} 1 & 2\varepsilon^2\alpha_1\gamma_2 & \varepsilon^2\alpha_1\gamma_2^2 & \alpha_1 \\ 0 & 1 + 2\varepsilon^2\alpha_2\gamma_2 & \gamma_2(1 + 2\varepsilon^2\alpha_2\gamma_2) & \alpha_2 \\ 2\alpha_1 & 2\varepsilon^2(\alpha_2 + \gamma_2\alpha_{12}^2) & 1 + \varepsilon^2\gamma_2(2\alpha_2 + \gamma_2\alpha_{12}^2) & \alpha_{12}^2 \\ 0 & 2\varepsilon^2\gamma_2 & \varepsilon^2\gamma_2^2 & 1 \end{pmatrix} \quad (36)$$

and $\alpha_{12}^2 = \alpha_1^2 + \varepsilon^2\alpha_2^2$.

If $\alpha_1(n) \equiv 0$, $\xi_1(n) = K$(constant) and this set degenerates into three linear recurrence relations between $\xi_2(n), \xi_3(n)$ and $\xi_4(n)$. Moreover, with the identification (32) for $\gamma_2(n)$ and $\alpha_2(n)$, it becomes, in the limit $h \to 0$, identical to the third order linear equation (4) for $\xi_4(x) \equiv (Ku(x))^{-1}$.

A discrete form of the equation (2) may then be obtained by eliminating $v(n)$ between the set of two equations (30-31). From the first of these we have a quadratic equation for $v(n)$ and we choose the root

$$v(n) = \frac{\varepsilon^2}{\gamma_2(n)} \left(-1 + \sqrt{\frac{u(n)}{u(n+1)} - \varepsilon^2 \gamma_2^2(n) u^2(n)} \right) \qquad (37)$$

to get the correct continuum limit. Substituting in the second one with the following identification

$$x = nh, \quad \alpha_2(n) = 2hf(x), \quad \gamma_2(n) = -\frac{h\varepsilon^2}{2}, \quad u(n) = u(x) \qquad (38)$$

we obtain for $y(x) = u(x)^{-1}$ the equation :

$$y(x+h)\,2 - h^2 f(x) \;=\; \sqrt{y(x+h)y(x) - \frac{h^2}{4}\varepsilon^2}$$

$$+ \;\; \sqrt{y(x+2h)y(x+h) - \frac{h^2}{4}\varepsilon^2}. \qquad (39)$$

The square roots may be eliminated by squaring twice. Introducing the compact notation:

$$\overline{\overline{y}} = y(x+2h), \quad \overline{y} = y(x+h), \quad \underline{y} = y(x-h), \quad \underline{\underline{y}} = y(x-2h) \qquad (40)$$

also valid for every function of x, we get the equation

$$\left[\overline{y} - y - \left(2 - h^2 f(x) \right)^2 \overline{y} \right]^2 - \left(2 - h^2 f(x) \right)^2 \left(4\overline{y}y - \varepsilon^2 h^2 \right) = 0. \qquad (41)$$

Setting $y(x) = \psi(x)\,\phi(x)$, $A(x) = \pm (2 - h^2 f(x))$, equation (41) becomes

$$\left(\overline{\overline{\psi}}\,\overline{\overline{\phi}} - \psi\phi - A(x)^2 \overline{\psi}\,\overline{\phi} \right)^2 - 4A(x)^2 \psi\overline{\overline{\psi}}\phi\overline{\overline{\phi}} = -h^2\varepsilon^2 A(x)^2. \qquad (42)$$

The left hand side of this relation can be identified with a perfect square if

$$\overline{\overline{\psi}}\,\overline{\overline{\phi}} - \psi\phi - A(x)^2 \overline{\psi}\,\overline{\phi} = A(x)\left(\overline{\psi}\phi + \overline{\phi}\psi \right) \qquad (43)$$

or

$$\overline{\overline{\psi}}\,\overline{\overline{\phi}} = \left(A(x)\overline{\psi} + \psi \right)\left(A(x)\overline{\phi} + \phi \right). \qquad (44)$$

Thus, the condition (43) introduced in (42) implies that

$$\left(\overline{\psi}\phi - \overline{\phi}\psi \right) = \pm ih\varepsilon \qquad (45)$$

while the condition (44) leads to

$$\overline{\overline{\psi}}\,\overline{\phi} - \overline{\overline{\phi}}\,\overline{\psi} = A(x)\left(k(x) - k(x)^{-1}\right)\overline{\psi}\,\overline{\phi} + k(x)\psi\overline{\phi} - k(x)^{-1}\phi\overline{\psi} \qquad (46)$$

where $k(x)$ is an arbitrary function. Those two last relations are compatible if one identifies $k(x)$ with -1. Therefore, the functions ψ and ϕ are two linear independent solutions of the second order difference equation

$$\Psi(x + 2h) + A(x)\Psi(x + h) + \Psi(x) = 0 \qquad (47)$$

with the Wronskian equal to $\pm ih\varepsilon$. Applying the linear transformation

$$\begin{pmatrix} \psi(x) \\ \phi(x) \end{pmatrix} \rightarrow \begin{pmatrix} a & b \\ c & d \end{pmatrix}\begin{pmatrix} \psi(x) \\ \phi(x) \end{pmatrix}, \qquad \begin{vmatrix} a & b \\ c & d \end{vmatrix} \neq 0$$

the solution of the difference equation (41) becomes

$$y(x) = \alpha\psi(x)^2 + 2\beta\psi(x)\,\phi(x) + \gamma\phi(x)^2 \qquad (48)$$

where α, β, γ related by $\alpha\gamma - \beta^2 = -\frac{1}{4}$ are expressible in terms of the constants a, b, c, d . Let us remark that the equation (41) and the nonlinear superposition principle for its solution (48) coincide with the results recently obtained by W.Schief [12].

Finally, we may show that in the continuum limit $h \rightarrow 0$, the equations (47) and (41) lead respectively to the second order linear equation (5) and the nonlinear equation (2). Making in (47) the identification

$$A(x) = -2 + h^2 f(x) \qquad (49)$$

we recover in the limit $h \rightarrow 0$, the linear differential equation (5). Moreover, setting in (35) $\alpha_1(n) = 0$ and expanding the coefficients of the three last recurrence relations to terms $O(h)$ we obtain in the variable $y(n) \equiv \xi_4(n)$

$$\mathcal{D}^3 y + 4q\mathcal{D}y + 2\frac{\overline{q} - q}{h}\overline{y} = 0, \text{ with } \mathcal{D}y = \frac{\overline{y} - y}{h} \qquad (50)$$

whose the continuum limit yields the third order linear equation (4).

On the other hand, expanding $\overline{\overline{y}}$ and $\overline{y}$ up to terms $O(h^2)$ we obtain

$$yy_{xx} - \frac{1}{2}y_x^2 - 2f(x)y^2 = \frac{1}{2}\varepsilon^2. \qquad (51)$$

Thus, the difference equation (41) deserves the name of " discrete Ermakov equation " for three reasons:

(i) it is linearisable into a third order linear equation equivalent to a subset of (35) which consist in three first order linear recurrence relations when $\alpha_1(n) = 0$,

(ii) it tends, in the limit $h \to 0$, to the nonlinear differential equation (2),

(iii) its general solution is related to the second order linear difference equation (47) by the quadratic form (48) as in the continuous case.

However, as conjectured in [5], there should exist a discretisation of Pinney equation of degree one, not two like (41). We will now give such an equation.

When one discretises the nonlinear equation (2) following the rules given in [5] one finds a candidate for discrete Ermakov equation linearisable into a third order linear difference equation

$$E \equiv y\mathcal{D}^2\underline{y} - \frac{1}{2}\mathcal{D}y\mathcal{D}\underline{y} + \left(\frac{\lambda_1}{2}(\overline{f} + \underline{f}) + \lambda_2 f\right)\left(\overline{y}\underline{y} + \overline{y}y + \underline{y}y + y^2\right) - c = 0 \quad (52)$$

with $\lambda_1 + \lambda_2 = 1/2$. Indeed, the difference $\mathcal{D}E = (\overline{E} - E)/h$ is given by

$$\mathcal{D}E = \frac{\overline{y} + y}{2}\left[\mathcal{D}^3\underline{y} + \frac{\lambda_1}{2}\left((\overline{\overline{y}} + \overline{y})(\overline{f} + f) - (y + \underline{y})(\overline{f} + \underline{f})\right)h^{-1}\right.$$
$$+ \lambda_2\left((\overline{\overline{y}} + \overline{y})\overline{f} - (y + \underline{y})f\right)h^{-1} = 0. \quad (53)$$

We will show here that we obtain the same result, starting from the form (11) in terms of the Schwarzian derivative of X. Although there are a number of discrete forms for this derivative with the property of invariance under the homographic transformation

$$X(x) \to \frac{aX(x) + b}{cX(x) + d}, \quad (54)$$

we will choose a particular form where the discrete analogue of the left hand side of (11) may be transformed into a discrete Riccati equation which is linearisable.

The Schwarzian derivative of the function $X(x)$ of the discrete variable $x = nh$ is defined by:

$$S(x) \equiv 4\frac{[X(x) - X(x - h)][X(x + h) - 3X(x) + 3X(x - h) - X(x - 2h)]}{h^2[X(x + h) - X(x - h)][X(x) - X(x - 2h)]}$$
$$- \frac{3}{2}\frac{[X(x + h) - 2X(x) + X(x - h)][X(x) - 2X(x - h) + X(x - 2h)]}{h^2[X(x + h) - X(x - h)][X(x) - X(x - 2h)]} \quad (55)$$

and corresponds to that given previously by Faddeev and Takhtajan [7]. It is easily seen that this has the correct continuum limit and very interestingly is related to the cross-ratio of four adjacent values of X, i.e.

$$S(x) = -\frac{2}{h^2}\left\{1 - 4\frac{[X(x - h) - X(x - 2h)][X(x + h) - X(x)]}{[X(x + h) - X(x - h)][X(x) - X(x - 2h)]}\right\} \quad (56)$$

so that $S(x)$ is obviously invariant under the homographic transformation of $X(x)$.

In analogy to the continuum case, the expression (55) is equivalent to the discrete Riccati equation

$$\overline{\Omega} - \Omega + \Omega\overline{\Omega} = -\tfrac{1}{2}S \tag{57}$$

where

$$\Omega \equiv -\frac{1}{h}\left(X - 2\underline{X} + \underline{X}\right) / \left(X - \underline{X}\right) \tag{58}$$

corresponding in the continuum limit to $\Omega(x) = -X_{xx}/(2X_x)$.

In the continuous case, the equation (2) may be written in the form (11) by the transformation (10). The discrete analogue of (11) is then from (56)

$$4\frac{(X - \underline{X})(X - \overline{X})}{(X - \underline{X})(X - \overline{X})} = 1 + h^2 f(x) \tag{59}$$

Making the substitution $y = \tfrac{h}{2}\sqrt{-2c}(X + \underline{X})/(X - \underline{X})$ in (59) we obtain

$$4y^2 + 2h^2 c - \left(1 + h^2 f(x)\right)\left(\overline{y} + y\right)\left(y + \underline{y}\right) = 0. \tag{60}$$

Equation (60) may also be written as

$$E \equiv y\mathcal{D}^2 \underline{y} - \frac{1}{2}\mathcal{D}y\mathcal{D}\underline{y} + \frac{f(x)}{2}\; y^2 + y(\overline{y} + \underline{y}) + \overline{y}\underline{y}\; - c = 0 \tag{61}$$

which corresponds to (52) for $\lambda_1 = 0$.

The equation (60) is our second type of discrete Ermakov equation. It also corresponds to a discretisation of equation (2) in three points but with the advantage over the previous form (41) to be of degree one. It is equivalent to the discrete Riccati equation (57) where S is identical to $2f(x)$ but a expression in a compact form of its general solution in terms of the solutions of a second order linear difference equation has not yet been derived [3], in contrast to the discrete equation (41), where this has been achieved.

4   Discrete Gambier equation

For $\alpha_1(n) = hA_1$, $\gamma_2(n) = -h\varepsilon^2/2$, $\alpha_2(n) = hA_2(x)$ the system (30-31) is linearisable into the set (35) of four recurrence relations which by expanding each element of the matrix (36) up to terms O(h) becomes

$$\begin{pmatrix} \xi_1(n+1) \\ \xi_2(n+1) \\ \xi_3(n+1) \\ \xi_4(n+1) \end{pmatrix} = \begin{pmatrix} 1 & 0 & 0 & hA_1 \\ 0 & 1 & -h\varepsilon^2/2 & hA_2(x) \\ 2hA_1 & 2h\varepsilon^2 A_2(x) & 1 & 0 \\ 0 & -h & 0 & 1 \end{pmatrix} \begin{pmatrix} \xi_1(n) \\ \xi_2(n) \\ \xi_3(n) \\ \xi_4(n) \end{pmatrix} \tag{62}$$

identical to the fourth order linear difference equation for $\xi_1(n)$

$$\mathcal{D}^4\xi_1 + (A_2 + \overline{A_2})\mathcal{D}^2\xi_1 + \frac{\overline{A_2} - A_2}{h}\mathcal{D}\xi_1 - \varepsilon^2 A_1^2\xi_1 = 0. \qquad (63)$$

The nonlinear equation for $y(x) = u(x)^{-1}$

$$\left[\overline{\overline{Y}}\left(\left(2 - \frac{h^2}{2}A_2(x)\right)^2\overline{y}^2 + \frac{\varepsilon^2 h^2}{4}\left(1 + \overline{Y}^2\right)\right) - \overline{y}\left(\overline{\overline{y}} + \overline{Y}\,\overline{\overline{Y}}y\right)\right]^2$$

$$= 4\overline{Y}\,\overline{\overline{Y}}\left(\overline{y}y - \frac{\varepsilon^2 h^2}{4}\overline{Y}\right)\left(\overline{\overline{y}}\,\overline{y} - \frac{\varepsilon^2 h^2}{4}\overline{\overline{Y}}\right), \text{ with } Y = 1 - hA_1 y \quad (64)$$

is of second degree in $\overline{\overline{y}}$ like (41) and is linearisable by the transformation $y(x) = (\xi_1(n+1) - \xi_1(n))/(hA_1\xi_1(n))$ into a fourth order linear difference equation.

Let us remark that in the continuum limit $h \to 0$ equations (63) and (64) become respectively

$$\xi_{1,xxxx} + 2A_2\xi_{1,xx} + A_{2,x}\xi_{1,x} - \varepsilon^2 A_1^2\xi_1 = 0, \qquad (65)$$

$$yy_{xx} - \frac{1}{2}y_x^2 + 2A_1y^2y_x + \frac{1}{2}A_1^2y^4 + A_2(x)y^2 - \frac{\varepsilon^2}{2} = 0, \qquad (66)$$

with $y = \xi_{1,x}/(A_1\xi_1)$ as it was given in the thesis of Gambier [9] .

5   Conclusion

This paper is an update of a previous contribution on the same subject [3]. The new results reported here are the explicit linearisation of the second order nonlinear difference equation (41) into a second order difference equation and the expression (64) of a discrete equation linearisable into a fourth order difference equation.

A group theory approach similar to ours is also applied by Winternitz (see these Proceedings) to derive a discretisation of the Pinney equation equivalent to (41).

Equations (2) and (19) represent particular cases of the Gambier equation XXVII which can be explicitly linearised. Grammaticos and Ramani [10] have proposed a discrete analogue of this last equation in the form of two discrete Riccati equations in " cascade". They use the singularity confinement criterion to generate a two-dimensional mapping but never write the explicit form of its general solution in terms of linear difference equations.

Acknowledgements

During this conference we have benefitted from an enlightening discussion with Dr W. Schief concerning the discretisation and linearisation of the Ermakov equation (2). We thank him for his advice. M.M. also thanks R. Conte for his interest in a discretisation of first degree.

A.K.C. and M.M. would like to thank the British Council and the National Fonds voor Wetenschappelijk Onderzoek for financial support for exchange visits during which much of this work was carried out.

E.H. would like to thank the government of the Islamic Republic of Iran for their sponsorship of his study at the University of Kent. M.M. acknowledges financial support from the project IUAP III funded by the Belgian government.

References

[1] Anderson R.L., Harnad J. and Winternitz P., *Physica* **D4** (1982) 164-182.

[2] Appell P., *Comptes Rendus* **91** (1880) 211-214.

[3] Common A.K., Hesameddini E. and Musette M., *J. Phys. A.* **29** (1996) 6343-6352.

[4] Common A.K., Hesameddini E. and Musette M., *Theory of nonlinear special functions " the Painlevé transcendents"*, eds L.Vinet and P.Winternitz (Springer, Berlin, 1997).

[5] Conte R. and Musette M., *Phys.Lett.* **224A** 101-110.

[6] Ermakov V.P., Univ.Izv.Kiev Series III, **9** (1880) 1-25.

[7] Faddeev L.D. and Takhtajan L.A., *Springer Lect.Notes Phys.* **246**(1986) 166-179.

[8] Gambier B., *Comptes Rendus* **142** (1906) 1403-1406.

[9] Gambier B., *Acta Mathematica* **33** (1910) 1-55.

[10] Grammaticos B. and Ramani A., *Physica A* **223** (1996) 125-136.

[11] Hesameddini E., Thesis (1996), University of Kent.

[12] Schief W.K., *A discrete Pinney equation*, to be published in Applied Mathematics Letters.

Generalized Bäcklund transformation and new explicit solutions of the two-dimensional Toda equation

Xing-Biao Hu[t,‡] and Peter A Clarkson[‡]
[t] State Key Laboratory of Scientific
and Engineering Computing,
Institute of Computational Mathematics
and Scientific Engineering Computing,
Academia Sinca, P.O. Box 2719, Beijing 100080, P.R. China
[‡] Institute of Mathematics and Statistics
University of Kent at Canterbury
Canterbury, Kent CT2 7NF, UK

Abstract

In this paper, we obtain a generalized Bäcklund transformation for the bilinear representation of the two-dimensional Toda lattice equation, in which Bäcklund parameters are functions of x, y, n. Furthermore, corresponding nonlinear superposition formula are derived and using these results, we obtain some new explicit solutions of the two-dimensional Toda lattice equation.

It is always of considerable interest in obtaining exact analytical solutions of nonlinear equations which have many important applications in several aspects of physics as well as other natural and applied sciences.

In this paper, we discuss the following nonlinear differential-difference equation

$$h_{n+1} + h_{n-1} = 2h_n + \frac{\partial^2}{\partial x \partial y}(\ln h_n) \tag{1}$$

which was first introduced by Darboux in 1915 [1]. By introducing a new variable Q_n related to h_n given by

$$h_n = \exp(Q_{n-1} - Q_n),$$

we can represent (1) in the form

$$\frac{\partial^2 Q_n}{\partial x \partial y} = \exp(Q_{n-1} - Q_n) - \exp(Q_n - Q_{n+1}). \qquad (2)$$

We shall refer to this equation as the two-dimensional Toda lattice equation. There have been several studies of equations (1) or (2) (see, e.g. [2-14]). Now, we consider corresponding bilinear equation of (1)

$$[D_x D_y - 4\sinh^2(\tfrac{1}{2}D_n) + 2H(x, y, n)]f_n \bullet f_n = 0 \qquad (3)$$

which is connected with (1) by the transformation

$$h_n = \frac{\partial^2}{\partial x \partial y}(\ln f_n) + 1 + H(x, y, n)$$

where $H(x, y, n) \equiv H_0(x, y)n + H_1(x, y)$ with $H_0(x, y)$, $H_1(x, y)$ functions of x and y.

In the following, we shall generalize the results in [9] and present a generalized Bäcklund transformation (BT) for the bilinear equation (3). Furthermore the corresponding nonlinear superposition formula is derived and some particular exact solutions of (3) are given.

Let f_n be a solution of (3) and f'_n be a solution of

$$[D_x D_y - 4\sinh^2(\tfrac{1}{2}D_n) + 2\hat{H}(x, y, n)]f'_n \bullet f'_n = 0 \qquad (4)$$

where $\hat{H}(x, y, n) \equiv \hat{H}_0(x, y)n + \hat{H}_1(x, y)$ with $\hat{H}_0(x, y)$, $\hat{H}_1(x, y)$ functions of x, y. If we can find two equations which relate f_n and f'_n and satisfy

$$P \equiv {f'_n}^2[D_x D_y - 4\sinh^2(\tfrac{1}{2}D_n) + 2H]f_n \bullet f_n$$
$$- f_n^2[D_x D_y - 4\sinh^2(\tfrac{1}{2}D_n) + 2\hat{H}]f'_n \bullet f'_n = 0$$

then this is a BT. Here we show that

$$[D_x + \lambda^{-1}\exp(-D_n) + \mu]f_n \bullet f'_n = 0 \qquad (5)$$
$$[D_y \exp(-\tfrac{1}{2}D_n) - \lambda\exp(\tfrac{1}{2}D_n) + \gamma\exp(-\tfrac{1}{2}D_n)]f_n \bullet f'_n = 0 \qquad (6)$$

indeed provides a BT connecting (3) and (4), where $\lambda = \lambda(x, y)$ is an arbitrary function of x, y, $\mu = \mu(x, y, n)$ satisfies $\mu_y = \hat{H}(x, y, n) - H(x, y, n)$ and $\gamma = n(\ln \lambda)_y + \bar{\gamma}(x, y)$ with $\bar{\gamma}(x, y)$ an arbitrary function.

In fact, making use of (A.1)-(A.3),(5) and (6), P can be rewritten as

$$P = 2D_y(D_x f_n \bullet f'_n) \bullet f_n f'_n + 2(H - \hat{H})f_n^2 {f'_n}^2$$
$$- 4\sinh(\tfrac{1}{2}D_n)[\exp(\tfrac{1}{2}D_n)f_n \bullet f'_n] \bullet [\exp(-\tfrac{1}{2}D_n)f_n \bullet f'_n]$$

$$= -2\lambda^{-1}D_y[\exp(-D_n)f_n \bullet f'_n] \bullet f_n f'_n + 2\frac{\lambda_y}{\lambda^2}f_n f'_n \exp(-D_n)f_n \bullet f'_n$$
$$-4\sinh(\tfrac{1}{2}D_n)[\exp(\tfrac{1}{2}D_n)f_n \bullet f'_n] \bullet [\exp(-\tfrac{1}{2}D_n)f_n \bullet f'_n]$$
$$= 4\sinh(\tfrac{1}{2}D_n)\left\{\left[\frac{1}{\lambda}D_y \exp(-\tfrac{1}{2}D_n) - \exp(\tfrac{1}{2}D_n)\right.\right.$$
$$\left.\left.+\frac{\gamma}{\lambda}\exp(-\tfrac{1}{2}D_n)\right]f_n \bullet f'_n\right\} \bullet [\exp(-\tfrac{1}{2}D_n)f_n \bullet f'_n]$$
$$= 0$$

Therefore, we have the following result:

Proposition 1. A BT connecting (3) and (4) is given by (5) and (6) with $\lambda = \lambda(x, y)$ being an arbitrary function of x, y, $\mu = \mu(x, y, n)$ satisfies $\mu_y = \hat{H}(x, y, n) - H(x, y, n)$ and $\gamma = n(\ln \lambda)_y + \bar{\gamma}(x, y)$, with $\bar{\gamma}(x, y)$ an arbitrary function.

Remark: When λ, μ and $\bar{\gamma}$ are arbitrary constants and $H = \hat{H} = 0$, Proposition 1 becomes a result in [9].

As an application of Proposition 1, we give some examples.

Example 1. If we choose $f_n = 1$, $\lambda = -1$, $\mu = \dfrac{n}{x-y}$, $\gamma = -\dfrac{2n+1}{2(x-y)}$ and $H = 0$, then it is easily verified that $f'_n = J_n(x-y)$ satisfies (5) and (6), where $\hat{H} = \dfrac{n}{(x-y)^2}$ and $J_n(z)$ is the standard Bessel function. Thus $J_n(x-y)$ satisfies

$$\left[D_x D_y - 4\sinh^2(\tfrac{1}{2}D_n) + \frac{2n}{(x-y)^2}\right]J_n(x-y) \bullet J_n(x-y) = 0$$

and

$$h_n = \frac{n}{(x-y)^2} + 1 + \frac{\partial^2}{\partial x \partial y}\ln J_n(x-y)$$

is a solution of (1).

Example 2. If we choose $f_n = 1$, $\lambda = -(x/y)^{1/2}$, $\mu = \dfrac{n}{2x}$, $\gamma = -\dfrac{2n+1}{4y}$ and $H = 0$, then it is easily verified that $f'_n = J_n(2\sqrt{xy})$ satisfies (5) and (6), where $\hat{H} = 0$ and J_n is the Bessel function.

Example 3. If we choose $f_n = 1$, λ is a constant, $\mu = -\lambda^{-1}$, $\gamma = \lambda$ and $H = 0$, then it is easily verified that $f'_n = n + \lambda^{-1}x + \lambda y + c$ (with c a constant) satisfies (5) and (6), and $\hat{H} = 0$.

In the following, we just consider the case $\mu_y = 0$, i.e.

$$\hat{H}(x, y, n) \equiv H(x, y, n).$$

In this case, Proposition 1 can be modified to be

Proposition 1'. A BT for (3) is

$$[D_x + \lambda^{-1} \exp(-D_n) + \mu]f_n \bullet f'_n = 0 \qquad (7)$$
$$[D_y \exp(-\tfrac{1}{2}D_n) - \lambda \exp(\tfrac{1}{2}D_n) + \gamma \exp(-\tfrac{1}{2}D_n)]f_n \bullet f'_n = 0 \qquad (8)$$

where $\lambda = \lambda(x, y)$, $\mu = \mu(x, n)$ are arbitrary functions and $\gamma = n(\ln \lambda)_y + \bar{\gamma}(x, y)$, with $\bar{\gamma}$ an arbitrary function.

We shall represent the transformation (7,8) symbolically by $f_n \xrightarrow{(\lambda, \mu, \gamma)} f'_n$. Thus from Examples 2 and 3, we have

$$1 \xrightarrow{(-(x/y)^{1/2}, \frac{n}{2x}, -\frac{2n+1}{4y})} J_n(2\sqrt{xy})$$

$$1 \xrightarrow{(\lambda, -\lambda^{-1}, \lambda)} n + \lambda^{-1}x + \lambda y + c$$

Henceforth, we denote $f_n(x, y) = f(n, x, y) = f(n) = f$ without confusion.

Proposition 2. Let f_0 be a solution of (3) and suppose that f_i, $i = 1, 2$, is a solution of (3), which is related by f_0 under BT (7,8) with $(\lambda_i, \mu_i, \gamma_i)$, i.e. $f_0 \xrightarrow{(\lambda_i, \mu_i, \gamma_i)} f_i$, $i = 1, 2$, $\lambda_1\lambda_2 \neq 0$, $f_j \neq 0$, $j = 0, 1, 2$. Then f_{12} defined by

$$\exp(-\tfrac{1}{2}D_n)f_0 \bullet f_{12} = k(x, y)[\lambda_1 \exp(-\tfrac{1}{2}D_n) - \lambda_2 \exp(\tfrac{1}{2}D_n)]f_1 \bullet f_2 \qquad (9)$$

where $k(x, y)$ is some function of x, y, is a new solution which is related by f_1 and f_2 under BT (7,8) with parameters $(\lambda_2, \mu_2 + \Delta, \gamma_2 + [\ln(k\lambda_1\lambda_2)]_y)$ and $(\lambda_1, \mu_1 + \Delta, \gamma_1 + [\ln(k\lambda_1\lambda_2)]_y)$, respectively, provided that

$$\frac{(k\lambda_1)_x}{k\lambda_1} = \mu_1(x, n) - \mu_1(x, n - 1) + \Delta \qquad (10)$$

$$\frac{(k\lambda_2)_x}{k\lambda_2} = \mu_2(x, n) - \mu_2(x, n - 1) + \Delta \qquad (11)$$

hold for some function $\Delta = \Delta(x, y, n)$.
Proof.
First, from

$$[(D_x + \lambda_1^{-1} \exp(-D_n) + \mu_1)f_0 \bullet f_1]f_2$$
$$-[(D_x + \lambda_2^{-1} \exp(-D_n) + \mu_2)f_0 \bullet f_2]f_1 = 0$$

we have

$$-D_x f_1(n) \bullet f_2(n) + (\mu_1 - \mu_2) f_1(n) f_2(n)$$
$$-\frac{1}{k\lambda_1\lambda_2} \exp(-D_n) f_0(n) \bullet f_{12}(n) = 0. \tag{12}$$

Second, from

$$\left\{ \exp\left(-\frac{\partial}{\partial n}\right) [\lambda_1 D_x + \exp(-D_n) + \lambda_1\mu_1] f_0 \bullet f_1 \right\} f_2(n)$$
$$- \left\{ \exp\left(-\frac{\partial}{\partial n}\right) [\lambda_2 D_x + \exp(-D_n) + \lambda_2\mu_2] f_0 \bullet f_2 \right\} f_1(n) = 0$$

we obtain, after some detailed calculations,

$$\frac{1}{2k} D_x f_0(n-1) \bullet f_{12}(n) - \tfrac{1}{2}\lambda_1 D_x f_1(n-1) \bullet f_2(n)$$
$$-\tfrac{1}{2}\lambda_2 D_x f_1(n) \bullet f_2(n-1) + \frac{(k\lambda_1)_x}{2k} f_1(n-1) f_2(n)$$
$$-\frac{(k\lambda_2)_x}{2k} f_1(n) f_2(n-1) + \lambda_1\mu_1(n-1) f_1(n-1) f_2(n)$$
$$-\lambda_2\mu_2(n-1) f_1(n) f_2(n-1) = 0. \tag{13}$$

Thus, using (10,11) and (13), we obtain

$$\{[\lambda_2 D_x + \exp(-D_n) + \lambda_2(\mu_2 + \Delta)] f_1 \bullet f_{12}\} f_0(n-1)$$
$$= \lambda_2 f_1(n)[\tfrac{1}{2} D_x f_0(n-1) \bullet f_{12}(n) - \tfrac{1}{2}k\lambda_1 D_x f_1(n-1) \bullet f_2(n)$$
$$-\tfrac{1}{2}k\lambda_2 D_x f_1(n) \bullet f_2(n-1) - \tfrac{1}{2}(k\lambda_1)_x f_1(n-1) f_2(n)$$
$$+\tfrac{1}{2}(k\lambda_2)_x f_1(n) f_2(n-1) + \Delta k\lambda_1 f_1(n-1) f_2(n) - \Delta k\lambda_2 f_1(n) f_2(n-1)$$
$$+k\lambda_1\mu_1(n) f_1(n-1) f_2(n) - k\lambda_2\mu_2(n) f_1(n) f_2(n-1)]$$
$$= \lambda_2 f_1(n)[\tfrac{1}{2} D_x f_0(n-1) \bullet f_{12}(n) - \tfrac{1}{2}k\lambda_1 D_x f_1(n-1) \bullet f_2(n)$$
$$-\tfrac{1}{2}k\lambda_2 D_x f_1(n) \bullet f_2(n-1) + \tfrac{1}{2}(k\lambda_1)_x f_1(n-1) f_2(n)$$
$$-\tfrac{1}{2}(k\lambda_2)_x f_1(n) f_2(n-1) + k\lambda_1\mu_1(n-1) f_1(n-1) f_2(n)$$
$$-k\lambda_2\mu_2(n-1) f_1(n) f_2(n-1)] = 0 \tag{14}$$

which implies that

$$[D_x + \lambda_2^{-1} \exp(-D_n) + (\mu_2(n) + \Delta)] f_1 \bullet f_{12} = 0.$$

Similarly we have

$$[D_x + \lambda_1^{-1} \exp(-D_n) + (\mu_1(n) + \Delta)] f_2 \bullet f_{12} = 0.$$

Finally, since f_1 and f_2 are two solutions of (3), then we have

$$
\begin{aligned}
0 &= f_2^2[D_x D_y - 4\sinh^2(\tfrac{1}{2}D_n) + 2H]f_1 \bullet f_1 \\
&\quad - f_1^2[D_x D_y - 4\sinh^2(\tfrac{1}{2}D_n) + 2H]f_2 \bullet f_2 \\
&= 2D_y(D_x f_1 \bullet f_2) \bullet f_1 f_2 \\
&\quad - 4\sinh(\tfrac{1}{2}D_n)[\exp(\tfrac{1}{2}D_n)f_1 \bullet f_2] \bullet [\exp(-\tfrac{1}{2}D_n)f_1 \bullet f_2] \\
&= -2D_y\left[\frac{1}{k\lambda_1\lambda_2}\exp(-D_n)f_0 \bullet f_{12}\right] \bullet f_1 f_2 \\
&\quad + \frac{4}{\lambda_2}\sinh(\tfrac{1}{2}D_n)\{[\lambda_1\exp(-\tfrac{1}{2}D_n) \\
&\quad - \lambda_2\exp(\tfrac{1}{2}D_n)]f_1 \bullet f_2\} \bullet [\exp(-\tfrac{1}{2}D_n)f_1 \bullet f_2] \\
&= -2\left(\frac{1}{k\lambda_1\lambda_2}\right)_y [\exp(-D_n)f_0 \bullet f_{12}]f_1 f_2 \\
&\quad + \frac{2}{k\lambda_1\lambda_2}f_0(n-1)f_1(n)\left[\exp\left(\tfrac{1}{2}\frac{\partial}{\partial n}\right)D_y\exp(-\tfrac{1}{2}D_n)f_2(n)\bullet f_{12}(n) \right. \\
&\quad \left. + \gamma_1(n+\tfrac{1}{2})f_2(n)f_{12}(n+1) - \lambda_1 f_2(n+1)f_{12}(n)\right] \\
&= \frac{2}{k\lambda_1\lambda_2}f_0(n-1)f_1(n)\exp\left(\tfrac{1}{2}\frac{\partial}{\partial n}\right)\{D_y\exp(-\tfrac{1}{2}D_n)f_2(n)\bullet f_{12}(n) \\
&\quad + \gamma_1(n)\exp(-\tfrac{1}{2}D_n)f_2(n)\bullet f_{12}(n) - \lambda_1\exp(\tfrac{1}{2}D_n)f_2(n)\bullet f_{12}(n) \\
&\quad - \left(\frac{1}{k\lambda_1\lambda_2}\right)_y k\lambda_1\lambda_2\exp(-\tfrac{1}{2}D_n)f_2(n)\bullet f_{12}(n)\}
\end{aligned} \tag{15}
$$

which implies that

$$
[D_y\exp(-\tfrac{1}{2}D_n) - \lambda_1\exp(\tfrac{1}{2}D_n) + \{\gamma_1(n) + [\ln(k\lambda_1\lambda_2)]_y\}\exp(-\tfrac{1}{2}D_n)]f_2 \bullet f_{12} = 0.
$$

Similarly, we can show that

$$
[D_y\exp(-\tfrac{1}{2}D_n) - \lambda_2\exp(\tfrac{1}{2}D_n) + \{\gamma_2(n) + [\ln(k\lambda_1\lambda_2)]_y\}\exp(-\tfrac{1}{2}D_n)]f_1 \bullet f_{12} = 0.
$$

Thus we have completed the proof of Proposition 2.

In what follows, we give some examples as an illustrative application of the above results.

Example 4. Setting $f_0 = 1$, $H = 0$ and λ_i, $i = 1, 2$, are constants. Then as shown in Example 3, we have

$$
1 \xrightarrow{(\lambda_i, -\lambda_i^{-1}, \lambda_i)} n + \lambda_i^{-1}x + \lambda_i y + c_i
$$

Furthermore, by use of (9), we obtain the following solution

$$
\begin{aligned}
f_{12} &= \lambda_1(n - 1 + \lambda_1^{-1}x + \lambda_1 y + c_1)(n + \lambda_2^{-1}x + \lambda_2 y + c_2) \\
&\quad - \lambda_2(n + \lambda_1^{-1}x + \lambda_1 y + c_1)(n - 1 + \lambda_2^{-1}x + \lambda_2 y + c_2). \tag{16}
\end{aligned}
$$

In general, continuing along this line, we can obtain a hierarchy of rational solutions.

Example 5. Setting $f_0 = 1$, $H = 0$, λ_1 is a constant and $\lambda_2 = -(x/y)^{1/2}$. Then we have

$$1 \xrightarrow{(\lambda_1, -\lambda_1^{-1}, \lambda_1)} n + \lambda_1^{-1}x + \lambda_1 y + c_1$$

$$1 \xrightarrow{(-(x/y)^{1/2}, \frac{n}{2x}, -\frac{2n+1}{4y})} J_n(2\sqrt{xy})$$

Furthermore if we set $k = 1$ and $\Delta = 0$. Then it is easily verified that (10) and (11) hold. By use of (9), we have a solution of (3) with $H = 0$:

$$f_{12} = \lambda_1(n - 1 + \lambda_1^{-1}x + \lambda_1 y + c_1)J_n(2\sqrt{xy})$$
$$+(x/y)^{1/2}(n + \lambda_1^{-1}x + \lambda_1 y + c_1)J_{n-1}(2\sqrt{xy})$$

Example 6. Setting $f_0 = 1$, $H = 0$, λ_1 is a constant and $\lambda_2 = -(x/y)^{1/2}$, then we have

$$1 \xrightarrow{(\lambda_1, -\lambda_1^{-1}, \lambda_1)} 1 + B \exp\left\{ An + \lambda_1^{-1}(e^A - 1)x + \lambda_1(1 - e^{-A})y \right\}$$

$$1 \xrightarrow{(-(x/y)^{1/2}, \frac{n}{2x}, -\frac{2n+1}{4y})} J_n(2\sqrt{xy})$$

where A and B are constants. Furthermore, if we set $k = 1$ and $\Delta = 0$, then it is easily verified that (10) and (11) hold. Thus, using (9), we have the following solution of (3) with $H = 0$:

$$\begin{aligned}
f_{12} = \ &\lambda_1\left(1 + B\exp\left\{A(n-1) + \lambda_1^{-1}(e^A - 1)x\right.\right.\\
&\left.+ \lambda_1(1 - e^{-A})y\right\}) J_n(2\sqrt{xy}) + (x/y)^{1/2}\left(1 + B\exp\left\{An\right.\right.\\
&\left.\left.+ \lambda_1^{-1}(e^A - 1)x + \lambda_1(1 - e^{-A})y\right\}\right) J_{n-1}(2\sqrt{xy}) \qquad (17)
\end{aligned}$$

In this paper, we have given a generalized BT of (3), in which Bäcklund parameters are functions of x, y, n. Furthermore, corresponding nonlinear superposition formula has been shown. Using these results, we have obtained some new explicit solutions of the two-dimensional Toda lattice equation (1).

Acknowledgement

This work of XBH was supported by National Natural Science Foundation of China.

Appendix

The following bilinear operator identities hold for arbitrary functions a, b, c and d:

$$(D_x D_y a \bullet a)b^2 - a^2 D_x D_y b \bullet b = 2D_y(D_x a \bullet b) \bullet ab \qquad (\text{A.1})$$

$$\begin{aligned}
[\sinh^2(\epsilon D_n)a \bullet a]b^2 &- a^2[\sinh^2(\epsilon D_n)b \bullet b] \\
&= \sinh(\epsilon D_n)[\exp(\epsilon D_n)a \bullet b] \bullet [\exp(-\epsilon D_n)a \bullet b]
\end{aligned} \qquad (\text{A.2})$$

$$D_y[\exp(2\epsilon D_n)a \bullet b] \bullet ab = \sinh(\epsilon D_n)[D_y \exp(\epsilon D_n)a \bullet b] \bullet [\exp(\epsilon D_n)a \bullet b] \qquad (\text{A.3})$$

References

1. Darboux G 1915 Lecons sur la theorie des surfaces vol **2**, 2nd edn (Paris:Gauthier-Villars)

2. Mikhailov A V 1978 Pisma Zh. Eksp. Teor. Fiz. **77** 24

3. Mikhailov A V 1979 JETP Lett. **30** 443

4. Fordy A P and Gibbons J 1980 Commun. Math. Phys. **77** 2

5. Nakamura A 1983 J. Phys. Soc. Japan **52** 380

6. Saitoh N, Takeno S and Takizawa E I 1985 J. Phys. Soc. Japan **54** 3701

7. Levi D and Winternitz P 1993 J. Math. Phys. **34** 3713

8. Matveev V B and Salle M A 1991 Darboux Transformations and Solitons (Berlin:Springer)

9. Hu X B 1994 J. Phys. A: Math. Gen. **27** 201

10. Kajiwara K and Satsuma J 1991 J. Math. Phys. **32** 506

11. Mcintosh I 1994 Nonlinearity **7** 85

12. Villarroel J and Ablowitz M J 1992 Phys. Lett. A **163** 293

13. Villarroel J and Ablowitz M J 1993 Physica D **65** 48

14. Villarroel J and Ablowitz M J 1994 J. Phys. A: Math. Gen **27** 931

Different Aspects of Relativistic Toda Chain

S. Kharchev[1], A. Mironov[2] and A. Zhedanov[3]
[1] ITEP, Bol. Cheremushkinskaya, 25,
Moscow, 117 259, Russia
[2] Theory Department, P.N. Lebedev Physics Institute,
Leninsky prospect, 53, Moscow, 117924, Russia
[3] Physics Department, Donetsk State University,
Donetsk, 340 055, Ukraine

Abstract

We demonstrate that the generalization of the relativistic Toda chain (RTC) is a special reduction of two-dimensional Toda Lattice hierarchy (2DTL). We also show that the RTC is gauge equivalent to the discrete AKNS hierarchy and the unitary matrix model. Relativistic Toda molecule hierarchy is also considered, along with the forced RTC. The simple approach to the discrete RTC hierarchy based on Darboux-Bäcklund transformation is proposed.

1 Introduction

Since the paper of Ruijsenaars [1], where has been proposed, the relativistic Toda chain (RTC) system was investigated in many papers [2]-[4]. This system can be defined by the equation:

$$
\begin{aligned}
\ddot{q}_n = \ & (1 + \epsilon\dot{q}_n)(1 + \epsilon\dot{q}_{n+1})\frac{\exp(q_{n+1} - q_n)}{1 + \epsilon^2\exp(q_{n+1} - q_n)} \ - \\
& - (1 + \epsilon\dot{q}_{n-1})(1 + \epsilon\dot{q}_n)\frac{\exp(q_n - q_{n-1})}{1 + \epsilon^2\exp(q_n - q_{n-1})}
\end{aligned}
\tag{1}
$$

which transforms to the ordinary (non-relativistic) Toda chain (TC) in the evident limit $\epsilon \to 0$. The RTC is integrable, which was discussed in different frameworks (see, for example, [2]-[4] and references therein). The RTC can be obtained as a limit of the general Ruijsenaars system [1].

In this paper we are going to review different Lax representations of the RTC, and to establish numerous relations of it with many well-known integrable systems like AKNS, unitary matrix model etc. It is also shown that the RTC hierarchy can be embedded to the 2DTL hierarchy [5].

Besides, we discuss the forced RTC hierarchy and its finite analog, the relativistic Toda molecule. At the end of this short paper we describe the simple approach to discrete evolutions of the RTC which is based on the notion of the Darboux-Bäcklund transformations and can be considered as a natural generalization of the corresponding notion in the usual Toda chain theory.

For more details we refer the reader to our lengthy paper [6].

2 Lax representation for RTC

Let us describe the Lax representation for the standard RTC equation. The usual procedure to obtain integrable non-linear equations consists of the two essential steps:

i) To find appropriate spectral problem for the Baker-Akhiezer function(s).

ii) To define the proper evolution of this function with respect to isospectral deformations.

Lax representation by three-term recurrent relation. In the theory of the usual Toda chain the first step implies the discretized version of the Schrödinger equation (see [7], for example). In order to get the relativistic extension of the Toda equations, one should consider the following "unusual" spectral problem

$$\Phi_{n+1}(z) + a_n \Phi_n(z) = z\{\Phi_n(z) + b_n \Phi_{n-1}(z)\} \ , \qquad n \in \mathbb{Z} \qquad (2)$$

representing a particular discrete Lax operator acting on the Baker-Akhiezer function $\Phi_n(z)$. This is a simple three-term recurrent relation (similar to those for the Toda and Volterra chains) but with "unusual" spectral dependence.

As for the second step, one should note that there exist *two* distinct integrable flows leading to the same equation (1). As we shall see below, the spectral problem (2) can be naturally incorporated into the theory of two-dimensional Toda lattice (2DTL) which describes the evolution with respect to two (infinite) sets of times $(t_1, t_2, ...)$, $(t_{-1}, t_{-2}, ...)$ (positive and negative times, in accordance with [5]). Here we describe the two particular flows (at the moment, we deal with them "by hands", i.e. introducing the corresponding Lax pairs by a guess) which lead to the RTC equations (1). The most simple evolution equation is that with respect to the first *negative* time and has the form

$$\frac{\partial \Phi_n}{\partial t_{-1}} = R_n \Phi_{n-1} \tag{3}$$

with some (yet unknown) R_n.

The compatibility condition determines R_n in terms of a_n and b_n $R_n = \frac{b_n}{a_n}$ and leads to the following equations of motion:

$$\frac{\partial a_n}{\partial t_{-1}} = \frac{b_n}{a_{n-1}} - \frac{b_{n+1}}{a_{n+1}}; \qquad \frac{\partial b_n}{\partial t_{-1}} = b_n \left(\frac{1}{a_{n-1}} - \frac{1}{a_n} \right) \tag{4}$$

In order to get (1), we should identify

$$a_n = \exp(-\epsilon p_n); \qquad b_n = -\epsilon^2 \exp(q_n - q_{n-1} - \epsilon p_n) \tag{5}$$

Note that in this parameterization the "Hamiltonian" R_n depends only on coordinates q_n's: $R_n = -\epsilon^2 \exp(q_n - q_{n-1})$.

Performing the proper rescaling of time in (4) we reach the RTC equation (1).

As we noted already, the evolution (3), which leads to the RTC equations is not the unique one. The other possible choice leading to the same equations is

$$\frac{\partial \Phi_n}{\partial t_1} = -b_n(\Phi_n - z\Phi_{n-1}) \tag{6}$$

The compatibility condition of (2) and (6) gives the equations

$$\frac{\partial a_n}{\partial t_1} = -a_n(b_{n+1} - b_n) \qquad \frac{\partial b_n}{\partial t_1} = -b_n(b_{n+1} - b_{n-1} + a_{n-1} - a_n) \tag{7}$$

This leads to the same RTC equation (1).

2 × 2 matrix Lax representation. The same RTC equation can be obtained from the matrix Lax operator depending on the spectral parameter [3] (generalizing the Lax operator for the TC [7]). Then the RTC arises as the compatibility condition for the following 2 × 2 matrix equations:

$$L_n^{(s)} \psi_n = \psi_{n+1} \quad , \quad \frac{\partial \psi_n}{\partial t} = A_n \psi_n \tag{8}$$

where

$$L_n^{(s)} = \begin{pmatrix} \zeta \exp(\epsilon p_n) - \zeta^{-1} & \epsilon \exp(q_n) \\ -\epsilon \exp(-q_n + \epsilon p_n) & 0 \end{pmatrix} ; \quad \psi_n = \begin{pmatrix} \psi_n^{(1)} \\ \psi_n^{(2)} \end{pmatrix} \tag{9}$$

$$A_n = \begin{pmatrix} \epsilon^2 \exp(q_n - q_{n-1} + \epsilon p_{n-1}) & -\epsilon \zeta^{-1} \exp(q_n) \\ \epsilon \zeta^{-1} \exp(-q_{n-1} + \epsilon p_{n-1}) & 1 - \zeta^{-2} + \epsilon^2 \end{pmatrix} \qquad (10)$$

One can easily reduce these equations to the system (2) and (3).

To conclude this section, we remark that L-operator (9), which determines the RTC is not unique; moreover, it is not the simplest one. Indeed, we shall see that there exists the whole family of the gauge equivalent operators, which contains more "natural" ones and includes, in particular, the well known operator generating the AKNS hierarchy. From general point of view, the whole RTC hierarchy is nothing but AKNS and vice versa.

3   RTC and unitary matrix model, AKNS, etc.

Now we are going to describe the generalized RTC hierarchy as well as its connection with some other integrable systems. We start our investigation from the framework of orthogonal polynomials

Unitary matrix model. It is well-known that the partition function τ_n of the unitary one-matrix model can be presented as a product of norms of the biorthogonal polynomial system [8]. Namely, let us introduce a scalar product of the form [1]

$$< A, B >= \oint \frac{d\mu(z)}{2\pi i z} \exp \sum_{m>0} (t_m z^m - t_{-m} z^{-m}) \; A(z) B(z^{-1}) \qquad (11)$$

Let us define the system of polynomials biorthogonal with respect to this scalar product

$$< \Phi_n, \Phi_k^\star >= h_n \delta_{nk} \qquad (12)$$

Then, the partition function τ_n of the unitary matrix model is equal to the product of h_n's:

$$\tau_n = \prod_{k=0}^{n-1} h_k \; , \qquad \tau_0 \equiv 1 \qquad (13)$$

The polynomials are normalized as follows:

$$\Phi_n(z) = z^n + \ldots + S_{n-1}, \qquad \Phi_n^\star(z) = z^n + \ldots + S_{n-1}^\star, \qquad S_{-1} = S_{-1}^\star \equiv 1 \qquad (14)$$

These polynomials satisfy the following recurrent relations:

$$\Phi_{n+1}(z) = z\Phi_n(z) + S_n z^n \Phi_n^\star(z^{-1})$$
$$\Phi_{n+1}^\star(z^{-1}) = z^{-1}\Phi_n^\star(z^{-1}) + S_n^\star z^{-n} \Phi_n(z) \qquad (15)$$

---

[1] The signs of positive and negative times are defined in this way to get the exact correspondence with the times introduced in [5].

and

$$\frac{h_{n+1}}{h_n} = 1 - S_n S_n^\star \qquad (16)$$

The above relations can be written in several equivalent forms. First, it can be presented in the form analogous to (2):

$$\Phi_{n+1} - \frac{S_n}{S_{n-1}}\Phi_n = z\left\{\Phi_n - \frac{S_n}{S_{n-1}}(1 - S_{n-1}S_{n-1}^\star)\Phi_{n-1}\right\} \qquad (17)$$

$$\Phi_{n+1}^\star - \frac{S_n^\star}{S_{n-1}^\star}\Phi_n^\star = z^{-1}\left\{\Phi_n^\star - \frac{S_n^\star}{S_{n-1}^\star}(1 - S_{n-1}S_{n-1}^\star)\Phi_{n-1}^\star\right\} \qquad (18)$$

From the first relation and using (2) and (5), one can immediately read off

$$\frac{S_n}{S_{n-1}} = -\exp(-\epsilon p_n); \qquad \frac{h_n}{h_{n-1}} = -\epsilon^2 \exp(q_n - q_{n-1}) \qquad (19)$$

Thus, the orthogonality conditions (12) lead exactly to the spectral problem for the RTC. We should stress that equations (17), (18) can be *derived* from the unitary matrix model.

Using the orthogonal conditions, it is also possible to obtain the equations which describe the time dependence of Φ_n, $\Phi_n^\star$. Differentiating (12) with respect to times t_1, t_{-1} gives the evolution equations:

$$\frac{\partial \Phi_n}{\partial t_1} = \frac{S_n}{S_{n-1}}\frac{h_n}{h_{n-1}}(\Phi_n - z\Phi_{n-1}); \qquad \frac{\partial \Phi_n}{\partial t_{-1}} = \frac{h_n}{h_{n-1}}\Phi_{n-1} \qquad (20)$$

$$\frac{\partial \Phi_n^\star}{\partial t_1} = -\frac{h_n}{h_{n-1}}\Phi_{n-1}^\star; \qquad \frac{\partial \Phi_n^\star}{\partial t_{-1}} = -\frac{S_n^\star}{S_{n-1}^\star}\frac{h_n}{h_{n-1}}(\Phi_n^\star - z^{-1}\Phi_{n-1}^\star) \qquad (21)$$

(see general evolution equations with respect to higher flows below). The compatibility conditions give the following nonlinear evolution equations:

$$\frac{\partial S_n}{\partial t_1} = S_{n+1}\frac{h_{n+1}}{h_n}; \qquad \frac{\partial S_n}{\partial t_{-1}} = S_{n-1}\frac{h_{n+1}}{h_n} \qquad (22)$$

$$\frac{\partial S_n^\star}{\partial t_1} = -S_{n-1}^\star\frac{h_{n+1}}{h_n}; \qquad \frac{\partial S_n^\star}{\partial t_{-1}} = -S_{n+1}^\star\frac{h_{n+1}}{h_n} \qquad (23)$$

As a consequence, in the polynomial case,

$$\frac{\partial h_n}{\partial t_1} = -S_n S_{n-1}^\star h_n; \qquad \frac{\partial h_n}{\partial t_{-1}} = S_{n-1}S_n^\star h_n \qquad (24)$$

These are exactly relativistic Toda equations written in somewhat different form. Indeed, from (24), (22), (23) and (16) one gets[2]

$$\frac{\partial^2}{\partial t_1^2} \log h_n = -\left(\frac{\partial}{\partial t_1} \log h_n\right)\left(\frac{\partial}{\partial t_1} \log h_{n+1}\right)\frac{\frac{h_{n+1}}{h_n}}{1 - \frac{h_{n+1}}{h_n}} + (n \to n-1)$$

(25)

On the other hand, the RTC is a particular case of the 2DTL hierarchy. Indeed, let us introduce the key objects in the theory of integrable systems - the τ-functions, which are defined through the relation $h_n = \tau_{n+1}/\tau_n$. Then, with the help of (22)-(24), one can show that the functions τ_n satisfy the first equation of the 2DTL:

$$\partial_{t_1}\partial_{t_{-1}} \log \tau_n = -\frac{\tau_{n+1}\tau_{n-1}}{\tau_n^2}$$

(26)

Therefore, it is natural to assume that the higher flows generate the whole set of non-linear equations of the 2DTL in spirit of [5]. This is indeed the case.

This completes the derivation of the RTC from the unitary matrix model.

RTC versus AKNS and "novel" hierarchies. Now let us demonstrate the correspondence between RTC and AKNS system. We have already seen that the orthogonality conditions naturally lead to the 2×2 formulation of the problem generated by the unitary matrix model:

$$L^{(U)}\begin{pmatrix}\Phi_n \\ \Phi_n^\star\end{pmatrix} = \begin{pmatrix}\Phi_{n+1} \\ \Phi_{n+1}^\star\end{pmatrix} \quad, \quad L^{(U)} = \begin{pmatrix} z & z^n S_n \\ z^{-n} S_n^\star & z^{-1}\end{pmatrix}$$

(27)

$$\frac{\partial}{\partial t_1}\begin{pmatrix}\Phi_n \\ \Phi_n^\star\end{pmatrix} = \begin{pmatrix} -S_n S_{n-1}^\star & z^n S_n \\ z^{1-n} S_{n-1}^\star & -z \end{pmatrix}\begin{pmatrix}\Phi_n \\ \Phi_n^\star\end{pmatrix}$$

(28)

$$\frac{\partial}{\partial t_{-1}}\begin{pmatrix}\Phi_n \\ \Phi_n^\star\end{pmatrix} = \begin{pmatrix} z^{-1} & -z^{n-1} S_{n-1} \\ -z^{-n} S_n^\star & S_{n-1} S_n^\star \end{pmatrix}\begin{pmatrix}\Phi_n \\ \Phi_n^\star\end{pmatrix}$$

(29)

(Equations (28), (29) follow from (20)-(21) and the original spectral problem (15)). Put $\Phi_n \equiv z^{n/2-1/4} F_n$, $\Phi_n^\star \equiv z^{-n/2+1/4} F_n^\star$. Then the spectral problem (27) can be rewritten in the matrix form

$$L_n^{(\mathrm{AKNS})} \mathcal{F}_n = \mathcal{F}_{n+1} \quad, \quad \mathcal{F}_n \equiv \begin{pmatrix} F_n \\ F_n^\star \end{pmatrix}$$

(30)

---

[2]The same equation holds for t_{-1}-flow.

where

$$L_n^{(\text{AKNS})} = \begin{pmatrix} \zeta & S_n \\ S_n^\star & \zeta^{-1} \end{pmatrix} , \quad \zeta \equiv z^{1/2} \tag{31}$$

This is the Lax operator for the discrete AKNS [9]. Obviously the evolution equations (28), (29) can be written in terms of F_n, $F_n^\star$ as

$$\frac{\partial F_n}{\partial t_1} = A_n^{(1)} F_n , \qquad A_n^{(1)} = \begin{pmatrix} -S_n S_{n-1}^\star & \zeta S_n \\ \zeta S_{n-1}^\star & -\zeta^2 \end{pmatrix} \tag{32}$$

$$\frac{\partial F_n}{\partial t_{-1}} = - A_n^{(-1)} F_n , \qquad A_n^{(-1)} = \begin{pmatrix} -\zeta^{-2} & \zeta^{-1} S_{n-1} \\ \zeta^{-1} S_n^\star & -S_{n-1} S_n^\star \end{pmatrix} \tag{33}$$

Note that after introducing the trivial flow

$$\frac{\partial F_n}{\partial t_0} = A_n^{(0)} F_n , \qquad A_n^{(0)} = \begin{pmatrix} 1 & 0 \\ 0 & -1 \end{pmatrix} \tag{34}$$

we get the difference non-linear Schrödinger system (DNLS) [9] (see also [7]) generated by the "mixed" flow

$$\frac{\partial F_n}{\partial T} \equiv \left(\frac{\partial}{\partial t_0} - \frac{\partial}{\partial t_{-1}} - \frac{\partial}{\partial t_1} \right) F_n = (A_n^{(0)} + A_n^{(-1)} - A_n^{(1)}) F_n \equiv$$
$$\equiv \begin{pmatrix} 1 + S_n S_{n-1}^\star - \zeta^{-2} & \zeta^{-1} S_{n-1} - \zeta S_n \\ \zeta^{-1} S_n^\star - \zeta S_{n-1}^\star & -1 - S_{n-1} S_n^\star + \zeta^2 \end{pmatrix} \tag{35}$$

Indeed, from the compatibility conditions for (31), (35) or, equivalently, just from (22) (along with the trivial evolution $\partial_{t_0} S_n = 2 S_n$, $\partial_{t_0} S_n^\star = -2 S_n^\star$) one gets the discrete version of the nonlinear Schrödinger equation:

$$\frac{\partial S_n}{\partial T} = -(S_{n+1} - 2 S_n + S_{n-1}) + S_n S_n^\star (S_{n+1} + S_{n-1}) \tag{36}$$

Note also that the "novel" hierarchy of [10] is equivalent to the RTC (and, therefore, to the AKNS hierarchy) as well. Namely, the Lax operator in [10], i.e.

$$L_n = \begin{pmatrix} z + u_n v_n & u_n \\ v_n & 1 \end{pmatrix} ; \qquad L_n \begin{pmatrix} \phi_n^{(1)} \\ \phi_n^{(2)} \end{pmatrix} = \begin{pmatrix} \phi_{n+1}^{(1)} \\ \phi_{n+1}^{(2)} \end{pmatrix} \tag{37}$$

defines the recurrent relation of the form (17):

$$\phi_{n+1}^{(1)} - \left(u_n v_n + \frac{u_n}{u_{n-1}} \right) \phi_n^{(1)} = z \left(\phi_n^{(1)} - \frac{u_n}{u_{n-1}} \phi_{n-1}^{(1)} \right) \tag{38}$$

thus revealing the connection with the RTC. Comparing (17) and (38) leads to the identification $u_n = S_n h_n$, $v_n = \frac{S_{n-1}^\star}{h_n}$, where h_n's satisfy (16). Moreover, from (37) and (15) it is easy to see that $\phi_n^{(1)} = \Phi_n$; $\phi_n^{(2)} = \frac{1}{h_n}\left(z^n \Phi_n^\star - S_{n-1}^\star \Phi_n\right)$ and, therefore, L_n can be obtained from $L_n^{(AKNS)}$ by the discrete gauge transformation:

$$L_n = U_{n+1} L_n^{(AKNS)} U_n^{-1}; \qquad U_n = z^{n/2 - 1/4} \begin{pmatrix} 1 & 0 \\ -\dfrac{S_{n-1}^\star}{h_n} & \dfrac{z^{1/2}}{h_n} \end{pmatrix}; \qquad z = \zeta^2$$

$$(39)$$

Evolution equations (22)- (24) in terms of new variables u_n, v_n have the form

$$\frac{\partial u_n}{\partial t_1} = u_{n+1} - u_n^2 v_n , \qquad \frac{\partial u_n}{\partial t_{-1}} = \frac{u_{n-1}}{1 + u_{n-1} v_n}$$

$$\frac{\partial v_n}{\partial t_1} = -v_{n-1} + u_n v_n^2 , \qquad \frac{\partial v_n}{\partial t_{-1}} = -\frac{v_{n+1}}{1 + u_n v_{n+1}}$$

$$(40)$$

and easily reproduce the usual AKNS equations in the continuum limit since

$$(\partial_{t_0} - \partial_{t_1} - \partial_{t_{-1}}) u_n = -(u_{n+1} - 2u_n + u_{n-1}) + (u_{n-1}^2 + u_n^2) v_n + \cdots$$

$$(\partial_{t_0} - \partial_{t_1} - \partial_{t_{-1}}) v_n = (v_{n+1} - 2v_n + v_{n-1}) - (v_n^2 + v_{n+1}^2) u_n + \cdots$$

$$(41)$$

We conclude with the remark that the operator $L_n^{(S)}$ in (9) is also gauge equivalent to $L_n^{(AKNS)}$ (see [6]).

Non-local Lax representation. There is another form of the recurrent relations which is non-local (i.e. contains all the functions with smaller indices) but instead expresses $\Phi_n(z)$ through themselves. This form is crucial for dealing with the RTC as a particular reduction of the 2DTL. Let us introduce the normalized functions

$$\mathcal{P}_n(z) \equiv \Phi_n(z) , \qquad \mathcal{P}_n^\star(z^{-1}) \equiv \frac{1}{h_n} \Phi_n^\star(z^{-1})$$

$$(42)$$

From (17), (18) one can show that in the forced and fast-decreasing cases some proper solutions satisfy the equations

$$z\mathcal{P}_n(z) = \mathcal{P}_{n+1}(z) - S_n h_n \sum_{k=-\infty}^{n} \frac{S_{k-1}^\star}{h_k} \mathcal{P}_k(z) \equiv \mathcal{L}_{nk} \mathcal{P}_k(z)$$

$$z^{-1}\mathcal{P}_n^\star(z^{-1}) = \frac{h_{n+1}}{h_n} \mathcal{P}_{n+1}^\star(z^{-1}) - S_n^\star \sum_{k=-\infty}^{n} S_{k-1} \mathcal{P}_k^\star(z^{-1}) \equiv \overline{\mathcal{L}}_{kn} \mathcal{P}_k^\star(z^{-1})$$

$$(43)$$

This expression is correct for general (non-polynomial) $\mathcal{P}_n$ and $\mathcal{P}_n^\star$ provided the sums run over all integer k. In the polynomial case, the sums automatically run over only non-negative k. The last representation of the spectral problem will be useful to determine the general evolution of the system. Indeed, these relations manifestly describe the embedding of the RTC into the 2DTL [5,11], which is given essentially by *two* Lax operators ($\mathcal{L}$ and $\overline{\mathcal{L}}$).

RTC as reduction of 2DTL. In order to determine the whole set of the evolution equations, one can use different tricks. For example, one can use embedding (43) of the system into the 2DTL, making use of the standard evolution of this latter [5]. Let us briefly describe the formalism of the 2DTL following [5]. In their framework, one introduces *two* different Baker-Akhiezer (BA) $\mathbb{Z} \times \mathbb{Z}$ matrices $\mathcal{W}$ and $\overline{\mathcal{W}}$. These matrices satisfy the linear system:

i) the matrix version of the spectral problem:

$$\mathcal{L}\mathcal{W} = \mathcal{W}\Lambda \ , \quad \overline{\mathcal{L}}\,\overline{\mathcal{W}} = \overline{\mathcal{W}}\Lambda^{-1} \tag{44}$$

ii) the matrix version of the evolution equations:

$$
\begin{aligned}
\frac{\partial \mathcal{W}}{\partial t_m} &= (\mathcal{L}^m)_+ \mathcal{W} \ , & \frac{\partial \overline{\mathcal{W}}}{\partial t_m} &= (\mathcal{L}^m)_+ \overline{\mathcal{W}} \\
\frac{\partial \mathcal{W}}{\partial t_{-m}} &= (\overline{\mathcal{L}}^{\,m})_- \mathcal{W} \ , & \frac{\partial \overline{\mathcal{W}}}{\partial t_{-m}} &= (\overline{\mathcal{L}}^{\,m})_- \overline{\mathcal{W}} \ , & m = 1, 2, \dots
\end{aligned}
\tag{45}
$$

where $\mathbb{Z} \times \mathbb{Z}$ matrices $\mathcal{L}$ and $\overline{\mathcal{L}}$ have (by definition) the following structure:

$$
\begin{aligned}
\mathcal{L} &= \sum_{i \le 1} \mathrm{diag}[b_i(s)]\Lambda^i \ ; \quad b_1(s) = 1 \\
\overline{\mathcal{L}} &= \sum_{i \ge -1} \mathrm{diag}[c_i(s)]\Lambda^i \ ; \quad c_{-1}(s) \ne 0
\end{aligned}
\tag{46}
$$

Here $\mathrm{diag}[b_i(s)]$ denotes an infinite diagonal matrix

$$\mathrm{diag}(\dots\ b_i(-1),\ b_i(0),\ b_i(1),\ \dots)$$

Λ is the shift matrix with the elements $\Lambda_{nk} \equiv \delta_{n,k-1}$ and for arbitrary infinite matrix $A = \sum_{i \in \mathbb{Z}} \mathrm{diag}[a_i(s)]\Lambda^i$ we set

$$(A)_+ \equiv \sum_{i \ge 0} \mathrm{diag}[a_i(s)]\Lambda^i \ , \quad (A)_- \equiv \sum_{i < 0} \mathrm{diag}[a_i(s)]\Lambda^i \tag{47}$$

i.e. $(A)_+$ is the upper triangular part of the matrix A (including the main diagonal) while $(A)_-$ is strictly the lower triangular part.

Note that (46) can be written in components as

$$
\begin{aligned}
\mathcal{L}_{nk} &= \delta_{n+1,k} + b_{k-n}(n)\theta(n-k) \ , \\
\overline{\mathcal{L}}_{nk} &= c_{-1}(n)\delta_{n-1,k} + c_{k-n}(n)\theta(k-n) \ ; \quad n, k \in \mathbb{Z}
\end{aligned}
\tag{48}
$$

The compatibility conditions imposed on (44),(45) give rise to the infinite set (hierarchy) of nonlinear equations for the operators $\mathcal{L}$, $\overline{\mathcal{L}}$ or, equivalently, for the coefficients $b_m(n)$, $c_m(n)$. This is what is called 2DTL hierarchy.

From (43), one gets two matrices

$$\mathcal{L}_{nk} = \delta_{n+1,k} - \frac{h_n}{h_k} S_n S_{k-1}^* \theta(n-k) \ , \quad k,n \in \mathbb{Z} \tag{49}$$

$$\overline{\mathcal{L}}_{nk} = \frac{h_n}{h_{n-1}} \delta_{n-1,k} - S_{n-1} S_k^* \theta(k-n) \ , \quad k,n \in \mathbb{Z} \tag{50}$$

which have exactly the form (48). Now using the technique developed in [5], one can get the whole evolution of the RTC hierarchy. However, it can be also easily obtained in the framework of the orthogonal polynomials.

Evolution and orthogonal polynomials. Differentiating the orthogonality conditions with respect to arbitrary times, one can obtain with the help of (43) the evolution of polynomials $\mathcal{P}_n$ and $\mathcal{P}_n^*$ [6]:

$$\frac{\partial \mathcal{P}_n}{\partial t_m} = -[(\mathcal{L}^m)_-]_{nk} \mathcal{P}_k; \qquad \frac{\partial \mathcal{P}_n}{\partial t_{-m}} = [(\overline{\mathcal{L}}^{\,m})_-]_{nk} \mathcal{P}_k$$

$$\frac{\partial \mathcal{P}_n^*}{\partial t_m} = -[(\mathcal{L}^m)_+]_{kn} \mathcal{P}_k^*; \qquad \frac{\partial \mathcal{P}_n^*}{\partial t_{-m}} = [(\overline{\mathcal{L}}^{\,m})_+]_{kn} \mathcal{P}_k^* \tag{51}$$

$$\frac{\partial h_n}{\partial t_m} = (\mathcal{L}^m)_{nn} h_n \ , \qquad \frac{\partial h_n}{\partial t_{-m}} = -(\overline{\mathcal{L}}^{\,m})_{nn} h_n$$

4   Forced RTC hierarchy

RTC-reduction of 2DTL. Let us formulate in some invariant terms what reduction of the 2DTL corresponds to the RTC hierarchy. Return again to the Lax representation (43) embedding the RTC into the 2DTL. Using (16), one can easily prove the following identities

$$\sum_{k=n}^{N} \frac{S_{k-1} S_{k-1}^*}{h_k} = \frac{1}{h_N} - \frac{1}{h_{N-1}}; \qquad \sum_{k=n}^{N} S_k S_k^* h_k = h_n - h_{N+1} \tag{52}$$

Because of these identities, the matrices $\mathcal{L}$ and $\overline{\mathcal{L}}^T$ have zero modes $\sim S_{k-1}$ and S_{k-1}^*/h_k respectively. Therefore, one could naively expect that they are not invertible and get (using (52)) that

$$(\mathcal{L}\overline{\mathcal{L}})_{nk} = \delta_{nk} - \frac{S_n S_k^* h_n}{h_{-\infty}}; \qquad (\overline{\mathcal{L}}\mathcal{L})_{nk} = \delta_{nk} - \frac{S_{n-1} S_{k-1}^* h_\infty}{h_k} \tag{53}$$

Since the reduction is to be described as an invariant condition imposed on $\mathcal{L}$ and $\overline{\mathcal{L}}$, these formulas might serve as a starting point to describe the reduction of the 2DTL to the RTC hierarchy only if their r.h.s. does not depend on the dynamical variables. It seems not to be the case.

However, these formulas require some careful treatment. Indeed, the formulation of the 2DTL in terms of infinite-dimensional matrices, although being correct as a formal construction requires some accuracy if one wants to work with the genuine matrices since the products of the infinite-dimensional matrices should be properly defined. In fact, this product exists for the "band" matrices, i.e. those with only a finite number of the non-zero diagonals, and in some other more complicated cases (of special divergency conditions). One can easily see from (43) that the RTC Lax operators do not belong to this class. Therefore, equations (53) just do not make sense in this case (this is why the interpretation of the general RTC hierarchy in invariant (say, Grassmannian) terms is a little bit complicated).

However, in the case of forced hierarchy, some of the indicated problems are removed since one needs to multiply only quarter-infinite matrices and, say, the product $\mathcal{L}\bar{\mathcal{L}}$ always exists. Certainly the inverse order of the multipliers is still impossible. Therefore, only the first formula in (53) becomes well-defined acquiring the form

$$(\mathcal{L}\bar{\mathcal{L}})_{nk} = \delta_{nk} \tag{54}$$

This formula can be already taken as a definition of the RTC-reduction of the 2DTL in the forced case as it does not depend on dynamical variables. Now we will show how this definition is reflected in different formulations of the 2DTL.

Determinant representation. One can show (see [6]) that τ-function of the forced hierarchy $\tau_n = 0$, $n < 0$ has the following determinant representation [12]

$$\tau_n(t) =$$
$$= \det \left[\partial^i_{t_1} (-\partial_{t_{-1}})^j \int_\gamma A(z,w) \exp \left\{ \sum_{m>0} (t_m z^m - t_{-m} w^{-m}) \right\} dz dw \right]\Bigg|_{i,j=0,\dots,n-1} \tag{55}$$

where the matrix A has the form $A(z,w) = \frac{\mu'(z)}{2\pi i z}\delta(z - w^{-1})$ [6].

Indeed, let us rewrite the orthogonality relation (12) in the matrix form. We define matrices D and D^* with the matrix elements determined as the coefficients of the polynomials $\Phi_n(z)$ and $\Phi^*_n(z)$

$$\Phi_i(z) \equiv \sum_j D_{ij} z^{j-1}, \quad \Phi^*_i(z) \equiv \sum_j D^*_{ij} z^{j-1} \tag{56}$$

Then, (12) looks like

$$D \cdot C \cdot D^{*T} = H \tag{57}$$

where superscript T means transposed matrix and H denotes the diagonal matrix with the entries $C_{ii} = h_{i-1}$ and C is the moment matrix with the matrix elements

$$C_{ij} \equiv \int_\gamma \frac{d\mu(z)}{2\pi i z} z^{i-j} \exp\left\{ \sum_{m>0} (t_m z^m - t_{-m} z^{-m}) \right\} \qquad (58)$$

Let us note that D ($D^{\star T}$) is the upper (lower) triangle matrix with the units on the diagonal (because of (14)). This representation is nothing but the Riemann-Hilbert problem for the forced hierarchy. Now taking the determinant of the both sides of (57), one gets

$$\det_{n\times n} C_{ij} = \prod_{k=0}^{n-1} h_k = \tau_n \qquad (59)$$

due to formula (13). The remaining last step is to observe that

$$C_{ij} = \partial_{t_1}^i (-\partial_{t_{-1}})^j \int_\gamma \frac{d\mu(z)}{2\pi i z} \exp\left\{ \sum_{m>0} (t_m z^m - t_{-m} z^{-m}) \right\} = \partial_{t_1}^i (-\partial_{t_{-1}})^j C_{11} \qquad (60)$$

i.e.

$$\tau_n(t) = \det_{n\times n} \left[\partial_{t_1}^i (-\partial_{t_{-1}})^j \int_\gamma \frac{d\mu(z)}{2\pi i z} \exp\left\{ \sum_{m>0} (t_m z^m - t_{-m} z^{-m}) \right\} \right] \qquad (61)$$

This expression coincides with (55). One can also remark that the moment matrix C_{ij} is Toeplitz matrix. This proves from the different approach that the RTC-reduction is defined by the Toeplitz matrices.

5   Relativistic Toda molecule

General properties. Let us consider further restrictions on the RTC which allows one to consider the *both* products in (53). Namely, in addition to the condition $\tau_n = 0$, $n < 0$ picking up forced hierarchy, we impose the following constraint

$$\tau_n = 0, \quad n > N \qquad (62)$$

for some N. This system should be called $N - 1$-particle relativistic Toda molecule, by analogy with the non-relativistic case and is nothing but RTC-reduction of the two-dimensional Toda molecule [13,14][3].

---

[3]Sometimes the Toda molecule is called non-periodic Toda [15]. It is an immediate generalization of the Liouville system.

$sl(N)$ Toda can be described by the kernel $A(z,w)$

$$A(z,w) = \sum_{k}^{N} f^{(k)}(z)g^{(k)}(w) \tag{63}$$

where $f^{(k)}(z)$ and $g^{(k)}(z)$ are arbitrary functions. From this description, one can immediately read off the corresponding determinant representation (55).

Indeed, equation (26) and condition (62) implies that $\log \tau_0$ and $\log \tau_N$ satisfy the free wave equation $\partial_{t_1}\partial_{t_{-1}} \log \tau_0 = \partial_{t_1}\partial_{t_{-1}} \log \tau_N = 0$. Since the relative normalization of τ_n's is not fixed, we are free to choose $\tau_0 = 1$. Then, $\tau_0(t) = 1$, $\tau_N(t) = \chi(t_1)\bar{\chi}(t_{-1})$, where $\chi(t_1)$ and $\bar{\chi}(t_{-1})$ are arbitrary functions. 2DTL with boundary conditions (5) was considered in [13]. The solution to (26) in this case is given by [14]:

$$\tau_n(t) = \det \partial_{t_1}^{i-1}(-\partial_{t_{-1}})^{j-1}\tau_1(t) \tag{64}$$

with

$$\tau_1(t) = \sum_{k=1}^{N} a^{(k)}(t)\bar{a}^{(k)}(t_{-1}) \tag{65}$$

where functions $a^{(k)}(t)$ and $\bar{a}^{(k)}(t_{-1})$ satisfy

$$\det \partial_{t_1}^{i-1}a^{(k)}(t) = \chi(t), \qquad \det(-\partial_{t_{-1}})^{i-1}\bar{a}^{(k)}(t_{-1}) = \bar{\chi}(t_{-1})$$

This result coincides with that obtained by substituting into (55) the kernel $A(z,w)$ of the form (63).

Lax representation. In all our previous considerations, we dealt with infinite-dimensional matrices. Let us note that the Toda molecule can be effectively treated in terms of $N \times N$ matrices like the forced case could be described by the quarter-infinite matrices. This allows one to deal with the *both* identities (53) since all the products of *finite* matrices are well-defined.

To see this, one can just look at the recurrent relation (43) and observe that there exists the finite-dimensional subsystem of (N) polynomials which is decoupled from the whole system. The recurrent relation for these polynomials can be considered as the finite-matrix Lax operator (which still does not depend on the spectral parameter, in contrast to (9)). Indeed, from (43) and condition (62), i.e. $h_n/h_{n-1} = 0$ as $n \geq N$ (the Toda molecule conditions in terms of S-variables read as $S_n = S_n^\star = 1$ for $n > N - 2$ or $n > 0$), one can see that

$$z\mathcal{P}_N(z) = \mathcal{P}_{N+1}(z) - \mathcal{P}_N(z), \quad z\mathcal{P}_{N+1}(z) = \mathcal{P}_{N+2}(z) - \mathcal{P}_{N+1}(z) \quad \text{etc.} \tag{66}$$

i.e. all the polynomials $\mathcal{P}_n$ with $n > N$ are trivially expressed through $\mathcal{P}_N$. Therefore, the system can be effectively described by the dynamics of only some first polynomials (i.e. has really finite number of degrees of freedom). Certainly, all the same is correct for the star-polynomials $\mathcal{P}_n^*$ although, in this case, it would be better to use the original non-singularly normalized polynomials Φ_n^*.

Now let us look at the corresponding Lax operators (49)-(50). They are getting quite trivial everywhere but in the left upper corner of the size $N \times N$ [6]. Therefore, one can restrict himself to the system of N polynomials $\mathcal{P}_n$, $n = 0, 1, \ldots, N - 1$ and the finite matrix Lax operators (of the size $N \times N$).

Now one needs only to check that this finite system still has the same evolution equations (51). It turns out to be the case only for the first $N - 1$ times. This is not so surprising, since, in the finite system with $N - 1$ degrees of freedom, only first $N - 1$ time flows are independent. Therefore, if looking at the finite matrix Lax operators, one gets the dependent higher flows. On the other hand, if one embeds this finite system into the infinite 2DTL, one observes that the higher flows can be no longer described inside this finite system. Just the finite system is often called relativistic Toda molecule.

To simplify further notations, we introduce, instead of S_n, S_n^*, the new variables $s_n \equiv (-)^{n+1} S_n$, $s_n^* \equiv (-)^{n+1} S_n^*$. Then, one can realize that the Lax operator can be constructed as the product of simpler ones $\mathcal{L} = \mathcal{L}_N \mathcal{L}_{N-1} \ldots \mathcal{L}_1$, where $\mathcal{L}_k$ is the unit matrix wherever but a 2×2-block:

$$\mathcal{L}_k \equiv \begin{pmatrix} 1 & \vdots & \\ \cdots & G_k & \cdots \\ & \vdots & 1 \end{pmatrix} \qquad G_k \equiv \begin{pmatrix} s_k & 1 \\ s_k s_k^* - 1 & s_k^* \end{pmatrix} \qquad (67)$$

Analogously, $\bar{\mathcal{L}} = \bar{\mathcal{L}}_1 \ldots \bar{\mathcal{L}}_{N-1} \bar{\mathcal{L}}_N$ with

$$\bar{\mathcal{L}}_k \equiv \begin{pmatrix} 1 & \vdots & \\ \cdots & \bar{G}_k & \cdots \\ & \vdots & 1 \end{pmatrix} \qquad \bar{G}_k \equiv \begin{pmatrix} s_k^* & -1 \\ 1 - s_k s_k^* & s_k \end{pmatrix} \qquad (68)$$

One can trivially see that $\mathcal{L}_k \bar{\mathcal{L}}_k = \bar{\mathcal{L}}_k \mathcal{L}_k = 1$, and, therefore, one obtains $\mathcal{L}\bar{\mathcal{L}} = \bar{\mathcal{L}}\mathcal{L} = 1$ (cf. (53)).

From these formulas, one obtains that $\det \mathcal{L} = \det \bar{\mathcal{L}} = 1$ which reminds once more of the $sl(N)$ algebra. More generally, the factorization property of the Lax operators opens the wide road to the group theory interpretation of the RTC molecule – see [6].

6   Discrete evolutions and limit to Toda chain

Darboux-Bäcklund transformations. Now we are going to discuss some discrete evolutions of the RTC given by the Darboux-Bäcklund transformations and their limit to the usual Toda chain. One can easily take the continuum limit of the formulas of this section to reproduce the TC as the limit of the RTC, both with the standard continuous evolutions.

The discrete evolution equations in the RTC framework were recently introduced by [3] in a little bit sophisticated way. Here we outline the simple approach based on the notion of the Darboux-Bäcklund transformation (DBT). More details will be presented in the separate publication [16].

Let discrete index i denote the successive DBT's. The spectral problem now can be written as follows:

$$\Phi_{n+1}(i|z) + a_n(i)\Phi_n(i|z) = z\ \Phi_n(i|z) + b_n(i)\Phi_{n-1}(i|z) \qquad (69)$$

Let us define the first forward DBT (treating it as a discrete analog of (3)):

$$\Phi_n(i+1|z) = \Phi_n(i|z) + \alpha_n^{(1)}(i)\Phi_{n-1}(i|z) \qquad (70)$$

where $\alpha_n^{(1)}(i)$ are some unknown functions. One requires that $\Phi_n(i+1)$ satisfies the same spectral problem as (69) but with the shifted value of $i \rightarrow i-1$. Then the compatibility condition gives the equations of the discrete RTC:

$$a_n(i+1) = a_{n-1}(i)\ \frac{a_n(i) - \alpha_{n+1}^{(1)}(i)}{a_{n-1}(i) - \alpha_n^{(1)}(i)};$$

$$b_n(i+1) = b_{n-1}(i)\ \frac{b_n(i) - \alpha_n^{(1)}(i)}{b_{n-1}(i) - \alpha_{n-1}^{(1)}(i)} \qquad (71)$$

$$z_i\frac{b_n(i)}{\alpha_n^{(1)}(i)} = z_i - a_n(i) + \alpha_{n+1}^{(1)}(i) \qquad (72)$$

where z_i are arbitrary constants.

One can also introduce the discrete analog of (6):

$$\Phi_n(i+1|z) = (1 - \alpha_n^{(2)}(i))\Phi_n(i|z) + z\alpha_n^{(2)}(i)\Phi_{n-1}(i|z) \qquad (73)$$

where $\alpha_n^{(2)}(i)$ are some new unknown functions of the corresponding discrete indices. We refer to this evolution as to the second forward DBT. Substitution of (73) to (69) gives quite different system of the discrete evolution equations:

$$a_n(i+1) = a_n(i)\frac{1 - \alpha_{n+1}^{(2)}(i)}{1 - \alpha_n^{(2)}(i)}; \quad b_n(i+1) = b_{n-1}(i)\frac{\alpha_n^{(2)}(i)}{\alpha_{n-1}^{(2)}(i)} \qquad (74)$$

$$a_n(i) + b_n(i) \frac{1 - \alpha_n^{(2)}(i)}{\alpha_n^{(2)}(i)} = z_i \frac{1}{1 - \alpha_{n+1}^{(2)}(i)} \tag{75}$$

This system is the discrete counterpart of the continuum system (7).

Actually, in [3], four different discrete systems of the RTC equations were written. From our point of view, the additional evolutional systems result from the *backward Darboux-Bäcklund transformations* which are complimentary to those described above [6].

Continuum limit. Introducing some discrete shift of time $\Delta > 0$, one can rewrite all the equations describing the first forward DBT as follows:

$$a_n(t + \Delta) = a_{n-1}(t) \frac{a_n(t) - \alpha_{n+1}^{(1)}(t)}{a_{n-1}(t) - \alpha_n^{(1)}(t)} \tag{76}$$

$$b_n(t + \Delta) = b_{n-1}(t) \frac{b_n(t) - \alpha_n^{(1)}(t)}{b_{n-1}(t) - \alpha_{n-1}^{(1)}(t)} \tag{77}$$

After the rescaling $z_i \to g\,\Delta$, one gets from (72) $\alpha_n^{(1)} \sim -g\Delta b_n/a_n$ and, in the continuum limit, $\Delta \to 0$ the last two equations lead directly to (4).

The analogous equations can be written for the second forward DBT but with z_i rescaled as $z_i \to \frac{1}{g\,\Delta}$. It is clear that, in the limit $\Delta \to 0$, one reproduces the continuum equations (7).

Limit to Toda chain. Now let us make the following expansion (compare with (5))

$$a_n(i) \simeq 1 - \epsilon p_n(i) \;\; ; \;\; b_n(i) \simeq -\epsilon^2 R_n(i)$$
$$z \simeq 1 + \epsilon\lambda \;\; ; \;\; z_i \simeq 1 + \epsilon\lambda_i \tag{78}$$

Introduce also functions $\Psi_n(i)$

$$\Phi_n(i) \simeq \epsilon^n \Psi_n(i) \tag{79}$$

It is easy to see that (70) leads to the forward DBT for the non-relativistic Toda chain if one identifies

$$\alpha_n^{(1)}(i) \simeq \epsilon A_n(i) \tag{80}$$

Indeed, in the limit $\epsilon \to 0$, one gets the standard Toda spectral and evolution equations.

There also exist some other interesting limits [6] leading to the modified discrete Toda equations [3].

7   Concluding remarks

From the point of view of studying the RTC hierarchy itself, the most promising representation is that describing the relativistic Toda hierarchy as a particular reduction of the two-dimensional Toda lattice hierarchy. However, even this quite large enveloping hierarchy is still insufficient. Loosely speaking, the Toda lattice is too "rigid" to reproduce both the continuous and discrete flows of the RTC. Therefore, one should embed the RTC into a more general system which admits more natural reductions.

This is done in the forthcoming publication [16], where we show that the RTC has a nice interpretation if considering it as a simple reduction of the two-component KP (Toda) hierarchy. It turns out that, in the framework of the 2-component hierarchy, the continuous AKNS system, Toda chain hierarchy *and* the discrete AKNS (which is equivalent to the RTC how we proved in this paper) can be treated on equal footing.

We acknowledge V. Fock, A. Marshakov and A. Zabrodin for the discussions. A.Z. is also grateful to V.Spiridonov, S.Suslov and L.Vinet for stimulating discussions. The work of S.K. is partially supported by grants RFFI-96-02-19085, INTAS-93-1038 and by Volkswagen Stiftung, that of A.M. – by grants RFFI-96-02-16210(a), INTAS-96-2058 and Volkswagen Stiftung.

References

1. S.N. Ruijsenaars, *Comm. Math. Phys.*, **133** (1990) 217-247

2. M. Bruschi and O. Ragnisco, **A129** (1988) 21-25; *Phys. Lett.*, **A134** (1989) 365-370; *Inverse Problems*, **5** (1989) 389-405

3. Yu. Suris, *Phys. Lett.*, **A145** (1990) 113-119; *Phys. Lett.*, **A156** (1991) 467-474; *Phys. Lett.*, **A180** (1993) 419-433; **solv-int/9510007**

4. Y. Ohta, K. Kajiwara, J. Matsukidaira, J. Satsuma, **solv-int/9304002**

5. K. Ueno and K. Takasaki, *Adv.Stud. in Pure Math.*, **4** (1984) 1-95

6. S. Kharchev, A. Mironov and A. Zhedanov, **hep-th/9606144**

7. L. Faddeev and L. Takhtadjan, *Hamiltonian methods in the theory of solitons*, Springer, Berlin, 1987

8. V. Periwal and D.i Shevits, *Phys. Rev. Lett.*, **64** (1990) 1326-1335 M.J. Bowick, A. Morozov and D. Shevits, *Nucl. Phys.*, **B354** (1991) 496-530

9. M.J. Ablowitz and J.E. Ladik, *J. Math. Phys.*, **17** (1976) 1011-1018

10. I. Merola, O. Ragnisco and Tu Gui-Zhang, **solv-int/9401005**
    A. Kundu and O. Ragnisco, **hep-th/9401066**

11. S. Kharchev and A. Mironov, *Int. J. Mod. Phys.*, **A17** (1992) 4803-4824

12. S. Kharchev, A. Marshakov, A. Mironov and A. Morozov, *Nucl. Phys.*,
    B397 (1993) 339-378; **hep-th/9203043**

13. R. Hirota, *J. Phys. Soc. Japan,* **57** (1987) 4285-4288
    R. Hirota, Y. Ohta and J. Satsuma, *Progr. Theor. Phys. Suppl.*, **94**
    (1988) 59-72

14. A. Leznov and M. Saveliev, *Physica,* **3D** (1981) 62

15. M. Olshanetsky and A. Perelomov, *Phys. Rep.*, **71** (1981) 71-313

16. S. Kharchev, A. Mironov and A. Zhedanov, *Discrete and Continuous
    Relativistic Toda in the Framework of 2-component Toda hierarchy,* to
    appear

Chapter 2

Integrable Mappings

Integrable Symplectic Maps

A.P. Fordy

Department of Applied Mathematical Studies and
Centre for Nonlinear Studies,
University of Leeds, Leeds LS2 9JT, UK
Email: allan@amsta.leeds.ac.uk

Abstract

We consider the construction of canonical transformations which preserve some given Hamiltonian function and thus the corresponding Hamiltonian flow. The methods employed depend upon whether or not we already know a Lax representation for the continuous system. The paper thus falls naturally into two parts:

1. In the case that we know a Lax representation for the continuous system we use a special similarity transformation. Our examples are finite reductions of integrable PDEs with the canonical transformation being a finite reduction of Darboux–Bäcklund transformation.

2. In the case that we are just given a Hamiltonian function, without any knowledge of other structures, then we use a novel form of the Hamilton-Jacobi equation to directly construct a canonical transformation. Examples include (a special case of) the McMillan map and Bäcklund transformations for some of the Painlevé equations.

1 Introduction

An important ingredient in the theory of *continuous* Hamiltonian dynamics is the *canonical transformation*. The interpretation is that we change to a new set of canonical co-ordinates, with the coefficients of the symplectic form being unchanged. An alternative interpretation is to *actively* change points in our symplectic manifold M^{2n}, $(q_i, p_i) \mapsto \bar{q}_i(\mathbf{q}, \mathbf{p}), \bar{p}_i(\mathbf{q}, \mathbf{p}))$, in such a way that

$$\sum_{i=1}^{n} d\bar{q}_i \wedge d\bar{p}_i = \sum_{i=1}^{n} dq_i \wedge dp_i.$$

In such a case we call this a symplectic map $\varphi : M^{2n} \to M^{2n}$, which can then be iterated. A given function $f : M^{2n} \to \mathcal{R}$ on phase space will change value under such a map and this defines a new function $\bar{f} : M^{2n} \to \mathcal{R}$, given by:

$$\bar{f}(\mathbf{q}, \mathbf{p}) = f(\bar{q}(\mathbf{q}, \mathbf{p}), \bar{p}(\mathbf{q}, \mathbf{p})).$$

If $\bar{f}(\mathbf{q}, \mathbf{p}) = f(\mathbf{q}, \mathbf{p})$, meaning that the functional form of f is unchanged by the transformation, then f is said to be an *invariant* of the symplectic map, analogous to a *first integral* in the continuous case.

In [9] Veselov proved a discrete version of the Arnol'd-Liouville theorem, which led him to the definition:

Definition 1.1 (Integrable Symplectic Map) *A symplectic map*

$$\varphi : M^{2n} \to M^{2n}$$

is said to be integrable *if it has n independent integrals in involution.*

In this paper we consider such integrable symplectic maps. Thus each of our maps is associated with the *continuous* Hamiltonian flows, generated by the map's integrals, but is not itself a discretisation of any of these, playing more the role of a Bäcklund transformation. This should be contrasted with the usual application of canonical transformations to solving the (nonlinear) equations of motion by linearisation.

We consider two different situations, depending upon whether or not we know a Lax representation for our continuous system. When we do, then we can use this machinery as a basis for constructing the corresponding Bäcklund transformation, which is canonical in these circumstances. When we have no knowledge of a Lax representation (or, for purposes of illustration, choose to ignore this information) we can sometimes directly construct the canonical transformation from a novel form of the Hamilton-Jacobi equation. In this paper we just present these ideas in the context of specific examples, leaving a more complete development for a later publication.

2   The Darboux Transformation

We now consider the classical Darboux transformation $L \mapsto \tilde{L}$, with:

$$
\begin{align}
L &= -\partial^2 + u = -(\partial + f)(\partial - f) \tag{1} \\
\tilde{L} &= -\partial^2 + v = -(\partial - f)(\partial + f) = -(\partial + g)(\partial - g), \tag{2}
\end{align}
$$

where $\partial = \frac{d}{dx}$. The KdV variable v is related to two *different* MKdV variables f and g, giving a differential relation between f and g:

$$f_x + g_x = f^2 - g^2, \tag{3}$$

together with its differential prolongations (infinite in number):

$$
\begin{aligned}
f_{xx} + g_{xx} &= 2ff_x - 2gg_x, \\
f_{xxx} + g_{xxx} &= 2(ff_{xx} + f_x^2) - 2(gg_{xx} + g_x^2),
\end{aligned}
\qquad (4)
$$
$$
\vdots \qquad\qquad \vdots
$$

Now recall that the isospectral flows of (1) and (2) are the members of the KdV hierarchy for u and v and the MKdV hierarchy for f and g. The stationary and similarity reductions of these PDEs give rise to ODEs of the form:

$$
f^{(n)} = \text{lower order terms,}
$$

which can be used to close the chain of relations (4) to a set depending only on a *finite* set of derivatives $(f, g, f_x, g_x, \cdots, f^{(n-1)}, g^{(n-1)})$.

We illustrate this with a number of examples. A more detailed discussion of the stationary case can be found in [5].

Example 2.1 (Stationary MKdV Equation) Consider the stationary reduction ($f_t = 0$) of the MKdV equation, which is Lagrangian,

$$
\mathcal{L}_c = \tfrac{1}{2}\left(f_x^2 + f^4 \right) + cf, \qquad f_{xx} = 2f^3 + c. \qquad (5)
$$

The Darboux transformation and its first prolongation is:

$$
f_x + g_x = f^2 - g^2, \qquad f_{xx} + g_{xx} = 2ff_x - 2gg_x.
$$

Using equation (5) and defining the canonical variables $q = f$, $p = f_x$, $Q = g$, $P = g_x$, the resulting equations can be *explicitly* solved for (Q_i, P_i) in terms of (q_i, p_i):

$$
Q = -q + \frac{c}{p - q^2}, \qquad P = -p + \frac{2cq}{p - q^2} - \frac{c^2}{(p - q^2)^2}, \qquad (6)
$$

which, when $c = 0$, reduces to minus the identity. This mapping has the integral:

$$
I_c(q, p) = \tfrac{1}{2}(p^2 - q^4) - cq.
$$

Remark 2.1 *The stationary MKdV equation (5) can be written as a bi-Hamiltonian system [2]. For this we need to consider c as a dynamical variable, which plays the role of the Casimir of a degenerate canonical bracket. The equation takes the form:*

$$
\mathbf{q}_x = J_1 \nabla h_0 = J_0 \nabla h_1,
$$

where $\mathbf{q} = (q, p, c)^T$, $h_0 = -\frac{1}{2}c^2$, $h_1 = \frac{1}{2}I_c$, and:

$$
J_0 = \begin{pmatrix} 0 & 1 & 0 \\ -1 & 0 & 0 \\ 0 & 0 & 0 \end{pmatrix}, \quad J_1 = \frac{1}{c}\begin{pmatrix} 0 & -q & -p \\ q & 0 & -2q^3 - c \\ p & 2q^3 + c & 0 \end{pmatrix}.
$$

Our map is not just canonical, preserving J_0, but also preserves the second Poisson matrix J_1, together with the hierarchy of commuting integrals (in this case only two).

Example 2.2 (Stationary 5th. Order MKdV Equation) We consider the stationary manifold for the flow corresponding to $H_2 + cH_{-1}$, giving the Lagrangian:

$$
\mathcal{L}_c = \tfrac{1}{2}f_{xx}^2 + 5f^2 f_x^2 + f^6 + cf,
$$

with flow:

$$
f_{xxxx} - 10f^2 f_{xx} - 10ff_x^2 + 6f^5 + c = 0. \tag{7}
$$

The Lagrangian $\mathcal{L}_c$ gives rise to canonical coordinates:

$$
q_1 = f, \quad q_2 = f_x, \quad p_2 = f_{xx}, \quad p_1 = 10f^2 f_x - f_{xxx}.
$$

and the Hamiltonian:

$$
h_1 = \tfrac{1}{2}p_2^2 + q_2 p_1 - 5q_1^2 q_2^2 - q_1^6 - cq_1,
$$

which, with the canonical bracket, generates (7). The Darboux transformation (3) takes the form:

$$
q_2 + Q_2 = q_1^2 - Q_1^2.
$$

We use this and its differential consequences, together with (7) (to write f_{xxxx} and g_{xxxx} in terms of canonical coordinates) and the resulting equations can be solved *explicitly* for Q_i, P_i in terms of q_i, p_i:

$$
\begin{aligned}
Q_1 &= -q_1 + \frac{c}{d}, \\
Q_2 &= -q_2 + \frac{2cq_1}{d} - \frac{c^2}{d^2}, \\
P_2 &= -p_2 + \frac{2c}{d}\left(2q_1^2 + q_2\right) - \frac{6c^2 q_1}{d^2} + \frac{2c^3}{d^3}, \\
P_1 &= -p_1 - \frac{2c}{d}\left(p_2 - 6q_1^3 - 4q_1 q_2\right) \\
&\quad - \frac{2c^2}{d^2}\left(11q_1^2 + q_2\right) + \frac{16c^3 q_1}{d^3} - \frac{4c^4}{d^4},
\end{aligned} \tag{8}
$$

where $d = p_1 + 2p_2 q_1 - 3q_1^4 - 4q_1^2 q_2 - q_2^2$.

Remark 2.2 *Once again, the mapping is not just canonical, but also preserves the second Poisson bracket of the stationary equation (7), together with all the commuting integrals.*

Example 2.3 (Bäcklund transformations for PII) The similarity solutions of the MKdV equation satisfy:

$$f_t = f_{xxx} - 6f^2 f_x \quad \text{where} \quad f(x,t) = \frac{y(z)}{t^{1/3}}, \quad z = \frac{x}{t^{1/3}}.$$

The function $y(z)$ satisfies PII:

$$y'' = 2y^3 - \tfrac{1}{3}zy + \gamma, \quad \gamma \quad \text{an arbitrary constant.} \tag{9}$$

The relation (3) and its first prolongation are just the same in the z coordinate:

$$y_i' + y_{i+1}' = y_i^2 - y_{i+1}^2, \quad y_i'' + y_{i+1}'' = 2y_i y_i' - 2y_{i+1}y_{i+1}', \tag{10}$$

where we have used y_i and y_{i+1} to denote the PII variables derived respectively from f and g. With y_i satifying (9) with $\gamma = \gamma_i$, we find:

$$2(y_i^3 + y_{i+1}^3) - \tfrac{1}{3}z(y_i + y_{i+1}) + \gamma_i + \gamma_{i+1} = 2y_i y_i' - 2y_{i+1}y_{i+1}'.$$

Using (10) to replace y_{i+1}', we get:

$$(y_i + y_{i+1})(2y_i' - 2y_i^2 + \tfrac{1}{3}z) = \gamma_i + \gamma_{i+1}. \tag{11}$$

The values of γ_i and γ_{i+1} are not independent. Differentiating (11), we find:

$$(y_i + y_{i+1})(2y_i' - 2y_i^2 + \tfrac{1}{3}z) = 2\gamma_i + \tfrac{1}{3}.$$

To be consistent, we have:

$$\gamma_{i+1} = \gamma_i + \tfrac{1}{3}. \tag{12}$$

Thus we derive the usual Bäcklund transformation [1], which takes the form of an explicit map from y_i to y_{i+1}, given by (11) with the condition (12). Taken in conjunction with the first of (10) this can be *explicitly* inverted to give:

$$(y_i + y_{i+1})(-2y_{i+1}' - 2y_{i+1}^2 + \tfrac{1}{3}z) = \gamma_i + \gamma_{i+1}.$$

Remark 2.3 *This calculation can easily be generalised to the similarity reductions of the higher order members of the MKdV hierarchy. Indeed, this (and a great deal more) has been done by Hone in [6, 7]. The similarity reductions of the Sawada–Kotera, fifth order KdV and Kaup–Kupershmidt equations lead to non-autonomous versions of the integrable cases of the Hénon–Heiles equations. For these, he has constructed Bäcklund transformations and rational solutions.*

2.1   Gauge Transformation

The transformation $L \mapsto \tilde{L}$, given by (2), can be represented as a gauge transformation between the corresponding 'modified' spectral problems. If we define $U(g)$ and $T(g)$ by:

$$U(g) = \begin{pmatrix} g & 1 \\ \lambda & -g \end{pmatrix}, \quad T(g) = \begin{pmatrix} 1 & 0 \\ g & 1 \end{pmatrix},$$

then, under the gauge transformation:

$$U \mapsto TUT^{-1} + T_x T^{-1},$$

we have:

$$U(g) \mapsto \begin{pmatrix} 0 & 1 \\ \lambda + v & 0 \end{pmatrix}, \quad \text{where} \quad v = g_x + g^2.$$

Thus, combining the two transformations:

$$U(-f) \mapsto \begin{pmatrix} 0 & 1 \\ \lambda + v & 0 \end{pmatrix} \mapsto U(g),$$

we have:

$$T(-f - g) : U(-f) \mapsto U(g),$$

where:

$$-f_x + f^2 = g_x + g^2.$$

This gauge transformation reduces to the identity when $f + g = 0$.

Any isospectral flow of $U \equiv U(g)$ satisfies a zero curvature condition:

$$U_t - V_x + [U, V] = 0,$$

which is invariant under the gauge transformation if V transforms as:

$$V \mapsto TVT^{-1} + T_t T^{-1}.$$

For stationary flows, this reduces to a similarity transformation.

For the stationary 5th order MKdV equation we use (7) to eliminate fourth order derivatives to obtain V (in canonical co-ordinates):

$$V^q = \begin{pmatrix} -16q_1\lambda^2 + 4\lambda(2q_1^3 - p_2) + c & 16\lambda^2 + 8\lambda(q_2 - q_1^2) - 2d_+(q) \\ 16\lambda^3 - 8\lambda^2(q_2 + q_1^2) + 2\lambda d_-(q) & 16q_1\lambda^2 + 4\lambda(p_2 - 2q_1^3) - c \end{pmatrix},$$

$$V^Q = \begin{pmatrix} 16Q_1\lambda^2 - 4\lambda(2Q_1^3 - P_2) - c & 16\lambda^2 - 8\lambda(Q_2 + Q_1^2) + 2d_-(Q) \\ 16\lambda^3 + 8\lambda^2(Q_2 - Q_1^2) - 2\lambda d_+(Q) & -16Q_1\lambda^2 - 4\lambda(P_2 - 2Q_1^3) + c \end{pmatrix},$$

where $d_\pm(q) = p_1 \pm 2q_1 p_2 \mp 3q_1^4 - 4q_1^2 q_2 \mp q_2^2$.

Under the trivial gauge transformation $f \mapsto -g$, $c \mapsto -c$, we have: $V^q \mapsto V^Q$. The similarity transformation:

$$V^Q = TV^qT^{-1}, \quad T = \begin{pmatrix} 1 & 0 \\ -q_1 - Q_1 & 1 \end{pmatrix}$$

leads to the set of equations:

$$
\begin{aligned}
(q_1 + Q_1)d_+(q) &= c, \\
q_2 + Q_2 &= q_1^2 - Q_1^2, \\
p_2 + P_2 &= 2(q_1 + Q_1)(q_2 + Q_1^2 - q_1Q_1), \\
d_+(q) &= -d_-(Q),
\end{aligned}
$$

which can be solved to give (8) with $d = d_+(q)$. Since we have $d_+(q) = -d_-(Q)$, these formulae can be *explicitly* inverted.

The first integrals of (7) are coefficients in:

$$det\, V^q = -256\lambda^5 - 32\lambda^2 h_1 + 8\lambda h_2 - c^2,$$

where

$$
\begin{aligned}
h_1 &= \tfrac{1}{2}p_2^2 + q_2p_1 - 5q_1^2q_2^2 - q_1^6 - cq_1, \\
h_2 &= \tfrac{1}{2}p_1^2 - 2q_1^2p_2^2 + 6q_1^5p_2 - \frac{9}{2}q_1^8 - 4q_1^2q_2p_1 + 2q_1q_2^2p_2 + 5q_1^4q_2^2 \\
&\quad - \tfrac{1}{2}q_2^4 + c(p_2 - 2q_1^3) \\
&= \tfrac{1}{2}d_+(q)d_-(q) + c(p_2 - 2q_1^3).
\end{aligned}
$$

These Poisson commute (with respect to *both* Poisson brackets) and are invariant under the similarity transformation. Thus the symplectic map (8) is integrable.

Remark 2.4 *It is interesting that the denominator $d = d_+(q)$ is a factor of the integral h_2 when $c = 0$.*

3   The Hamilton-Jacobi Equation

In this section we use a novel form of the Hamilton-Jacobi equation to construct a canonical transformation which preserves a given Hamiltonian function.

In the standard theory of canonical transformations (generally time dependant) between (q_i, p_i) and (Q_i, P_i), we have a generating function $S(q_i, \cdots, q_n, Q_i, \cdots, Q_n, t)$ satisfying:

$$\sum_{i+i}^{n} P_i dq_i - H dt = \sum_{i=1}^{n} P_i dQ_i - K dt + dS, \tag{13}$$

implying:

$$p_i = \frac{\partial S}{\partial q_i}, \quad P_i = -\frac{\partial S}{\partial Q_i}, \tag{14}$$

$$K = H + \frac{\partial S}{\partial t}, \tag{15}$$

where $H(q_1, \cdots, q_n, p_1, \cdots, p_n, t)$ and $K(Q_1, \cdots, Q_n, P_1, \cdots, P_n, t)$ are the corresponding Hamiltonian functions. We can, of course, similarly define generating functions depending upon other choices of variable (such as (q_i, P_i, t)) by manipulating the differentials in (13) [3].

In the context of completely integrable Hamiltonian systems, it is usual to choose K to be a simple function of just P_i, so that these momenta are constants of motion, while Q_i are just linear in t. The Hamilton-Jacobi equation (15) is then a first order, nonlinear PDE for $S(q_i, P_i, t)$ as a function of (q_i, t) with P_i entering as parameters. The outcome is to solve the *equations of motion* for $q_i(t)$ as functions of t, up to a quadrature coming from the canonical transformations (14).

However, our purpose is different. Rather than solve the equation of motion, we construct a Bäcklund transformation for Hamilton's equations, in the form of a canonical transformation. Instead of specifying some simple form for K, we require that $K(Q_i, P_i)$ has the *same* functional form as $H(q_i, p_i)$. In the autonomous case these have *exactly* the same functional form, but to include non-autonomous systems in our framework we must allow our Hamiltonian to depend upon a (possibly some) parameter(s) α. In this case the Hamiltonian is denoted $H_\alpha(q_i, p_i, t)$. We then require the *transformed* Hamiltonian to be of the *same* functional form, but with parameter $\bar\alpha$ (generally different). Thus, the Hamilton-Jacobi equation (15) becomes:

$$H_{\bar\alpha}\left(Q_i, -\frac{\partial S}{\partial Q_i}, t\right) = H_\alpha\left(q_i, \frac{\partial S}{\partial q_i}, t\right) + \frac{\partial S}{\partial t}, \tag{16}$$

where we have written

$$K_{\bar\alpha}\left(Q_i, -\frac{\partial S}{\partial Q_i}, t\right) \equiv H_{\bar\alpha}\left(Q_i, -\frac{\partial S}{\partial Q_i}, t\right).$$

In this paper we just present a number of examples. Although at present the method is very adhoc, it appears to be very effective.

Example 3.1 (Stationary MKdV Equation) Here we reproduce the transformation (6) obtained via the Darboux transformation. We have:

$$H_b(q, p) = p^2 - q^4 - 2bq,$$

so (16) gives:
$$S_Q^2 - Q^4 - 2\bar{b}Q = S_q^2 - q^4 - 2bq,$$

since $S_t = 0$ for an autonomous system. To remove the quartic terms we define $F(q, Q)$ by:
$$S(q, Q) = \tfrac{1}{3}(q^3 + Q^3) + F(q, Q),$$

where $F(q, Q)$ satisfies:
$$F_Q^2 - F_q^2 + 2Q^2 F_Q - 2q^2 F_q = 2(\bar{b}Q - bq).$$

Any nontrivial solution of this equation will give us a generating function. We can choose:
$$F_Q = F_q \quad \text{and} \quad \bar{b} = b,$$

to obtain
$$F_Q = \frac{b}{q + Q} = F_q \quad \Rightarrow \quad F(q, Q) = b\ln(q + Q).$$

Thus, we have:
$$S(q, Q) = \tfrac{1}{3}(q^3 + Q^3) + b\ln(q + Q),$$

leading to the transformation (6).

Remark 3.1 *In this example we found that the parameters b and $\bar{b}$ were equal, as seems to be always the case for autonomous systems.*

Example 3.2 (The McMillan Map) Here we start with the Hamiltonian function:
$$H_\alpha(q, p) = (1 + q^2)p^2 + q^2 + 2aqp$$

and *derive* the McMillan map as the canonical transformation which preserves this function. The Hamilton-Jacobi equation (16) (with $\bar{\alpha} = \alpha$) implies:
$$Q^2 S_Q^2 - q^2 S_q^2 + S_Q^2 - q^2 + Q^2 - S_q^2 = 2\alpha(QS_Q + qS_q).$$

When $\alpha = 0$ this equation has the symmetry $(q, p) \mapsto (-P, Q)$, with generating function $S(q, Q) = qQ$. We therefore try a first order perturbation in α:
$$S(q, Q) = qQ + \alpha F(q, Q), \tag{17}$$

leading to two equations for F (the coefficients of α and α^2):
$$q(1 + q^2)F_Q - Q(1 + q^2)F_q = 2qQ,$$
$$(1 + Q^2)F_Q^2 - (1 + q^2)F_q^2 = 2(QF_Q + qF_q).$$

Once again, *any* solution of these equations will suffice. Choosing $F_Q = 0$ leads to:

$$F_q = -\frac{2q}{1+q^2} \quad \text{so} \quad F(q,Q) = -\ln(1+q^2).$$

With this solution for F, the generating function (17) gives:

$$p = Q - \frac{2\alpha q}{1+q^2}, \quad P = -q,$$

so, labelling (q,p) and (Q,P) respectively with indices n and $n+1$, we have the McMillan map:

$$q_{n+1} + q_{n-1} = \frac{2\alpha q_n}{1+q_n^2}.$$

Remark 3.2 *A generating function of the form:*

$$S(q,P) = qP - \alpha \ln\left((1+q^2)(1+P^2)\right)$$

can also be constructed, leading to the map:

$$q_{n+1} = q_n - \frac{2\alpha p_{n+1}}{1+p_{n+1}^2}, \quad p_n = p_{n+1} - \frac{2\alpha q_n}{1+q_n^2}.$$

Example 3.3 (Painlevé II) Hamiltonian formulations were written down for all the Painlevé equations by Okamoto [8]. However, since PII is just a non-autonomous version of the MKdV equation, we prefer to use the Hamiltonian:

$$H_\alpha(q,p,t) = \tfrac{1}{2}(p^2 - q^4 - tq^2) - \alpha q,$$

so that (16), but now with $S_t \neq 0$, gives:

$$\tfrac{1}{2}(S_Q^2 - Q^4 - tQ^2) - \bar{\alpha}Q = \tfrac{1}{2}(S_q^2 - q^4 - tq^2) - \alpha q + S_t.$$

As before, we define $F(q,Q)$ by:

$$S(q,Q,t) = \tfrac{1}{3}(q^3 + Q^3) + \tfrac{1}{2}t(q+Q) + F(q,Q),$$

where $F(q,Q)$ satisfies:

$$(2Q^2 + t)F_Q + F_Q^2 = (2q^2 + t)F_q + F_q^2 + (2\bar{\alpha} + \tfrac{1}{2})Q + (\tfrac{1}{2} - 2\alpha)q.$$

Since $F(q,Q)$ does not explicitly depend upon t, we must have:

$$F_Q = F_q = \frac{(4\bar{\alpha} - 1)Q + (1 - 4\alpha)q}{4(Q^2 - q^2)}.$$

Since $F(q, Q)$ must be a function of the single variable $q + Q$, we deduce:

$$\bar{\alpha} = \alpha - \tfrac{1}{2} \quad \text{and} \quad F_Q = F_q = \frac{\alpha + \bar{\alpha}}{2(Q + q)},$$

giving the generating function:

$$S(q, Q, t) = \tfrac{1}{3}(q^3 + Q^3) + \tfrac{1}{2}t(q + Q) + \tfrac{1}{2}(\alpha + \bar{\alpha})\ln(q + Q), \quad \bar{\alpha} = \alpha - \tfrac{1}{2}.$$

This gives us the *explicitly invertible* map:

$$Q = -q + \frac{\alpha - \tfrac{1}{4}}{p - q^2 - \tfrac{1}{2}t}, \quad q = -Q - \frac{\bar{\alpha} + \tfrac{1}{4}}{P + Q^2 + \tfrac{1}{2}t},$$

where $\bar{\alpha} = \alpha - \tfrac{1}{2}$. This is the same as (11).

Example 3.4 (Painlevé IV) The equation PIV can be generated by the Hamiltonian function:

$$H_{(\alpha,\beta)} = \tfrac{1}{2}q\left(p^2 - (q + 2t)^2\right) + 2\alpha q + \frac{\beta}{q},$$

so that (16), with $S_t \neq 0$, gives

$$\tfrac{1}{2}\left(S_Q^2 - (Q + 2t)^2\right) + 2\bar{\alpha}Q + \frac{\bar{\beta}}{Q} = \tfrac{1}{2}\left(S_q^2 - (q + 2t)^2\right) + 2\alpha q + \frac{\beta}{q} + S_t.$$

We can write S as:

$$S(q, Q, t) = \tfrac{1}{2}(q^2 + Q^2) + 2t(q + Q) + F(q, Q), \tag{18}$$

where $F(q, Q)$ satisfies:

$$\tfrac{1}{2}QF_Q(F_Q + 2Q + 4t) + 2\bar{\alpha}Q + \frac{\bar{\beta}}{Q} = \tfrac{1}{2}qF_q(F_q + 2q + 4t) + 2\alpha q + \frac{\beta}{q} + 2(q + Q).$$

Since $F(q, Q)$ does not explicitly depend upon t, we must have:

$$QF_Q - qF_q = 0 \quad \text{so that} \quad F(q, Q) = G(\xi) \quad \text{where} \quad \xi = qQ.$$

The function $G(\xi)$ must satisfy:

$$\tfrac{1}{2}(q - Q)(\xi G_\xi^2 - 2\xi G_\xi) + 2(\bar{\alpha} - 1)Q - 2(\alpha + 1)q + \frac{\bar{\beta}}{Q} - \frac{\beta}{q} = 0.$$

For this to be an equation whose coefficients depend *only* upon ξ, we must have $\bar{\alpha} = \alpha + 2$ and $\bar{\beta} = \beta$, so that

$$(G_\xi - 1)^2 = 1 + \frac{4(\alpha + 1)}{\xi} - \frac{2\beta}{\xi^2} \tag{19}$$

and, from (18),

$$S(q, Q, t) = \tfrac{1}{2}(q^2 + Q^2) + 2t(q + Q) + G(qQ).$$

The formula $\dot{q}/q = p = S_q = q + 2t + QG_\xi$ implies:

$$G_\xi = \frac{\dot{q} - q^2 - 2tq}{qQ},$$

which, when taken with (19), gives the *explicitly invertible* Bäcklund transformation:

$$Q = \frac{(\dot{q} - q^2 - 2tq)^2 + 2\beta}{2q(\dot{q} - q^2 - 2tq) + 2(\alpha + 1)}, \quad \bar{\alpha} = \alpha + 2, \quad \bar{\beta} = \beta. \tag{20}$$

This is just (the inverse of) the transformation (2.23) of [4].

4   Conclusions

In this paper we have been concerned with constructing symplectic maps which preserve some given set of Poisson commuting Hamiltonian functions. The presentation fell naturally into two parts. In the presence of a Lax matrix for the continuous system, it is possible to construct the corresponding symplectic map through an appropriate similarity transformation. On the other hand, given a function on phase space, it is sometimes possible to directly solve a novel form of the Hamilton-Jacobi equation to construct this symplectic map.

In the examples given, it was necessary to make some ad hoc steps in the calculation. For one degree of freedom, the method seems to be effective. However, for more degrees of freedom, with an appropriate number of integrals in involution, the corresponding Hamilton-Jacobi equation is too difficult to solve in all but the simplest examples. The problem of removing these ad hoc steps and generalising to higher number of degrees of freedom is the subject of future research.

Acknowledgements

The first part of this paper is based on joint work with A.B. Shabat and A.P. Veselov. The second part grew out of general discussions with F.W. Nijhoff about integrable maps.

References

[1] H. Airault. Rational solutions of Painlevé equations. *Stud. Appl. Math.*, 61:31–53, 1979.

[2] M. Antonowicz, A.P. Fordy, and S. Wojciechowski. Integrable stationary flows : Miura maps and bi-Hamiltonian structures. *Phys. Letts. A*, 124:143–50, 1987.

[3] V.I. Arnol'd. *Mathematical Methods of Classical Mechanics*. Springer-Verlag, Berlin, 1978.

[4] A.P. Bassom, P.A. Clarkson, and A.C. Hicks. Bäcklund transformations and solution hierarchies for the fourth Painlevé equation. *Stud. Appl. Math.*, 95:1–71, 1995.

[5] A.P. Fordy, A.B. Shabat, and A.P. Veselov. Factorisation and Poisson correspondences. *Theor. Math. Phys.*, 105(2):225–45, 1995. Translated journal: *TMP*, 105:1369–86, Plenum, 1996.

[6] A.N.W. Hone. Non-autonomous Hénon-Heiles systems. In *Proceedings of the First Non-Orthodox School of Nonlinearity and Geometry, Warsaw.* 1995.

[7] A.N.W. Hone. *Integrable Systems and their Finite-Dimensional Reductions*. PhD thesis, University of Edinburgh, 1996.

[8] K. Okamoto. Polynomial Hamiltonians associated with Painlevé equations. I. *Proc. Japan Acad.*, 56(A):264–8, 1980.

[9] A.P. Veselov. Integrable maps. *Russ. Math.i Surveys*, 46, N5:1–51, 1991.

An iterative process on quartics and integrable symplectic maps

J.P. Francoise[†] and O. Ragnisco

† Laboratoire de Geométrie Differentielle et Appliquée
Université Pierre et Marie Curie
Place Jussieu,Paris, France
Email: jpf@ccr.jussieu.fr

‡ Dipartimento di Fisica, Università di Roma Tre
Via Vasca Navale 84, Roma, Italy
Email: ragnisco@fis.uniroma3.it

Abstract

An iterative process on closed curves is considered, which is relevant in the context of the Dirichlet problem for the Wave equation. Such iterative process is symplectic and integrable, being related to a special case of the integrable symplectic map denoted as *Discrete Garnier System*. Its explicit solution in terms of Weierstrass $\mathcal{P}$ functions is derived.

1   Introduction

In this short note, we will rederive a classical result concerning the Dirichlet Problem on a closed curve [1], [2] taking advantage of some recent findings in the area of integrable symplectic maps.

Accordingly, Section 2 is devoted to recall how, in a natural way, the Dirichlet problem on a suitable region D in the plane can be associated with an iterative process T on its boundary Γ.

In Section 3, we will introduce an integrable symplectic map that obtains as a "degenerate" (in a sense that will be clarified later) case of a discrete analog of the Garnier System [3], and will elucidate its geometric meaning.

In Section 4, an iterative process on a closed curve in the plane will be obtained as a "reduced system" from the above symplectic map, and its explicit solution in terms of elliptic functions will be derived.

2   An iterative process on closed curves

2.1   Dirichlet problem for the wave equation

Suppose that D is a region on the plane, convex in the coordinate directions, i.e. its boundary $\Gamma = \partial D$ intersects each line $x = c$, $y = c$ at not more than two points.

Given a function $f : \Gamma \to \mathbf{R}$, the Dirichlet problem for the equation

$$\frac{\partial^2 u}{\partial x \partial y} = 0$$

on D consists in finding on D a function u such that

$$\frac{\partial^2 u}{\partial x \partial y} = 0$$

is satisfied with the condition $u|_\Gamma = f$.

One may impose various requirements of smoothness on f. By the change of variables $t = x + y$, $X = x - y$, the differential equation becomes the one-dimensional wave equation

$$\frac{\partial^2 u}{\partial t^2} - \frac{\partial^2 u}{\partial X^2} = 0.$$

The Dirichlet problem is not always solvable and, if it is solvable, not always uniquely. It has been the object of a series of papers [1], [2], [4], [5].

We associate with the boundary Γ an iterative process defined by a mapping T. Suppose that T_1 is the transformation carrying the point $m \in \Gamma$ into the point $T_1 m$ of Γ with the same coordinate x and that T_2 is the transformation carrying the point $m \in \Gamma$ into the point $T_2 m$ of Γ with the same coordinate y. We define T as $T_2 \circ T_1$. We obviously have:

$$T_1^2 = T_2^2 = Id, \quad T^{-1} = T_1 \circ T_2 \tag{2.1}$$

having denoted by Id the identity map.

Theorem 1 (F.John [1]) *Let Γ be such that for some point $m_0 \in \Gamma$, the orbit $T^n m_0$ is dense. Then, the Dirichlet problem for Γ cannot have more than one continuous solution.*

If Γ has bounded curvature, then it is C^2. If the rotation number μ is irrational, a theorem of Denjoy (see again [2]) ensures that the Dirichlet problem can have only one continuous solution.

2.2   Iteration on a quartic

We consider a closed curve of the plane, which is defined as:

$$Q(x, y) = a(x)y^2 + b(x)y + c(x) = \alpha(y)x^2 + \beta(y)x + \gamma(y) = 0 \tag{2.2}$$

where $a, b, c, \alpha, \beta, \gamma$ are polynomials of degree two.

Proposition 1 *The iterative process T defined on the quartic curve (2.2) is either periodic or all its orbits are dense on Q.*

PROOF: As already observed this results follows from the Denjoy's theorem. However, we shall give here an alternative proof, whose interest is related to what will follow in this paper. We observe that the iterative process T is defined by:

$$x + x' = -\frac{\beta(y)}{\alpha(y)}, \tag{2.3}$$

$$y + y' = -\frac{b(x)}{a(x)}, \tag{2.4}$$

It extends as a mapping of the plane $F = T_2 \circ T_1$:

$$T_1 : \begin{pmatrix} x \\ y \end{pmatrix} \Rightarrow \begin{pmatrix} x' = -x - \beta(y)/\alpha(y) \\ y' = y \end{pmatrix} \tag{2.5}$$

$$T_2 : \begin{pmatrix} x \\ y \end{pmatrix} \Rightarrow \begin{pmatrix} x' = x \\ y' = -y - b(x)/a(x) \end{pmatrix} \tag{2.6}$$

The mapping F is symplectic for $\omega = dx \wedge dy$. By construction, it has an invariant function $Q(x, y)$. We know [6] that there exists (at least locally) a canonical coordinate system which linearizes the mapping F. This implies the result.

3   Discrete Analog of Garnier System

Moser [7] was the first who noticed that the Neumann system [8] is intimately related to the Hill's equation with the so-called N-gap potential. Since then, a considerable number of papers have been published where integrable finite-dimensional hamiltonian systems have been derived by restricting the solution manifold of a given hierarchy of integrable evolution equations to some finite-dimensional invariant submanifold. There are a few variants of this technique, which are denoted as "restricted flows" technique [9] and "non-linearization" technique [10]. It is out of the scope of the present paper to give a detailed analysis of such approaches, and to discuss their relationships with the perhaps more familiar one relying on stationary flows, introduced in [11], and applied for the first time in a discrete context in [12]. For other interesting finite-dimensional reductions of discrete equations see [13]. By a discrete variant of the "nonlinearization technique" [3], [14], or of the "restricted flow" technique [15], discrete-time analogs of some celebrated integrable finite-dimensional systems, like Garnier and Neumann systems were derived. Other integrable discretisations of those systems have been constructed, through different approaches, in [16] and [17]. In [3] by taking N replicas of the Toda-lattice spectral problem and restricting the "potentials" to the N-gap sector, it was obtained the following symplectic process:

$$F : \begin{pmatrix} p \\ q \end{pmatrix} \to \begin{pmatrix} p' \\ q' \end{pmatrix} = \begin{pmatrix} \pm \rho q \\ \pm \rho^{-1}(\Lambda q - p - <q, q > q) \end{pmatrix} \quad (3.1)$$

where

$$\rho = (<q, \Lambda q> - <q, p> - <q, q>^2)^{1/2} \quad (3.2)$$

In formula (3.1), p and q belong to $\mathbf{R}^N$, $\Lambda = diag(\lambda_1, \cdots, \lambda_N)$ and $< \cdot, \cdot >$ denotes the usual euclidean inner product. It turns out [3] [15] that the functions:

$$K_j = \sum_{r \neq j}^{N} \frac{(p_j q_r - p_r q_j)^2}{\lambda_j - \lambda_r} + (\lambda_j - <q, q>)p_j q_j + p_j^2 \quad (3.3)$$

are invariant under F and are in involution for the standard symplectic form $\omega = \sum_{j=1}^{N} dp_j \wedge dq_j$.

Hence, when no degeneracy occurs (i.e. $\lambda_j \neq \lambda_k$ for $j \neq k$), F is completely integrable in the Arnold-Liouville sense.

The iterative process (3.1) has a simple geometrical interpretation. Indeed, it can be factorized as follows:

$$F = T_1 \circ T_2 \circ T_3, \quad (3.4)$$

where

$$T_3 : \begin{pmatrix} p \\ q \end{pmatrix} \to \begin{pmatrix} p' \\ q' \end{pmatrix} = \begin{pmatrix} q \\ p \end{pmatrix} \quad (3.5)$$

$$T_2 : \begin{pmatrix} p \\ q \end{pmatrix} \to \begin{pmatrix} p' \\ q' \end{pmatrix} = \begin{pmatrix} p \\ \Lambda p - q - <p, p> p \end{pmatrix} \quad (3.6)$$

$$T_3 : \begin{pmatrix} p \\ q \end{pmatrix} \to \begin{pmatrix} p' \\ q' \end{pmatrix} = \begin{pmatrix} p <p, q>^{1/2} \\ q <p, q>^{-1/2} \end{pmatrix}. \quad (3.7)$$

Let us introduce the family of quartic hypersurfaces:

$$Q(p, q) := - <\Lambda p, q> + <q, q><p, q> + <p, p> = C. \quad (3.8)$$

We first notice that $Q = \sum_{j=1}^{N} K_j$ (see (3.3)) so that the family (3.8) is invariant under F. Moreover, we recall that, if a given map $F : \mathbf{R}^k \to \mathbf{R}^k$ factorizes as $F = F_1 \circ F_2$, any function $Q : \mathbf{R}^k \to \mathbf{R}^k$ transforms as $Q' = F^*Q = (F_2^* \circ F_1^*)Q$. Then, the factorization (3.4) has a simple geometrical interpretation. Namely, by T_1 we get the symmetric quartic:

$$Q^*(p, q) := - <\Lambda p, q> + <q, q><p, p> + <q, q> = C \quad (3.9)$$

Then, the *shear transformation* T_2 amounts to take the next point of intersection of the hyperplane $p = const$ with the quartic $Q^* = C$. Finally, T_3 is just the symmetry with respect to the hyperplane $p = q$, and we get the image of (p, q) by F, which is a pair on $Q = C$.

From the decomposition (3.1) we can easily derive the following

Lemma 1 *Let $l_{jk} := p_j q_k - p_k q_j$. Then*

$$F^* l_{jk} = l_{jk} + (\lambda_j - \lambda_k) q_j q_k. \tag{3.10}$$

It is then natural to focus attention on the case of maximal degeneracy, when $\Lambda = \lambda I$, I denoting the $N \times N$ identity matrix. In this case, the functions K_j are no longer defined (only their sum exists and equals Q), but the l_{jk} themselves are preserved. Their Poisson brackets generate the Lie algebra $so(N)$, and so one can build up out of them $N - 1$ functionally independent invariants in involution, given for instance by:

$$I_j = \sum_{r \neq j}^{N} \frac{(p_j q_r - p_r q_j)^2}{b_j - b_r} \tag{3.11}$$

where $b_1, \cdots b_N$ are real constants such that $b_j \neq b_k$ for $j \neq k$. Moreover, as in the nondegenerate case, we have:

$$\{Q, I_j\} = 0, \quad j = 1, \cdots, N - 1 \tag{3.12}$$

Therefore the mapping remains completely integrable in the Arnold-Liouville sense.

4 The reduced system

Through the identification $T^* \mathbf{R}^N \simeq \mathbf{R}^N \times \mathbf{R}^N$, the symplectic process F (3.1) can be viewed as a mapping from a pair of vectors $(p, q) \in \mathbf{R}^N \times \mathbf{R}^N$ to a new pair $(p', q') \in \mathbf{R}^N \times \mathbf{R}^N$. In the case of maximal degeneracy, both p' and q' belong to the linear span of p and q:

$$p' = \pm \rho q \tag{4.1}$$
$$q' = \pm \rho^{-1}[(\lambda - <q, q>)q - p] \tag{4.2}$$

with $\rho = (\lambda - <q, q>) <q, q> - <p, q>$

To investigate the behaviour of (4.1), (4.2) it is convenient to look at the evolution of the three functions:

$$x := <q, p> \tag{4.3}$$
$$y := <q, q> \tag{4.4}$$
$$z := <p, p> \tag{4.5}$$

under the action of F.

We get:

$$x' = -x + \lambda y - y^2 \tag{4.6}$$
$$y' = (x')^{-1}[y(\lambda - y)^2 - 2x(\lambda - y) + z] \tag{4.7}$$
$$z' = x'y \tag{4.8}$$

By using the invariant function (see (3.8)):

$$Q(p, q) = Q(x, y, z) = -x(\lambda - y) + z = C \tag{4.9}$$

z can be eliminated in favour of x and y, yielding the two-dimensional map, *symplectic* for $dx \wedge dy$:

$$T : \begin{pmatrix} x \\ y \end{pmatrix} \to \begin{pmatrix} x' \\ y' \end{pmatrix} = \begin{pmatrix} -x + y(\lambda - y) \\ \lambda - y - \dfrac{C}{-x + y(\lambda - y)} \end{pmatrix} \tag{4.10}$$

Theorem 2 *The mapping (4.10) defines an iterative process of the type described in subsection (2.1) on the quartic:*

$$\mathcal{G} = -x^2 + x(\lambda - y)y - Cy = G \tag{4.11}$$

PROOF: The invariance under F of the functions l_{jk} (Lemma 1) entails the invariance of $\frac{1}{2}\sum_j \sum_k l_{jk}^2 = < p,p >< q,q > - < p,q >^2$. By reduction, we get that $yz - x^2$ is preserved by (4.10), and thus $\mathcal{G}$ is preserved by T. From Proposition 1 it then follows that the orbits of T are either periodic or dense on $\mathcal{G} = G$.

Actually, by exploiting the procedure outlined for instance in [6] for linearizing integrable symplectic maps, we can ever exhibit the explicit solutions for the iterated of T in terms of elliptic functions. They read:

$$x^{(n)} = \mathcal{P}(\nu) - \mathcal{P}(n\nu + a) \tag{4.12}$$
$$y^{(n)} = \frac{\lambda}{2} - \frac{\mathcal{P}'(\nu) - \mathcal{P}'(n\nu + a)}{\mathcal{P}(\nu) - \mathcal{P}(n\nu + a)} \tag{4.13}$$

In formulae (4.12),(4.13) $\mathcal{P}, \mathcal{P}'$, denote the Weierstrass elliptic function and its derivative; the invariants g_2, g_3 are given by:

$$g_2 = 4(G - \frac{\lambda}{2}C) + \frac{\lambda^4}{12} \tag{4.14}$$
$$g_3 = -\frac{2}{3}\frac{\lambda^6}{12} - \frac{1}{3}\lambda^2 \left(G - \frac{\lambda}{2} \right) - C^2 \tag{4.15}$$

and ν is defined on the fundamental parallelogram (corresponding to the periods $2\omega_1, 2\omega_2$) by: $\mathcal{P}(\nu) = \frac{\lambda^2}{12}$. Observing that $C = \mathcal{P}'(\nu)$, we immediately get that $C = 0$ implies $\nu = \frac{\omega_i}{2}$ so that T is periodic of period 2, while $G = 0$ implies $\nu = \frac{2\omega_i}{3}$, entailing periodicity of period 3.

The integration of the map F in the case $\Lambda = \lambda I$ is then reduced to the solution of a system of 2 linear first order difference equations, whose coefficients are given functions of the discrete variable n.

References

[1] F. John, "The Dirichlet problem for a hyperbolic equation" *Amer. J. Math.* **63**, 141–154 (1941).

[2] V.I. Arnold, "Small denominators I, mappings of the circumference into itself", *Amer. Math. Soc. Transl.* **46**, 213–284.

[3] O. Ragnisco, "A simple method to generate integrable symplectic maps", in: *Solitons and chaos* I.Antoniou, F.J. Lambert eds., 227–231 (Springer, 1991).

[4] D.G. Bargin, R.Duffin, "The Dirichlet problem for the vibrating string equation", *Bull. Amer. Math. Soc.* **45**, 851–859 (1939).

[5] A. Huber, "Die erste Randwertaufgabe fur geschlossene Bereiche bei der Gleichung $\frac{\partial^2 z}{\partial x \partial y} = f(x,y)$", *Monatgh. Math. Phys.* **39**, 79–100 (1939).

[6] M. Bruschi, O. Ragnisco, P. Santini, and G. Tu, " Integrable symplectic maps" *Physica D* **49**, 273–294 (1991); M. Maeda, *Acta Math. Japon.* **25** , 405–420 (1980).

[7] J. Moser, "Geometry of quadrics and spectral theory" *Chern Symposium* 1979, 147–188, Springer 1981.

[8] C. Neumann, " De problemate quodam mechanica, quod ad primam integralium ultraellipticorum classem revocatur", *J. Reine Angew. Math.* **56**, 46–69 (1859).

[9] M. Antonowicz, S. Rauch-Wojciechowski, *Phys. Lett. A* **147** , 455 (1990); M. Antonowicz, S.Rauch-Woijciechowski, *J. Phys. A: Math.Gen* **24**, 5043 (1992).

[10] Cao Ce Wen, "Classical integrable systems generated through nonlinearization of eigenvalue problems" in *Nonlinear Physics*, Gu C et al. eds., Springer Verlag, Berlin, Heidelberg 1990;
Li Yi Shen and Y Zeng, *J.Math.Phys* **30** 1679 (1989).

[11] O.I. Bogoyavlenski, S.P. Novikov, *Funct. Anal. Appl.* **13**, 6 (1976).

[12] G.R.W. Quispel, J.A. Roberts, C.J. Thompson. *Physica D* **34**, 183–192 (1990).

[13] H.W. Capel, F.W. Nijhoff, V.G. Papageorgiou, *Phys. Lett. A* **147** 106 (1990); H.W. Capel, F.W. Nijhoff, V.G. Papageorgiou, "Lattice Equations and Integrable Mappings", in *"Nonlinear Evolution Equations and Dynamical System"s*, S. Carillo and O. Ragnisco eds., 182, Springer 1990.

[14] O. Ragnisco, C. Cao, Y. Wu, "On the relation of the stationary Toda equation and the symplectic maps", *J. Phys. A: Math. and Gen.* **28**, 573–588 (1995).

[15] O. Ragnisco, "A discrete Neumann system" *Phys. Lett. A* **167**, 165–171 (1992); O. Ragnisco, S. Rauch-Wojciechowski "Integrable maps for the Garnier and for the Neumann systems", *J. Phys. A: Math. Gen* **29**, 1115–1124 (1996); O. Ragnisco, Yu.B. Suris, "On the r-matrix structure of the Neumann system and its discretizations", in *Algebraic aspects of integrable systems: in memory of Irene Dorfman* A. Fokas and I.M. Gel'fand eds, vol.**26**, Birkhauser 1996.

[16] Yu.B. Suris, "A discrete-time Garnier system" *Phys. Lett. A* **189**, 281–289 (1994); "A family of integrable standard-like maps related to symmetric spaces" *Phys. Lett. A* **192**, 9–16 (1994).

[17] A.P. Veselov " Integrable systems with discrete time and difference operators" *Funct. Anal. Appl.* **22**, 1–13 (1988).

Integrable Mappings of KdV type and Hyperelliptic Addition Theorems

F.W. Nijhoff[†] and V.Z.Enolskii[‡]
† Department of Applied Mathematical Studies,
University of Leeds, Leeds LS2 9JT, UK
‡ Institute of Magnetism of NASU
36 Vernadskii str., 252142 Kiev, Ukraine

1 Introduction

Since a few years integrable discrete-time systems have attracted considerable attention, [29]. They have appeared in various forms: partial difference equations (lattice equations), [2, 22, 17, 31, 33, 18], integrable mappings, [34, 11, 36, 28, 18, 26, 35, 37], and discrete Painlevé equations, [21]. They have intriguing new applications, for instance in discrete geometry, [9], quantum theory of conformal models, [20], and cellular automata, [1, 12, 8].

In [32, 16], cf. also [29, 30], integrable multidimensional mappings were considered as arising from finite-dimensional reductions of lattice equations. This reduction from what are partial difference analogues of the KdV equation naturally led to the establishment of the important integrability characteristics: Lax pairs, classical r-matrix structures including a Lagrangian formalism. The Liouville integrability of the resulting mappings, in the spirit of [36], was established in [16] in the sense that the discrete-time flow is the iterate of a canonical transformation, preserving a suitable symplectic structure, possessing invariants which are in involution with respect to this symplectic form. According to [36], and in complete analogy with the continuous-time situation, it follows that the discrete-time flow can thus be linearized on a hypertorus which is the intersection of the level sets of the invariants. However, the explicit parametrization of the map represents in itself a separate problem which was not pursued to the end.

In this note we address this latter problem. We concentrate on the multidimensional rational mappings studied in [32, 16], which form the multidimensional generalizations of the well-known McMillan map [27], and show that they can be parametrized (uniformized) by hyperelliptic abelian functions. This amounts to the exact integration of the discrete flow by means of the technique of finite-gap integration, [7]. In principle, by virtue of the fact that these mappings admit a Lax representation whose invariant variety is a hyperelliptic curve, it is clear that they have to be parametrized in terms of the hyperelliptic abelian functions. Nevertheless, there are two aspects of the application of the finite-gap technique that are in our view special to the discrete situation: on the one hand the problem of the identification of the discrete-time flow with an interpolating continuous-time flow, on the other hand the development of a direct procedure to integrate the map in terms of an appropriate class of Abelian functions.

It is this latter question that leads us to consider among the hyperelliptic Abelian functions the class of Kleinian functions. It is known that there are a number of different realizations of hyperelliptic functions, (see e.g. [24]). The Kleinian theory of hyperelliptic Abelian functions, [23], was widely studied in the previous century by people like Burkhardt [15], Wiltheiss [38], Bolza [10], and Baker [3], and was carefully documented in Baker's monographs [4, 5]. The modern exposition of the hyperelliptic Kleinian function theory as well new results are given in the recent papers [13, 14].

Our approach is based on the employment of special addition theorems for hyperelliptic abelian functions, which can be interpreted, analogously to the elliptic case, as multidimensional rational maps. Conversely, we expect that the consideration of wider classes of integrable maps, [30] might have an application in establishing effective addition theorems for higher-genus Abelian functions.

2   KdV mappings

The example of a concrete integrable family of mappings that exhibit the structure outlined above, is the mapping of the KdV type (i.e. mappings arising from the periodic initial value problem of lattice versions of the KdV equation [32, 16]). These are rational mappings $\mathbb{R}^{2P} \rightarrow \mathbb{R}^{2P} : (\{v_j\}) \mapsto (\{v_j'\})$ of the form

$$v_{2j-1}' \; = \; v_{2j} \; , \quad v_{2j}' \; = \; v_{2j+1} + \frac{a}{v_{2j}} - \frac{a}{v_{2j+2}} \quad j = 1, \cdots, P, \qquad (2.1)$$

where $a = p^2 - q^2$, $p, q \in \mathbb{R}$ are parameters of the lattice and the prime denotes the discrete time-shift corresponding to a translation in the second lattice direction. Imposing the periodicity condition $v_{i+2P} = v_i$. The mapping (2.1) has the Casimirs (w.r.t. the relevant Poisson bracket)

$$\sum_{j=1}^{P} v_{2j} \; = \; \sum_{j=1}^{P} v_{2j-1} \; = \; c \, , \qquad (2.2)$$

where c is chosen to be invariant under the mapping, in which case we obtain a $(2P-2)$-dimensional generalization of the McMillan mapping [27].

The mapping (2.1) admits the Lax representation,

$$L'_n(x) \cdot M_n(x) = M_{n+1}(x) \cdot L_n(x) , \qquad (2.3)$$

in which x is a spectral parameter, L_n is the lattice translation operator at site n,

$$L_j = \mathcal{L}_{2j} \cdot \mathcal{L}_{2j-1} , \quad M_j = \begin{pmatrix} w_j & 1 \\ \Lambda_{2j} & 0 \end{pmatrix}, \mathcal{L}_i = \begin{pmatrix} v_i & 1 \\ \Lambda_i & 0 \end{pmatrix}, \qquad (2.4)$$

in which $\Lambda_n = \Lambda_n(x) = x + \omega_n$, $\omega_{2k} = p^2 - q^2 = a$, $\omega_{2k+1} = 0, k = 1, 2, \ldots$. In fact, from the condition (2.3) one obtains $w_j = v_{2j-1} - a/v_{2j}$ as well as the mapping (2.1).

3   Discrete dynamics of auxiliary spectrum

The discrete map (2.1) is a completely integrable map for which the Lagrangian and Hamiltonian formalism were developed, [16]. The map admits a discrete Poisson structure and a quadratic r-matrix algebra, [29].

Introduce the monodromy matrix

$$T_n(x) \equiv \overset{\longleftarrow}{\prod_{j=n}^{2P+n-1}} \mathcal{L}_j(x) = \begin{pmatrix} a_n(x) & b_n(x) \\ c_n(x) & d_n(x) \end{pmatrix} , \qquad (3.1)$$

for the entries of which we have (for $\Lambda_n \neq 0$)

$$
\begin{aligned}
a_{n+1}(x) &= v_n b_n(x) + d_n(x), \quad c_{n+1}(x) = \Lambda_n(x) b_n(x) , \\
b_{n+1}(x) &= \frac{1}{\Lambda_n(x)}[(a_n(x) - d_n(x))v_n - b_n(x)v_n^2 + c_n(x)], \\
d_{n+1}(x) &= a_n(x) - v_n b_n(x) .
\end{aligned}
\qquad (3.2)
$$

The time-dependent part of the map is described by operator

$$T'_{2j-1} = M_j T_{2j-1} M_j^{-1} , \quad j = 1, \ldots, P \qquad (3.3)$$

for the odd sites, whereas for the even sites one needs

$$T'_{2j+1} = V'_{2j} T'_{2j} V'^{-1}_{2j} = M_{j+1} T_{2j+1} M_{j+1}^{-1} , \qquad (3.4)$$

what leads to analogous to (3.2) formulae.

Denote by $T_{n,m}(x)$ the monodromy operator depending on both spatial and time variables. Then the equations

$$\det(T_{n,m}(x) - \tilde{y}E) = 0, \quad n = 1, \ldots, P, \tag{3.5}$$

where E is 2×2 unit matrix, define the hyperelliptic curve of genus $g = P-1$, which branching points are independent in m and n.

$$V(y,x): \quad y^2 = \sum_{i=0}^{2g+1} x^i \lambda_i = \lambda_{2g+1} \prod_{i=i}^{2g+1} (x - a_j) = R(x),$$

$$y = \tilde{y} - \frac{1}{2}(a_{n,m}(x) + b_{n,m}(x)), \tag{3.6}$$

where $a_i, i = 1, \ldots, 2g+1$ are branching points and λ_i are symmetric functions of order i build from them.

The mapping (2.1) induces the mapping $(V)^g \to (V)^g$ of auxiliary spectrum $(x_1(n,m), \ldots x_g(n,m))$, where $(x_i(n,m)$, are roots of the polynomials

$$b_{n,m}(x) = c \prod_{i=1}^{g} (x - x_i(n,m)), \tag{3.7}$$

where c is the Casimir element (2.2). To shorten the notation we shall omit below the second from the variables (n,m), where it does not come to contradiction.

Proposition 3.1. *The mapping $(V)^g \to (V)^g$ is described as follows*

$$\Lambda_n(x_i(n))b_{n+1}(x_i(n)) - \Lambda_{n-1}(x_i(n))b_{n-1}(x_i(n))$$
$$= \kappa_n v_n \sqrt{R(x_i(n))}, \tag{3.8}$$

where

$$v_n = \frac{\kappa_n}{c}\left(\sqrt{A_{n+1}} - \sqrt{A_n}\right), \quad \kappa_n^2 = 1$$

$$A_{n+1} = \lambda_1 + \sum_{j=1}^{P-1}(x_j(n) + x_j(n+1)) - \omega_n. \tag{3.9}$$

Proof. Solving a_n in terms of $\tau(x) = \text{tr}(T_n(x))$ and $\Delta(x) = \det(T_n(x))$, we obtain

$$a_n(x) = \frac{1}{2}\tau(x) + \frac{\kappa_n}{2}\sqrt{R(x) - 4\Lambda_{n-1}(x)b_n(x)b_{n-1}(x)}, \tag{3.10}$$

where $\kappa_n = (-1)^n \sigma$, where $\sigma = \pm 1$ is the sign associated with the choice of sheet of the curve. Using (3.2), we obtain from (3.10)

$$\Lambda_n(x)b_{n+1}(x) - \Lambda_{n-1}(x)b_{n-1}(x) + v_n^2 b_n(x)$$

$$= v_n \kappa_n \sqrt{R(x) - 4\Lambda_{n-1}(x) b_n(x) b_{n-1}(x)}. \qquad (3.11)$$

By substituting to (3.11) $x = x_i(n)$ we obtain (3.8).

To find v_n we use (3.10) and (3.2)

$$v_n = \frac{\kappa_{n+1} \sqrt{R(x) - 4 b_{n+1}(x) c_{n+1}(x)} + \kappa_n \sqrt{R(x) - 4 b_n(x) c_n(x)}}{2 b_n(x)}. \qquad (3.12)$$

Considering the leading term in the asymptotic expansion as $x \to \infty$, we obtain (3.8). □

We remark, that the function v_n given by the equation (3.12) is four-valued whilst the initial equation of the map (2.1) admits only the involution $v_i \to -v_i$. By fixing the opposite signs of the square roots by the prescription $\kappa_n = (-1)^n \sigma$ we obtain the required two-valued solution.

4   Integration

To complete the integration we must solve the Jacobi inversion problem and to describe the dynamics by the restriction to the discrete linear winding of the Jacobi variety of the curve.

4.1   Kleinian σ–functions

To solve the Jacobi inversion problem we shall use the apparatus of Kleinian σ-functions (see [3, 4, 5] and also [7, 13, 14]).

Write a genus two curve with $2g + 2$ branching points $a_1, \ldots, a_{2g+1}, a = \infty$ in the form (3.6). Equip the curve V by the canonical homology basis $(\mathfrak{a}_1, \ldots, \mathfrak{a}_g; \mathfrak{b}_1, \ldots, \mathfrak{b}_g)$ as it is shown for the case g=5 on the Figure 1.

Define on V the canonical sets of holomorphic differentials,

$$d\boldsymbol{u}^T = (du_1, \ldots, du_g), \quad du_k = \frac{x^{k-1} dx}{y} \qquad (4.1)$$

and differentials of the second kind with pole in infinity

$$d\boldsymbol{r}^T = (dr_1, \ldots, dr_g),$$

$$dr_j = \sum_{k=j}^{2g+1-j} (k+1-j) \lambda_{k+1+j} \frac{x^k dx}{4y}, \quad j = 1, \ldots, g. \qquad (4.2)$$

Introduce $g \times 2g$ periods matrices, $(2\omega, 2\omega')$ and $(2\eta, 2\eta')$ of their periods. In the considered case $\det \omega \neq 0$ and the matrix $\tau = \omega^{-1}\omega$ belongs to *Siegel upper half space* of degree 2, i.e. the matrix τ is symmetric ant its imaginary part is positively defined. Define also the matrix $\varkappa = \eta(2\omega)^{-1}$.

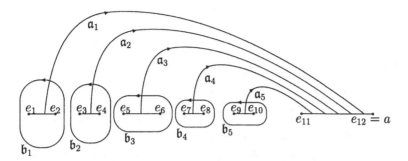

Figure 1: A homology basis on a Riemann surface of the hyperelliptic curve of genus 5 with the real branching points $e_1, \ldots, e_{12} = a$ (upper sheet). The cuts are drawn from e_{2i-1} to e_{2i} for $i = 1, \ldots, 6$. The $\mathfrak{a}$–cycles are completed on the lower sheet (the picture on lower sheet is just flipped horizontally).

The fundamental hyperelliptic σ–function is defined for the by the formula

$$\sigma(u) = \sqrt{\frac{\pi^g}{\det 2\omega}} \frac{1}{\sqrt[4]{\prod_{i \neq j}(a_i - a_j)}} \exp\{u^T \varkappa u\} \theta[\varepsilon]((2\omega)^{-1} u | \tau), \qquad (4.3)$$

where $[\varepsilon] = \begin{bmatrix} \varepsilon'^T \\ \varepsilon^T \end{bmatrix}$, is the characteristic of the vector of Riemann constants, and $\theta[\varepsilon](v|\tau)$ is the standard theta function with characteristic,

$$\theta[\varepsilon](v|\tau) = \sum_{m \in \mathbb{Z}^g} \exp\{\pi \imath (m + \varepsilon')^T \tau (m + \varepsilon') + 2\pi \imath (n + \varepsilon')^T (v + \varepsilon)\} \quad (4.4)$$

The fundamental Kleinian σ–function is a natural generalization of the Weierstrass σ function to the hyperelliptic case and inherits its principal property: *is automorphic with respect to the transformations from the group* $\mathrm{Sp}(2g, \mathbb{Z})$.

Introduce Kleinian $\wp$–functions through second logarithmic derivative of the Kleinian σ–function,

$$\wp_{ij}(u) = -\frac{\partial^2 \ln \sigma(u)}{\partial u_i \partial u_j}, \quad \wp_{ijk}(u) = -\frac{\partial^3 \ln \sigma(u)}{\partial u_i \partial u_j \partial u_k}, \quad i, j, k = 1, \ldots, g \quad \text{etc}$$

$$(4.5)$$

The Abel map $\mathfrak{A} : (V)^g \to \mathrm{Jac}(V)$ of the symmetrised product $V \times \cdots \times V$ to the Jacobi variety $\mathrm{Jac}(V) = \mathbb{C}^g / \omega \otimes \omega'$ of the curve V is defined by the equations

$$\int_{(0,a_1)}^{(y_1,x_1)} \frac{x_1^{i-1}dx}{y_1} + \ldots + \int_{(0,a_g)}^{(y_g,x_g)} \frac{x_g^{i-1}dx}{y_g} = u_i, \quad i = 1, \ldots, g. \qquad (4.6)$$

We omit below the first coordinate of the curve in the integral bounds.

The Jacobi inversion problem is formulated as the problem of inversion of the Abel map and is solved in terms of the Kleinian σ functions as follows: The Abel preimage of the point $u \in \text{Jac}(V)$ is given by the set $\{(y_1, x_1), \ldots, (y_g, x_g)\} \in (V)^g$, where $\{x_1, \ldots, x_g\}$ are the zeros of the polynomial

$$\mathcal{P}(x; u) = x^g - x^{g-1}\wp_{g,g}(u) - x^{g-2}\wp_{g,g-1}(u) - \ldots - \wp_{g,1}(u), \qquad (4.7)$$

and $\{y_1, \ldots, y_g\}$ are given by

$$y_k = -\left.\frac{\partial\mathcal{P}(x; u)}{\partial u_g}\right|_{x=x_k}, \qquad (4.8)$$

4.2   Discrete linear winding of Jacobi variety

In the standard KdV theory the restriction of the Jacobi inversion problem to special linear windings is referred to as *Dubrovin equations* [19]. We inherit here this terminology and call the equations (3.8) *discrete Dubrovin equations*.

The following proposition describes the windings

Proposition 4.1. *Let $x_1(n_0, m_0), \ldots, x_g(n_0, m_0)$ is arbitrary point on $(V)^g$. Then for arbitrary n, m the following formulae are valid*

$$\sum_{i=1}^{g} \int_{x_i(n_0,m_0)}^{x_i(n,m)} du = -(U + V)\left(\left[\frac{n - n_0}{2}\right] + \left[\frac{m - m_0}{2}\right]\right)$$
$$- 2U\left(\left\{\frac{n - n_0}{2}\right\} + \left\{\frac{m - m_0}{2}\right\}\right), \qquad (4.9)$$

where $[\cdot]$ and $\{\cdot\}$ mean integer and fractional parts of a number correspondingly and the winding vectors U and V are given as

$$U = \int_{\infty}^{0} du, \quad V = \int_{\infty}^{a} du, \quad a = p^2 - q^2. \qquad (4.10)$$

Proof. Write the eigenfunction Ψ_n for the spectral problem $T_n(x)\Psi_n = y\Psi_n$ in two equivalent forms

$$\Psi_n = \begin{pmatrix} 1 \\ \frac{y - a_n(x)}{b_n(x)} \end{pmatrix} \varphi_n = \begin{pmatrix} 1 \\ \frac{c_n(x)}{y - d_n(x)} \end{pmatrix} \varphi_n$$

Then because of the equation $\Psi_{n+1} = V_n \Psi_n$, the function

$$f_n(x) = \frac{\varphi_{n+1}(x)}{\varphi_n(x)}$$

can be presented in two equivalent forms

$$f_n(x) = \frac{\Lambda_n(x) b_{n+1}(x)}{y - a_{n+1}(x)} = \frac{y - d_{n+1}(x)}{b_n(x)}$$

and hence it is meromorphic on V and has $g+1$ simple zeros in the points $(d(x_1(n+1)), x_1(n+1)) \ldots , (d(x_g(n+1)), x_g(n+1)), Q_n = (d(\omega_n), \omega_n)$, and $g+1$ simple poles in the points $(d(x_1(n)), x_1(n)), \ldots , (d(x_g(n)), x_g(n)), \frac{1}{\xi} = \sqrt{x} = \infty$. The Abel theorem then states that

$$\sum_{i=1}^{g} \int_{x_n}^{x_{n+1}} d\boldsymbol{u} + \int_{\infty}^{Q_n} d\boldsymbol{u} = 0. \tag{4.11}$$

The behavior of φ_n under the discrete-time flow is different for even and odd cites. For odd sites we have from the equations $T_{2n-1}(x)\Psi_{2n-1} = y\Psi_{2n-1}$ and $\Psi'_{2n-1} = M_n(x)\Psi_{2n-1}$

$$\frac{\varphi'_{2n-1}(x)}{\varphi_{2n-1}(x)} = \frac{y - d'_{2n-1}(x)}{b_{2n-1}(x)} = \frac{\Lambda_{2n}(x) b'_{2n-1}(x)}{y - a'_{2n-1}(x)}. \tag{4.12}$$

corresponding to

$$\sum_{j=1}^{g} \int_{x_j^0}^{x_j'(2n-1)} d\boldsymbol{u} = \sum_{j=1}^{g} \int_{x_j^0}^{x_j(2n-1)} d\boldsymbol{u} - \int_{\infty}^{Q_{2n}} d\boldsymbol{u}. \tag{4.13}$$

To find the evolution law for odd cites we find from $T_{2n}(x)'\Psi'_{2n} = y\Psi'_{2n}$ and $\Psi'_{2n+1} = V'_n\Psi'_{2n}$ the relation

$$\frac{\varphi'_{2n}(x)}{\varphi'_{2n+1}(x)} = \frac{b'_{2n}(x)}{y - d_{2n+1}(x)} \tag{4.14}$$

Using (4.12) we find

$$\frac{\varphi'_{2n}(x)}{\varphi'_{2n+1}(x)} = \frac{b'_{2n}(x)}{b_{2n+1}(x)} \tag{4.15}$$

and therefore the Abel theorem gives

$$\sum_{j=1}^{g} \int_{x_j^0}^{x_j'(2n)} d\boldsymbol{u} = \sum_{j=1}^{g} \int_{x_j^0}^{x_j(2n+1)} d\boldsymbol{u}. \tag{4.16}$$

The compatibility of the spatial and time evolutions follows from (4.11, 4.13, 4.16), which enable to check, that the vector standing on the left-hand side of (4.9) in terms of the winding vectors (4.10). is independent of the staircase path in the lattice that was used in [32].   □

We are ready now to recover the potential v_n.

Theorem 4.1. *The map (2.1) is given in terms of Kleinian hyperelliptic functions*

$$v_n = \sqrt{\wp_{gg}(\boldsymbol{u}_{n+1}) + \wp_{gg}(\boldsymbol{u}_n) + \lambda_1 - \omega_n}$$
$$- \sqrt{\wp_{gg}(\boldsymbol{u}_{n-1}) + \wp_{gg}(\boldsymbol{u}_n) + \lambda_1 - \omega_n}, \qquad (4.17)$$

where

$$\boldsymbol{u}_n = \boldsymbol{u}_0 + (\boldsymbol{U} + \boldsymbol{V})\left(\left[\frac{n}{2}\right] + \left[\frac{m}{2}\right]\right) + 2\boldsymbol{U}\left(\left\{\frac{n}{2}\right\} + \left\{\frac{m}{2}\right\}\right),$$

where $\boldsymbol{u}_0$ being arbitrary point in Jac(V) the winding vectors $\boldsymbol{U}$ and $\boldsymbol{V}$ given in (4.10) and $[\cdot],\{\cdot\}$ mean integer and fractional part of the number correspondingly.

Proof. The proof is straightforward and based on (3.9) and the formulae (4.11), (4.13), (4.16).   □

5   Addition theorems for hyperelliptic functions

In the case when the associated algebraic curve is an elliptic curve the mapping was given by McMillan [27] and described by the formula

$$x_{n+1} + x_{n-1} = \frac{ax_n}{1 - bx_n^2} \qquad (5.1)$$

which presents the direct implementation of the addition theorem for Jacobi elliptic functions,

$$\text{sn}(u + v) = \frac{\text{sn}u\,\text{cn}v\,\text{dn}v + \text{sn}v\,\text{cn}u\,\text{dn}u}{1 - k^2\text{sn}^2u\,\text{sn}^2v},$$

where snu, etc. are the standard Jacobi elliptic function with module k[6].

The formula (4.17) being derived represents in the combination with the definition of the map (2.1) and solution of the Jacobi inversion problem

(4.7,4.8) a special addition theorems for Kleinian hyperelliptic functions $\wp_{gg}(u), \wp_{ggg}(u), \ldots, \wp_{gg1}(u)$, defined on subvarieties

$$\{u|u_0 + pU + qV, \quad p, q \in \mathbb{N}\} \subset \text{Jac}(V). \tag{5.2}$$

These equations can be also interpreted as an integrable discrete map of Kummer surface $\text{Kum}(V)$ associated with the curve V. To demonstrate the evidence of this statement we use the description of Jacobi and Kummer varieties as algebraic varieties given in [13, 14].

We introduce (cf. [25]) new functions h_{ik} defined by the formula

$$h_{ik} = 4\wp_{i-1,k-1} - 2\wp_{k,i-2} - 2\wp_{i,k-2}$$
$$+ \frac{1}{2}\left(\delta_{ik}(\lambda_{2i-2} + \lambda_{2k-2}) + \delta_{k,i+1}\lambda_{2i-1} + \delta_{i,k+1}\lambda_{2k-1}\right), \tag{5.3}$$

where the indices $i, k \in 1, \ldots, g+2$. We assume that $\wp_{nm} = 0$ if n or m is < 1 and $\wp_{nm} = 0$ if n or m is $> g$. It is evident that $h_{ij} = h_{ji}$. We shall denote the matrix of h_{ik} by H. It was proved in [13, 14] that the Jacobi variety $\text{Jac}(V)$ is embedded into the complex space

$$\mathbb{C}^{g+\frac{g(g+1)}{2}} = \{\wp_{ggi}(u), \wp_{jk}(u) = \wp_{kj}(u) \quad u \in \text{Jac}(V), i, j, k = 1, \ldots, g\}$$

as algebraic variety given by intersection of $\frac{g(g+1)}{2}$ cubics

$$4\wp_{ggi}\wp_{ggk} = -\det H\left[\begin{smallmatrix} i, g+1, g+2 \\ k, g+1, g+2 \end{smallmatrix}\right], \quad 1 \leq i \leq j \leq g. \tag{5.4}$$

The rank of any from $\frac{g(g-2)}{2}$ matrices

$$\det H \left[\begin{smallmatrix} i,j,g+1,g+2 \\ k,l,g+1,g+2 \end{smallmatrix}\right] \tag{5.5}$$

does not exceed 3; these last conditions define the quartic Kummer surface $\text{Kum}(V)$

We shall use this description of Jacobi and Kummer varieties to describe the considered discrete map in terms of Kleinian hyperelliptic functions.

Proposition 5.1. *The discrete Dubrovin equations are equivalent to the set addition theorem for hyperelliptic Kleinian functions restricted to the discrete windings*

$$\frac{v_n \kappa_n}{c}\wp_{g,g,i}(u_n) = \wp_{g,i}(u_n)\left[\wp_{g,g}(z_{n+1}) - \wp_{g,g}(u_{n-1}) + a\right]$$
$$+\wp_{g,g-i}(u_{n-1}) - \wp_{g,g-i}(u_{n+1}) + a\wp_{g,i}(u_{n-1}), \quad i = 1, \ldots, g. \tag{5.6}$$

Proof. In follows from (3.8), and the expressions (4.8) that the points $x_i(n)$, $i = 1, \ldots, g$ are the roots of the polynomial

$$\sum_{i=1}^{g} x^i \left(\wp_{g,i}(u_{n-1}) - \wp_{g,i}(u_{n+1})\right)$$

$$+ \; a\left(\wp_{g,i+1}(\boldsymbol{u}_{n-1}) - \delta_{i,g} - \frac{v_n \kappa_n}{c}\wp_{g,g,i+1}(\boldsymbol{u}_n)\right) = 0, \qquad (5.7)$$

where two and three index symbols $\wp_{nm}$ and $\wp_{kln}$ equals zero if the subscript $i < 1$ or $i > g$. Applying (6) we obtain (5.6).   □

The equations (5.6) represents addition formulae for hyperelliptic Kleinian functions restricted to the special subvarieties in $\mathrm{Jac}(V)$ (5.2), (4.10).

6   Discrete map of Kummer surface

Consider the case of genus two, when the following expressions are valid

$$\wp_{22}(\boldsymbol{u}) = x_1 + x_2, \quad \wp_{12}(\boldsymbol{u}) = -x_1 x_2, \quad \wp_{11}(\boldsymbol{u}) = \frac{F(x_1, x_2) - 2y_1 y_2}{4(x_1 - x_2)^2},$$

where

$$F(x_1, x_2) = \sum_{k=0}^{k=2} x_1^k x_2^k (2\lambda_{2k} + \lambda_{2k+1}(x_1 + x_2)) \qquad (6.1)$$

Only two from the functions $\wp_{ij}(\boldsymbol{u})$ are independent. Three functions $\wp_{22}(\boldsymbol{u})$, $\wp_{21}(\boldsymbol{u})$ and $\wp_{11}(\boldsymbol{u})$ are connected by a quartic relation which is the remarkable Kummer surface $\mathcal{K}$ in $\mathbb{C}^3$, which equation is given in the form

$$\mathcal{K}: \quad \det K = 0, \qquad (6.2)$$

where K is the 4×4 matrix

$$K = \begin{pmatrix} -\Lambda_0 & \frac{1}{2}\lambda_1 & 2\wp_{11} & -2\wp_{12} \\ \frac{1}{2} & -\lambda_2 - 4\wp_{11} & \frac{1}{2}\lambda_3 + 2\wp_{12} & 2\wp_{22} \\ 2\wp_{11} & \frac{1}{2}\lambda_3 + 2\wp_{12} & -\lambda_4 - 4\wp_{22} & 2 \\ -2\wp_{12} & 2\wp_{22} & 2 & 0 \end{pmatrix}. \qquad (6.3)$$

The three indexed symbols satisfy to the equations

$$y_1 = x_1 \wp_{222}(\boldsymbol{u}) + \wp_{221}(\boldsymbol{u}), \quad y_2 = x_2 \wp_{222}(\boldsymbol{u}) + \wp_{221}(\boldsymbol{u}) \qquad (6.4)$$

The equations (5.6) read in this case

$$\frac{v_n \kappa_n}{c}\wp_{222}(\boldsymbol{u}_n) = \wp_{22}(\boldsymbol{u}_n)\left[\wp_{22}(\boldsymbol{u}_{n+1}) - \wp_{22}(\boldsymbol{u}_{n-1}) + a\right]$$
$$- \left[\wp_{12}(\boldsymbol{u}_{n+1}) - \wp_{12}(\boldsymbol{u}_{n-1}) + a\right] + a\wp_{22}(\boldsymbol{u}_{n-1}),$$
$$\frac{v_n \kappa_n}{c}\wp_{122}(\boldsymbol{u}_n) = \wp_{12}(\boldsymbol{u}_n)\left[\wp_{22}(\boldsymbol{u}_{n+1}) - \wp_{22}(\boldsymbol{u}_{n-1}) + a\right] + a\wp_{22}(\boldsymbol{z}_{n-1}),$$

The odd Kleinian functions are related to even functions as follows

$$\wp_{222}^2 = 4\wp_{11} + \lambda_3\wp_{22} + 4\wp_{22}^3 + 4\wp_{12}\wp_{22} + \lambda_4\wp_{22}^2 + \lambda_2,$$
$$\wp_{221}^2 = \lambda_0 - 4\wp_{11}\wp_{12} + \lambda_4\wp_{12}^2 + 4\wp_{22}\wp_{12}^2, \qquad (6.5)$$

where $\wp_{11}$ is expressible in terms of $\wp_{22}$ and $\wp_{12}$ by the aid of the equation of Kummer surface to which it enters quadraticaly. To take square powers of the last equations we come to the discrete map of the Kummer surface being induced by the KdV discrete map.

7 Acknowledgement

FWN is grateful to A.R. Its and V.G. Papageorgiou for stimulating discussions at the early stages of this work. VZE acknowledges support from the Royal Society for his visit to the University of Leeds.

References

[1] M J Ablowitz, J M Keiser and L A Takhtajan. Class of stable multistate time-reversible cellular automata with rich particle content. *Phys. Rev. A*, 44:6909–6912, 1991.

[2] M J Abowitz and F J Ladik. On the solution of a class of nonlinear partial difference equations. *Stud. Appl. Math.*, 55:1–12, 1977.

[3] H F Baker. *Abels theorem and the allied theory including the theory of Theta functions*. Cambridge Univ. Press, Cambridge, 1897.

[4] H F Baker. On the hyperelliptic sigma functions. *Amer. Journ. Math.*, 20:301–384, 1898.

[5] H F Baker. *Multiply Periodic Functions*. Cambridge Univ. Press, Cambridge, 1907.

[6] H Bateman and A Erdelyi. *Higher Transcendental Functions*, volume 2. McGraw-Hill, New York, 1955.

[7] E D Belokolos, A I Bobenko, V Z Enolskii, A R Its, and V B Matveev. *Algebro Geometrical Aproach to Nonlinear Integrable Equations*. Springer, Berlin, 1994.

[8] A I Bobenko, M Bordemann, C Gunn and U Pinkall. On two integrable cellular automata. *Commun. Math. Phys.*, 158:127–134, 1993.

[9] A I Bobenko and U Pinkall. Discrete surfaces with constant negative Gaussian curvature and the Hirota equation. *J. Diff. Geometry*, 43(3):527–611, 1985.

[10] O Bolza. On the first and second derivatives of hyperelliptic σ–functions. *Amer. Journ. Math.*, 17:11, 1895.

[11] M Bruschi, O Ragnisco, P M Santini, and G Z Tu. Integrable symplectic maps. *Physica D*, 49:273–294, 1991.

[12] M Bruschi, P M Santini and O Ragnisco. Integrable cellular automata. *Phys. Lett. A*, 169:151, 1992.

[13] V M Buchstaber, V Z Enolskii, and D V Leykin. *Hyperelliptic Kleinian functions and applications*, volume 179, pages 1–34. Solitons, Geometry and Topology: on Crossroad Advances in the Math. Sciences, AMS Trans. ser.2, Moscow State University and University of Maryland, College Park, 1997.

[14] V M Buchstaber, V Z Enolskii, and D V Leykin. *Hyperelliptic Kleinian functions, hyperelliptic Jacobians and applications*, Reviews in Mathematics and Mathematical Physics, volume 10, pages 1–115. Gordon and Breach, London, 1997.

[15] H Burkhardt. Beiträge zur Theorie der hyperelliptische Sigmafunctionen. *Math. Ann.*, 32:381–442, 1888.

[16] H W Capel, F W Nijhoff, and V G Papageorgiou. Complete integrability of Lagrangian mappings and lattices of KdV type. *Phys. Lett*, 155A:377–387, 1991.

[17] E Date, M Jimbo, and T Miwa. Method for generating discrete soliton equations, i-iv. *J. Phys. Soc. Japan*, 51:4116–4131, 1982, 52:388–393; 761–771, 1983.

[18] P A Deift, L C Li, and C Tomei. Matrix factorizations and integrable systems. *Commun. Pure Appl. Math.*, 42:443–521, 1989.

[19] B A Dubrovin, V B Matveev, and S P Novikov. Nonlinear equations of the KdV type, finite gap linear operators and abelian varieties. *Uspekhi Matem. Nauk*, 31:1:55–135, 1976.

[20] L D Faddeev and A Yu Volkov. Abelian current algebra and the Virasoro algebra on the lattice. *Phys. Lett. B*, 315:311–318, 1993.

[21] B Grammaticos, F W Nijhoff, and A Ramani. Discrete Painlevé equations. Lectures delivered at the Cargèse school : The Painlevé property. One century later, 131p, 1996.

[22] R Hirota. Nonlinear partial difference equations i–iii. *J. Phys. Soc. Japan*, 43:1424–1433; 2074–2089, 1977.

[23] F Klein. Ueber hyperelliptishe Sigmafunctionen. *Math. Ann.*, 32:351–380, 1888.

[24] A Krazer and W Wirtinger. *Abelsche Funktionen und allgemeine Thetafunctionen*, Encyklopädie der Mathematische Wissenshaften II(2), Heft 7, pages 603–882. Teübner, 1915.

[25] D V Leykin. On Weierstrass cubic for hyperelliptic functions. *Uspekhi Matem. Nauk*, 50(6): 191–192, 1995.

[26] V B Matveev and M A Salle. *Darboux transformations and solitons*. Springer Ser. Nonlinear Dynamics, Berlin, Heidelberg, 1991.

[27] E M McMillan. A problem in the stability of periodic systems. In W E Brittin and H Odabasi, editors, *Topics in Modern Physics. A Tribute to E U Condon*, pages 219–244, Boulder, 1971. Colorado Associated Univ. Press.

[28] J Moser and A P Veselov. Discrete versions of some integrable systems and factorisation of matrix polynomials. *Commun. Math. Phys.*, 139:217–243, 1991.

[29] F W Nijhoff, V G Papageorgiou, and H W Capel. Integrable time-discrete systems: lattices and mappings. In P.P.Kulish, editor, *Proceedings of the International Workshop on Quantum groups, Lectures Notes Math.*, volume 1510, pages 312–325. Springer, 1992.

[30] F W Nijhoff, V G Papageorgiou, H W Capel, and G R W Quispel. The lattice Gel'fand–Dikii hierarchy. *Inv. Problems*, 6:597–621, 1992.

[31] F W Nijhoff, G R W Quispel, and H W Capel. Direct linearization of nonlinear difference-difference equations. *Phys. Lett.*, 97A:125–128, 1983.

[32] V G Papageorgiou, F W Nijhoff, and H W Capel. Integrable mappings and nonlinear integrable lattice equations. *Phys. Lett.*, 147A:106–114, 1990.

[33] G R W Quispel, F W Nijhoff, H W Capel, and van der Linden. Linear integral equations and nonlinear difference–difference equations. *Physica*, 125A:344–380, 1984.

[34] G R W Quispel, J A G Roberts, and C J Thompson. Integrable mappings and soliton equations. *Phys. Lett*, 126A:419–421, 1989.

[35] Yu.B. Suris, *In this volume*.

[36] A P Veselov. Integrable mappings. *Russ. Math. Surv.*, 46:1–51, 1991.

[37] C. Viallet, *In this volume*.

[38] E Wiltheiss. Ueber die potenzreihen der hyperelliptishen thetafunctionen. *Math. Ann.*, 31:410–423, 1888.

R–matrix hierarchies, integrable lattice systems, and their integrable discretizations

Yuri B. Suris

Centre for Complex Systems and Visualization,
University of Bremen,
Universitätsallee 29, 28359 Bremen, Germany
Email: suris@cevis.uni-bremen.de

1 General framework

This paper is devoted to an old problem: how to discretize a given integrable system, retaining its integrability? An approach to this problem will be presented, recently elaborated by the present author [1], [2], [3], [4] and based on the notion of r–matrix hierarchies. This approach is by no means unique, neither does it pretend to be the best possible one (though it has many advantageous features). Because of the lack of place we will not discuss here other approaches. The references may be found in the above mentioned papers.

We recall some basic notions about integrable hierarchies and their r–matrix theory. Our formulations result from the observations on the large "experimental material" collected in the last decades of research in this area.

Let $\mathcal{P}$ be a Poisson manifold; in fact we consider here only the simplest case of a symplectic space $\mathbf{R}^{2N}(x, p)$ with canonically conjugated coordinates $x = (x_1, \ldots, x_N)^T$ and $p = (p_1, \ldots, p_N)^T$, so that the Poisson brackets of the coordinate functions read:

$$\{x_k, x_j\} = \{p_k, p_j\} = 0, \quad \{x_k, p_j\} = \delta_{kj}. \tag{1.1}$$

Let $H(x, p)$ be a completely integrable Hamiltonian, i.e. suppose that the Hamiltonian system

$$\dot{x} = \{x, H\}, \quad \dot{p} = \{p, H\} \tag{1.2}$$

posesses N functionally independent integrals in involution. Then (1.2) usually (probably always) admits a *Lax representation*, i.e. there exist two maps

$T : \mathcal{P} \mapsto g$ and $M : \mathcal{P} \mapsto g$ into some Lie algebra g such that (1.2) is equivalent to

$$\dot{T} = [T, M]. \tag{1.3}$$

In the cases we are dealing with in this paper, g is the algebra $gl(N)$ or a loop algebra over $gl(N)$. It carries some additional structures. In particular, it is an assosiative algebra with respect to the usual matrix multiplication, and admits a non–degenerate scalar product $\langle \cdot, \cdot \rangle$.

An important observation is that usually

$$H(x, p) = \varphi(T),$$

where $\varphi(T)$ is an Ad–invariant function on g. This is related to the fact that integrable systems appear not separately, but are organized in *hierarchies*. Namely, to every Ad–invariant function $\varphi(T)$ there corresponds a Lax equation of the form (1.3). Moreover, the matrices M often may be presented in the following form:

$$M = R(\nabla \varphi(T)), \tag{1.4}$$

where $R : g \mapsto g$ is a linear operator, called hereafter an *R–operator governing the corresponding hierarchy*. (Recall that the gradient $\nabla \varphi(T)$ of a smooth function φ on an algebra g with a scalar product $\langle U, V \rangle$ is defined by the relation

$$\langle \nabla \varphi(T), U \rangle = \frac{d}{d\epsilon} \varphi(T + \epsilon U) \Big|_{\epsilon=0}, \quad \forall U \in g,$$

and for an Ad–invariant φ one has $[\nabla \varphi(T), T] = 0$).

An *r–matrix theory* [7] provides a sort of explanation of the relation (1.4). Namely, the formula (1.4) is usually a consequence of a more fundamental fact, namely that $T = T(x, p)$ form a Poisson submanifold in g equipped with a *linear r–matrix bracket*:

$$\{\varphi, \psi\}(T) = \langle T, [R(\nabla \varphi), \nabla \psi] + [\nabla \varphi, R(\nabla \psi)] \rangle. \tag{1.5}$$

A sufficient condition for (1.5) to define a Poisson bracket is given by the *modified Yang–Baxter equation* [7]. As a consequence of (1.5), the Hamiltonian equation $\dot{T} = \{T, H\}$ on g with an Ad–invariant function $\varphi(T)$ has the Lax form (1.3) with M given in (1.4).

2   R–operators from splitting $g = g_+ \oplus g_-$

A very important class of the R–operators is given by the following construction. Let g be a Lie algebra of some associative algebra, equipped with a non–degenerate invariant scalar product. Let g, as a linear space, be a direct sum of its two subalgebras:

$$g = g_+ \oplus g_-,\tag{2.1}$$

and let π_+, π_- denote the projections from g onto corresponding subspaces:

$$T = \pi_+(T) + \pi_-(T), \quad \pi_\pm(T) \in g_\pm, \quad \forall T \in g.\tag{2.2}$$

Then it is easy to check that the operator

$$R = \frac{1}{2}(\pi_+ - \pi_-)\tag{2.3}$$

satisfies the modified Yang–Baxter equation, which allows to define a linear bracket (1.5) on g.

The corresponding Hamiltonian equations with Ad–invariant Hamiltonian functions read: $\dot{T} = [T, R(\nabla\varphi(T))]$, which in the present case may be presented in two equivalent forms:

$$\dot{T} = [T, \pi_+(\nabla\varphi(T))] = -[T, \pi_-(\nabla\varphi(T))].\tag{2.4}$$

It is very remarkable that in such a general setting the equation (2.4) admits an explicit solution in terms of a factorization problem in a Lie group. The following construction is due to Adler, Kostant, Symes, Reyman, Semenov–Tian–Shansky (see [7]).

Let G be a Lie group with the Lie algebra g, and let G_+ and G_- be its two subgroups having g_+ and g_-, respectively, as Lie algebras. Then in a certain component Ω of the group unity I a following factorization is uniquely defined:

$$T = \Pi_+(T)\Pi_-(T), \quad \Pi_\pm(T) \in G_\pm, \forall T \in \Omega \subset G.\tag{2.5}$$

Theorem 1 *Let $f : g \mapsto g$ be an Ad–covariant function. Then the solution of the differential equation*

$$\dot{T} = [T, \pi_+(f(T))] = -[T, \pi_-(f(T))].\tag{2.6}$$

with an initial condition $T(0) = T_0$ is given, at least for t small enough, by

$$T(t) = \Pi_+^{-1}\left(e^{tf(T_0)}\right) T_0 \Pi_+\left(e^{tf(T_0)}\right) = \Pi_-\left(e^{tf(T_0)}\right) T_0 \Pi_-^{-1}\left(e^{tf(T_0)}\right).\tag{2.7}$$

This construction admits also a discrete–time formulation. Namely, for an arbitrary flow of the type (2.6) one can write down a *difference* equation, satisfied by every trajectory of the flow. The next Theorem goes back to the work of Symes on the Toda lattice [8].

Theorem 2 *Let $F : g \mapsto G$ be an Ad–covariant function. Then the solution of the difference equation*

$$T_{n+1} = \Pi_+^{-1}\left(F(T_n)\right) T_n \Pi_+\left(F(T_n)\right) = \Pi_-\left(F(T_n)\right) T_n \Pi_-^{-1}\left(F(T_n)\right) \qquad (2.8)$$

with an initial condition T_0 is given by

$$T_n = \Pi_+^{-1}\left(F^n(T_0)\right) T_0 \Pi_+\left(F^n(T_0)\right) = \Pi_-\left(F^n(T_0)\right) T_0 \Pi_-^{-1}\left(F^n(T_0)\right). \qquad (2.9)$$

Comparing the formulas (2.9), (2.7), we see that the map (2.8) is the time h shift along the trajectories of the flow (2.6) with

$$f(T) = h^{-1}\log(F(T)).$$

Although we have seen that the equation (2.6) admits a Hamiltonian interpretation, this structure is not needed to formulate or to prove the Theorem 1. In this sense it is a purely kinematic result. However, the Hamilonian property provides us with some usuful additional information. The flows (2.6) are Hamiltonian with the Hamiltonian functions $\varphi(T)$, where

$$\nabla\varphi(T) = f(T)$$

(the existence of such φ is obvious, if $f(T)$ is an analytic function on matrices, which is usualy the case in applications). Hence the maps (2.8) are Poisson. If the set of Lax matrices $T(\mathcal{P})$ for the system at hand forms a Poisson submanifold for the bracket (1.5), then this manifold is invariant under flows (2.6), as well as under maps (2.8).

Analogs of the above Theorems hold also for the equations governed by R–operators different from (2.3). They need only to satisfy the modified Yang–Baxter equation.

3   Recipe for integrable discretization

The results of the previous Section lead to the following prescription. Given an integrable system admitting the Lax representation of the form (2.6), take as its integrable discretization the difference equation (2.8) with the same Lax matrix T and with some

$$F(T) = I + hf(T) + o(h).$$

Of course, this makes sense only if the corresponding factors $\Pi_\pm(F(T))$ admit more or less explicit expressions, allowing to write down the corresponding difference equations in a more or less closed form. The choice of $F(T)$ is a transcendent problem, which however turns out to be solvable for many of (hopefully, for the majority of or even for all) the known integrable systems. In particular, for the examples discussed below, one can take simply $F(T) = I + hf(T)$.

Let us stress once more the advantages of this approach to the problem of integrable discretization.

- It is universally applicable to any system with an R-operator satisfying the mYB.

- The discretizations obtained in this way possess the same integrals of motion as their continuous time ancestors.

- If the continuous time system is Hamiltonian with respect to the Poisson bracket (1.5), then our discretizations have the Poisson property with respect to the same bracket.

- The initial value problem for our discrete time equations can be solved in terms of the same factorization in a Lie group as the initial value problem for the continuous time system.

4 Toda lattice and its integrable discretization

Consider the following set consisting of tri-diagonal matrices depending on the variables $(a, b) \in \mathbb{R}^{2N}$ and an additional (spectral) parameter λ:

$$T(a, b, \lambda) = \sum_{k=1}^{N} b_k E_{kk} + \lambda \sum_{k=1}^{N} E_{k+1,k} + \lambda^{-1} \sum_{k=1}^{N} a_k E_{k,k+1}. \qquad (4.1)$$

Here E_{jk} stands for the matrix whose only nonzero entry on the intersection of the jth row and the kth column is equal to 1. In the periodic case all the subscripts belong to $\mathbb{Z}/N\mathbb{Z}$, i.e. $E_{N,N+1} = E_{N,1}$, $E_{N+1,N} = E_{1,N}$, and $a_0 = a_N$. In the open-end case we set $E_{N+1,N} = E_{N,N+1} = 0$, which allows to get rid of the spectral parameter (more precisely, in this case the similarity transformation with the matrix $\text{diag}(1, \lambda, \ldots, \lambda^{N-1})$ removes λ from the matrix T; therefore in the open-end case we simply set $\lambda = 1$).

It is natural to consider these matrices as belonging to the following algebras.

1) For the *open-end case* we set $g = gl(N)$, the algebra of $N \times N$ matrices. The (nondegenerate, invariant) scalar product is choosen as $\langle U, V \rangle = \text{tr}(U \cdot V)$. There holds (2.1), where g_+ is the set of lower triangular matrices, and g_- is a set of strictly upper triangular matrices. The Lie group G corresponding to g is $GL(N)$, G_+ is a subgroup consisting of lower triangular nondegenerate

matrices, and G_- is subgroup consisting of upper triangular matrices with unities on the diagonal.

2) For the *periodic case* $g \subset gl(N)[\lambda, \lambda^{-1}]$ is a twisted subalgebra of the loop algebra $gl(N)[\lambda, \lambda^{-1}]$, consisting of formal semiinfinite Laurent series over $gl(N)$ satisfying

$$T(\omega\lambda) = \Omega T(\lambda)\Omega^{-1}, \quad \Omega = \mathrm{diag}(1, \omega, \dots, \omega^{N-1}), \quad \omega = \exp(2\pi i/N). \quad (4.2)$$

The scalar product is choosen as $\langle U(\lambda), V(\lambda) \rangle = \mathrm{tr}(U(\lambda) \cdot V(\lambda))_0$, the subscript 0 denoting the free term of the formal Laurent series. Again, there holds (2.1), where g_+ consists of Laurent series with the non–negative powers of λ, and g_- consists of Laurent series with negative powers of λ. The group G consists of $GL(N)$–valued functions of the complex parameter λ, satisfying (4.2) and regular in $\mathbb{C}P^1 \backslash \{0, \infty\}$. The elements of the subgroup G_+ are regular in the neighbourhood of $\lambda = 0$, and the elements of G_- are regular in the neighbourhood of $\lambda = \infty$ and take the value I in this point.

Define the operator R as in (2.3), and construct the Poisson bracket (1.5) on g.

Proposition 1 [9] *The set of matrices $T(a, b, \lambda)$ forms a Poisson submanifold in g. The coordinate representation of the restriction of the bracket (1.5) to this set is:*

$$\{a_k, b_k\} = -\{a_k, b_{k+1}\} = a_k . \quad (4.3)$$

The Toda hierarchy consists of Hamiltonian flows on g with Hamiltonian functions of the form $H(a, b) = \varphi(T(a, b, \lambda))$ with Ad–invariant functions φ. The simplest flow of the Toda hierarchy (Toda lattice proper) corresponds to

$$\varphi(T) = \frac{1}{2}\mathrm{tr}(T^2)_0 = \frac{1}{2}\sum_{k=1}^{N} b_k^2 + \sum_{k=1}^{N} a_k.$$

In coordinates this flow may be presented as

$$\dot{a}_k = a_k(b_{k+1} - b_k), \quad \dot{b}_k = a_k - a_{k-1}. \quad (4.4)$$

(Open–end boundary conditions correspond to $a_0 = a_N = 0$, and the periodic ones correspond to $a_0 \equiv a_N$, $b_{N+1} \equiv b_1$).

We want now to represent this flow in a different form of Newtonian equations of motion. This form arises, if we parametrise the variables (a, b) with the Poisson brackets (4.3) by means of canonically conjugated variables (x, p) with the Poisson brackets (1.1).

The most standard way to do this in the theory of the Toda lattice is:

$$a_k = \exp(x_{k+1} - x_k), \quad b_k = p_k.$$

We consider here a different parametrization, which is in a sense dual to the previous one:

$$a_k = \exp(p_k), \quad b_k = x_k - x_{k-1}. \tag{4.5}$$

This duality played an important role in the process of discovering the Toda lattice [10].

The Hamiltonian function $\varphi(T)$ takes the form

$$H_{\mathrm{T}}(x,p) = \sum_{k=1}^{N} \exp(p_k) + \frac{1}{2}\sum_{k=1}^{N}(x_k - x_{k-1})^2, \tag{4.6}$$

which generates the canonical equations of motion

$$\begin{aligned}
\dot{x}_k &= \partial H_{\mathrm{T}}/\partial p_k = \exp(p_k), \\
\dot{p}_k &= -\partial H_{\mathrm{T}}/\partial x_k = x_{k+1} - 2x_k + x_{k-1}.
\end{aligned}$$

These equations of motion imply the Newtonian ones

$$\ddot{x}_k = \dot{x}_k\left(x_{k+1} - 2x_k + x_{k-1}\right). \tag{4.7}$$

To obtain integrable discretizations of these equations, we apply the recipe formulated in the Sect. 3 with $F(T) = I + hf(T) = I + hT$.

Proposition 2 [1] *The factor* $\mathbf{B}(a,b,\lambda) = \Pi_+(I + hT(a,b,\lambda))$ *has the form*

$$\mathbf{B}(a,b,\lambda) = \sum_{k=1}^{N}\beta_k E_{kk} + h\lambda\sum_{k=1}^{N}E_{k+1,k}. \tag{4.8}$$

Here the coefficients β_k *are uniquely defined by the relations*

$$\beta_k = 1 + hb_k - \frac{h^2 a_{k-1}}{\beta_{k-1}}, \qquad \beta_k = 1 + O(h). \tag{4.9}$$

The discrete time Lax equation

$$\tilde{T} = \mathbf{B}^{-1}T\mathbf{B}, \tag{4.10}$$

is equivalent to the following map in the coordinates (a,b):

$$\tilde{a}_k = a_k\frac{\beta_{k+1}}{\beta_k}, \quad \tilde{b}_k = b_k + h\left(\frac{a_k}{\beta_k} - \frac{a_{k-1}}{\beta_{k-1}}\right). \tag{4.11}$$

In the open–end case, due to $a_0 = 0$, we obtain from (4.9) the expressions for β_k in terms of finite continued fractions. In the periodic case the recurrent relations in (4.9) uniquely define β_k's as N-periodic infinite continued fractions, and it can be proved that for h small enough these continued fractions converge and their values satisfy the asymptotic relations in (4.9). In each case, the relations (4.9) imply the following sharpened asymptotics:

$$\beta_k = 1 + hb_k + O(h^2), \tag{4.12}$$

which makes it evident that the map (4.11) approximates the flow (4.4).

According to the general tehory, the map (4.11) is Poisson with respect to the bracket (4.3). We want to find out how this map looks like in the canonically conjugated coordinates (x, p).

Proposition 3 *In the parametrization (4.5) the equations of motion (4.11) may be presented in the form of the following two equations:*

$$h \exp(p_k) = (\tilde{x}_k - x_k)\Big(1 + h(x_k - \tilde{x}_{k-1})\Big), \tag{4.13}$$

$$h \exp(\tilde{p}_k) = (\tilde{x}_k - x_k)\Big(1 + h(x_{k+1} - \tilde{x}_k)\Big), \tag{4.14}$$

which imply the Newtonian equations of motion

$$\frac{\tilde{x}_k - x_k}{x_k - \underset{\sim}{x}_k} = \frac{1 + h(x_{k+1} - x_k)}{1 + h(x_k - \underset{\sim}{\tilde{x}}_{k-1})}. \tag{4.15}$$

The auxiliary variables β_k are given by

$$\beta_k = 1 + h(x_k - \tilde{x}_{k-1}). \tag{4.16}$$

5 Relativistic Toda lattice and its integrable discretization

Our next example will be the relativistic Toda lattice. See [2] for the references.

The Lax matrix of the relativistic Toda lattice may be taken as

$$T(c, d, \lambda) = L(c, d, \lambda)U^{-1}(c, d, \lambda), \tag{5.1}$$

where

$$L(c, d, \lambda) = \sum_{k=1}^{N} d_k E_{kk} + \lambda \sum_{k=1}^{N} E_{k+1,k}, \quad U(c, d, \lambda) = \sum_{k=1}^{N} E_{kk} - \lambda^{-1} \sum_{k=1}^{N} c_k E_{k,k+1}. \tag{5.2}$$

These matrices depend on the variables $(c, d) \in \mathbb{R}^{2N}$ and on the spectral parameter λ. They may be considered as belonging to the same algebra g as defined in the previous section. Define also the operator R and the bracket (1.5) as in the previous section.

Proposition 4 [11] *The set of matrices $T(c, d, \lambda)$ forms a Poisson submanifold in g. The coordinate representation of the restriction of the bracket (1.5) to this set is:*

$$\{c_k, d_{k+1}\} = -c_k, \quad \{c_k, d_k\} = c_k, \quad \{d_k, d_{k+1}\} = c_k. \tag{5.3}$$

Now the simplest flow of the relativistic Toda hierarchy corresponds to the Hamiltonian function

$$\varphi(T) = \frac{1}{2}\mathrm{tr}(T^2)_0 = \frac{1}{2}\sum_{k=1}^{N}(d_k + c_{k-1})^2 + \sum_{k=1}^{N}(d_k + c_{k-1})c_k.$$

and reads:

$$\dot{d}_k = d_k(c_k - c_{k-1}), \quad \dot{c}_k = c_k(d_{k+1} + c_{k+1} - d_k - c_{k-1}). \tag{5.4}$$

It may be considered under open–end boundary conditions ($d_{N+1} = c_0 = c_N = 0$), or under periodic ones (all the subscripts are taken (mod N), so that $d_{N+1} \equiv d_1$, $c_0 \equiv c_N$, $c_{N+1} \equiv c_1$).

The Poisson brackets (5.3) also may be parametrized by canonically conjugated variables (x, p), see [5]. Here we chose another parametrization, dual to the one used there:

$$d_k = x_k - x_{k-1} - \exp(p_{k-1}), \quad c_k = \exp(p_k). \tag{5.5}$$

Under this parametrization the Hamiltonian function of the relativistic Toda lattice becomes

$$H_{\mathrm{RT}} = \sum_{k=1}^{N}\exp(p_k)(x_k - x_{k-1}) + \frac{1}{2}\sum_{k=1}^{N}(x_k - x_{k-1})^2. \tag{5.6}$$

Correspondingly, the flow (5.4) takes the form of canonical equations of motion:

$$\dot{x}_k = \partial H_{\mathrm{RT}}/\partial p_k = \exp(p_k)(x_k - x_{k-1}),$$
$$\dot{p}_k = -\partial H_{\mathrm{RT}}/\partial x_k = x_{k+1} - 2x_k + x_{k-1} + \exp(p_{k+1}) - \exp(p_k).$$

As an immediate consequence of these equations one gets the Newtonian equations of motion

$$\ddot{x}_k = \dot{x}_k\left(x_{k+1} - 2x_k + x_{k-1}\right) + \left(\frac{\dot{x}_{k+1}\dot{x}_k}{x_{k+1} - x_k} - \frac{\dot{x}_k\dot{x}_{k-1}}{x_k - x_{k-1}}\right). \tag{5.7}$$

Applying the recipe of the Sect.3 with $F(T) = I + hf(T) = I + hT$ to discretize the above equations, we get:

Proposition 5 [2] *The factor* $\mathbf{A} = \Pi_+(I + hT(c, d, \lambda))$ *has the structure:*

$$\mathbf{A}(c, d, \lambda) = \sum_{k=1}^{N} a_k E_{kk} + h\lambda \sum_{k=1}^{N} E_{k+1,k}, \qquad (5.8)$$

where the quantities a_k *are defined by the relations*

$$a_{k+1} = 1 + hd_{k+1} + \frac{hc_k}{a_k}, \qquad a_k = 1 + O(h). \qquad (5.9)$$

The matrix difference equation

$$\tilde{T} = \mathbf{A}^{-1}T\mathbf{A} \qquad (5.10)$$

is equivalent to the following map in the variables (c, d):

$$\tilde{d}_k = d_k \frac{a_{k+1} - hd_{k+1}}{a_k - hd_k}, \qquad \tilde{c}_k = c_k \frac{a_{k+1} + hc_{k+1}}{a_k + hc_k}. \qquad (5.11)$$

The auxiliary functions a_k, as in the case of the Toda lattice, have in the open–end case expressions in the form of finite continued fractions, and in the periodic case – in the form of periodic continued fractions which converge for h small enough. In each case they satisfy the folowing sharpened asymptotic relations:

$$a_k = 1 + h(d_k + c_{k-1}) + O(h^2), \qquad (5.12)$$

which shows once more that the map (5.11) approximates the flow (5.4). The map (5.11) is Poisson with respect to the bracket (5.3).

Switching on the parametrization (5.5), we have the following results.

Proposition 6 *In the parametrization* (5.5) *the equations of motion* (5.11) *may be presented in the form of the following two equations:*

$$h\exp(p_k) = \frac{(\tilde{x}_k - x_k)}{(x_k - \tilde{x}_{k-1})} \left(1 + h(x_k - \tilde{x}_{k-1})\right), \qquad (5.13)$$

$$h\exp(\tilde{p}_k) = \frac{(\tilde{x}_k - x_k)}{(\tilde{x}_k - \tilde{x}_{k-1})} \frac{(\tilde{x}_{k+1} - \tilde{x}_k)}{(x_{k+1} - \tilde{x}_k)} \left(1 + h(x_{k+1} - \tilde{x}_k)\right). \qquad (5.14)$$

which imply also the Newtonian equations of motion

$$\frac{\tilde{x}_k - x_k}{x_k - \underset{\sim}{x}_k} = \frac{(x_{k+1} - x_k)}{(x_{k+1} - x_k)} \frac{(x_k - \tilde{x}_{k-1})}{(x_k - x_{k-1})} \frac{\left(1 + h(x_{k+1} - x_k)\right)}{\left(1 + h(x_k - \tilde{x}_{k-1})\right)}. \qquad (5.15)$$

The auxiliary functions a_k *have the expressions*

$$a_k = 1 + h(x_k - \tilde{x}_{k-1}). \qquad (5.16)$$

6   Bruschi–Ragnisco lattice
and its integrable discretization

It is interesting to note that the equations of motion (5.7) have a remarkable feature: their right–hand side may be seen as a *sum* of the right–hand sides of the usual Toda lattice (4.7) and of another well–known integrable lattice, namely the Bruschi–Ragnisco one [12]:

$$\ddot{x}_k = \frac{\dot{x}_{k+1}\dot{x}_k}{x_{k+1} - x_k} - \frac{\dot{x}_k\dot{x}_{k-1}}{x_k - x_{k-1}}. \tag{6.1}$$

We will demonstrate that the situation becomes even more bizarre when looking at the discrete time counterparts of these systems. To this end we need first to recall the construction of the Bruschi–Ragnisco hierarchy, and to derive the discretization of its simplest flow.

Consider the $N \times N$ rank 1 Lax matrix $T(b, c)$ with the entries given by:

$$T_{kj} = b_j \prod_{i=k}^{j-1} c_i , \quad k \le j; \qquad T_{kj} = b_j \left(\prod_{i=j}^{k-1} c_i \right)^{-1} , \quad k > j . \tag{6.2}$$

Proposition 7 [13] *The rank 1 matrices (6.2) form a Poisson submanifold in $g = gl(N)$ equipped with the standard Lie–Poisson structure. The coordinate representation of the restriction of this structure to the above set is:*

$$\{c_k, b_k\} = -\{c_k, b_{k+1}\} = c_k. \tag{6.3}$$

A somewhat exceptional feature of the Bruschi–Ragnisco hierarchy is that its Hamiltonian functions are given not by the Ad–invariant functions of T (all such functions are dependant on the set of rank 1 matrices), but by *linear* functionals $\mathrm{tr}(M^m T)$. Here

$$M = \sum_{k=1}^{N-1} E_{k+1,k} \qquad \text{or} \qquad M = \sum_{k=1}^{N-1} E_{k+1,k} + C E_{1,N} \tag{6.4}$$

in the open–end or in the periodic case, respectively (in the latter case C is the value of the Casimir function $C = c_1 \ldots c_N$ of the bracket (6.3)).

The corresponding Lax equations of motion have the form

$$\dot{T} = [T, M^m]. \tag{6.5}$$

They are linear and can easily be solved:

$$T(t) = \exp(-tM^m)T(0)\exp(tM^m). \tag{6.6}$$

In particular, for $m = 1$ we get the Hamiltonian function

$$\varphi(T) = \mathrm{tr}(MT) = \sum_{k=1}^{N} b_{k+1}c_k.$$

The equations of motion:

$$\dot{b}_k = b_{k+1}c_k - b_k c_{k-1}, \quad \dot{c}_k = c_k(c_k - c_{k-1}). \tag{6.7}$$

They may be considered either under open–end boundary conditions ($c_0 = b_{N+1} = 0$), or under periodic ones (all the subscripts are taken (mod N), so that $c_0 \equiv c_N$, $b_{N+1} \equiv b_1$).

Observe now that the bracket (6.3) for the Bruschi–Ragnisco lattice formally is identical with that for the Toda lattice (4.3). This suggests the parametrization

$$c_k = \exp(p_k), \quad b_k = x_k - x_{k-1}. \tag{6.8}$$

The corresponding Hamiltonian function H_{BR} takes the form

$$H_{\mathrm{BR}}(x,p) = \sum_{k=1}^{N} \exp(p_k)(x_{k+1} - x_k), \tag{6.9}$$

which generates the canonical equations of motion

$$\begin{aligned}\dot{x}_k &= \partial H_{\mathrm{BR}}/\partial p_k = \exp(p_k)(x_{k+1} - x_k),\\ \dot{p}_k &= -\partial H_{\mathrm{BR}}/\partial x_k = \exp(p_k) - \exp(p_{k-1}).\end{aligned}$$

These equations of motion imply the Newtonian ones (6.1).

Since the Lax equations for the Bruschi–Ragnisco hierarchy do not have the form (2.4), the recipe of the Sect.3 can not be applied literally. However, according to its spirit, the discrete time Bruschi–Ragnisco lattice must belong to the same hierarchy, i.e. to share the Lax matrix with the continuous time one, and its explicit solution should be given by

$$T(nh) = (I + hM)^{-n}T(0)(I + hM)^n. \tag{6.10}$$

Hence the corresponding discrete Lax equation should have the form

$$\tilde{T} = (I + hM)^{-1}T(I + hM). \tag{6.11}$$

Proposition 8 *The matrix equation (6.11) in the coordinates (b,c) has the form:*

$$\tilde{b}_k(1 + h\tilde{c}_{k-1}) = b_k + hb_{k+1}c_k, \quad \tilde{c}_k = c_k \frac{1 + h\tilde{c}_k}{1 + h\tilde{c}_{k-1}}. \tag{6.12}$$

By construction, this map is Poisson with respect to the bracket (6.3). We want now to represent this map in the canonically conjugated variables (x, p).

Proposition 9 *In the coordinates* (x, p) *given by* (6.8) *the equations of motion* (6.12) *may be presented in the form of the following two equations:*

$$h \exp(p_k) = \frac{(\tilde{x}_k - x_k)}{(x_{k+1} - x_k)} \frac{(x_k - x_{k-1})}{(x_k - \tilde{x}_{k-1})},$$ (6.13)

$$h \exp(\tilde{p}_k) = \frac{(\tilde{x}_k - x_k)}{(x_{k+1} - \tilde{x}_k)},$$ (6.14)

which imply also the Newtonian equations of motion

$$\frac{\tilde{x}_k - x_k}{x_k - \underset{\sim}{x}_k} = \frac{(x_{k+1} - x_k)}{(x_{k+1} - \underset{\sim}{x}_k)} \frac{(x_k - \tilde{x}_{k-1})}{(x_k - x_{k-1})}.$$ (6.15)

So, the right–hand side of the equations of motion of the discrete time relativistic Toda lattice (5.15) may be seen as a *product* of the right–hand sides of the equations of motion of the discrete time Toda lattice (4.15) and the discrete time Bruschi–Ragnisco lattice (6.15).

The relations we established between three well known lattice systems, may be symbolically presented as

$$\text{RTL} = \text{TL} + \text{BRL}, \qquad \text{dRTL} = \text{dTL} \times \text{dBRL}.$$

They are, in my opinion, rather beautiful and unexpected. However, they probably are nothing more than a pure curiosity, which demonstrates once more: the field of integrable systems of classical mechanics, even in its best studied parts, is far from being exhausted. Many important and less important but funny findings await us in this fascinating world.

7 Second flow of the Toda hierarchy and its integrable discretization

We want to demonstrate now that our procedure of integrable discretization leads to reasonable results not only for the *simplest* flows of various hierarchies. To this end we turn to the *second* flow of the Toda hirarchy descibed by the Lax matrix (4.1) and the Hamiltonian function

$$\psi(T) = \frac{1}{3}\text{tr}(T^3)_0 = \frac{1}{3}\sum_{k=1}^{N} b_k^3 + \sum_{k=1}^{N} a_k(b_k + b_{k+1}).$$

The coordinate representation of this flow is:

$$\dot{a}_k = a_k(b_{k+1}^2 - b_k^2 + a_{k+1} - a_{k-1}), \quad \dot{b}_k = a_k(b_{k+1} + b_k) - a_{k-1}(b_k + b_{k-1}). \quad (7.1)$$

In the coordinates (x, p) introduced by (4.5) the Hamiltonian function $\psi(T)$ takes the form

$$H_{T2}(x, p) = \sum_{k=1}^{N} \exp(p_k)(x_{k+1} - x_{k-1}) + \frac{1}{3}\sum_{k=1}^{N}(x_k - x_{k-1})^3, \quad (7.2)$$

which generates the canonical equations of motion

$$\begin{aligned}
\dot{x}_k &= \partial H_{T2}/\partial p_k = \exp(p_k)(x_{k+1} - x_{k-1}), \\
\dot{p}_k &= -\partial H_{T2}/\partial x_k = \exp(p_{k+1}) - \exp(p_{k-1}) + (x_{k+1} - x_k)^2 - (x_k - x_{k-1})^2.
\end{aligned}$$

These equations of motion imply the Newtonian ones

$$\ddot{x}_k = \dot{x}_k\left((x_{k+1} - x_k)^2 - (x_k - x_{k-1})^2\right)$$

$$+\dot{x}_{k+1}\dot{x}_k\left(\frac{1}{x_{k+2} - x_k} + \frac{1}{x_{k+1} - x_{k-1}}\right) - \dot{x}_k\dot{x}_{k-1}\left(\frac{1}{x_{k+1} - x_{k-1}} + \frac{1}{x_k - x_{k-2}}\right).$$

$$(7.3)$$

Applying the recipe of the Sect. 3 with $F(T) = I + hf(T) = I + hT^2$, we have:

Proposition 10 *The factor* $\mathbf{C}(a, b, \lambda) = \Pi_+(I + hT^2(a, b, \lambda))$ *has the form*

$$\mathbf{C}(a, b, \lambda) = \sum_{k=1}^{N}\gamma_k E_{kk} + h\lambda\sum_{k=1}^{N}\delta_k E_{k+1,k} + h\lambda^2\sum_{k=1}^{N}E_{k+2,k}. \quad (7.4)$$

Here the coefficients γ_k, δ_k *are uniquely defined by the relations*

$$\gamma_k = 1 + h(b_k^2 + a_k + a_{k-1}) - \frac{h^2 a_{k-1}\delta_{k-1}^2}{\gamma_{k-1}} - \frac{h^2 a_{k-1}a_{k-2}}{\gamma_{k-2}}, \qquad \gamma_k = 1 + O(h), \quad (7.5)$$

$$\delta_k = b_{k+1} + b_k - \frac{h a_{k-1}\delta_{k-1}}{\gamma_{k-1}}, \qquad \delta_k = O(1). \quad (7.6)$$

The discrete time Lax equation

$$\tilde{T} = \mathbf{C}^{-1}T\mathbf{C}, \quad (7.7)$$

is equivalent to the following map in the coordinates (a, b):

$$\tilde{a}_k = a_k\frac{\gamma_{k+1}}{\gamma_k}, \quad \tilde{b}_k = b_k + h\left(\frac{a_k\delta_k}{\gamma_k} - \frac{a_{k-1}\delta_{k-1}}{\gamma_{k-1}}\right). \quad (7.8)$$

This time the sharpened asymptotics for the auxiliary functions γ_k, δ_k read:

$$\gamma_k = 1 + h(b_k^2 + a_k + a_{k-1}) + O(h^2), \quad \delta_k = b_{k+1} + b_k + O(h), \quad (7.9)$$

which make it obvious that the map (7.8) approximates the flow (7.1).

In the canonically conjugated variables (x, p) the map (7.8) allows the following formulation.

Proposition 11 *In the parametrization* (4.5) *the equations of motion* (7.8) *may be presented in the form of the following two equations:*

$$h \exp(p_k) = \frac{(\tilde{x}_k - x_k)(x_k - \tilde{x}_{k-2})\left(1 + h(x_k - \tilde{x}_{k-1})^2\right)}{(x_{k+1} + x_k - \tilde{x}_k - \tilde{x}_{k-1})(x_k + x_{k-1} - \tilde{x}_{k-1} - \tilde{x}_{k-2})}, \quad (7.10)$$

$$h \exp(\tilde{p}_k) = \frac{(\tilde{x}_k - x_k)(x_{k+2} - \tilde{x}_k)\left(1 + h(x_{k+1} - \tilde{x}_k)^2\right)}{(x_{k+1} + x_k - \tilde{x}_k - \tilde{x}_{k-1})(x_{k+2} + x_{k+1} - \tilde{x}_{k+1} - \tilde{x}_k)}, \quad (7.11)$$

which imply the Newtonian equations of motion

$$\frac{\tilde{x}_k - x_k}{x_k - \underset{\sim}{x}_k} = \frac{\left(x_{k+2} - x_k\right)\left(x_k + x_{k-1} - \tilde{x}_{k-1} - \tilde{x}_{k-2}\right)\left(x_{k+1} + x_k - \tilde{x}_k - \tilde{x}_{k-1}\right)}{\left(x_{k+2} + x_{k+1} - \underset{\sim}{x}_{k+1} - x_k\right)\left(x_k - \tilde{x}_{k-2}\right)\left(x_{k+1} + x_k - \underset{\sim}{x}_k - x_{k-1}\right)}$$

$$\times \frac{\left(1 + h(x_{k+1} - x_k)^2\right)}{\left(1 + h(x_k - \tilde{x}_{k-1})^2\right)} \quad (7.12)$$

The auxiliary variables γ_k, δ_k *are given by*

$$\gamma_k = \frac{(x_{k+1} - \tilde{x}_{k-1})(x_k - \tilde{x}_{k-2})}{(x_{k+1} + x_k - \tilde{x}_k - \tilde{x}_{k-1})(x_k + x_{k-1} - \tilde{x}_{k-1} - \tilde{x}_{k-2})}\left(1 + h(x_k - \tilde{x}_{k-1})^2\right),$$

$$\delta_k = x_{k+1} - \tilde{x}_{k-1}. \quad (7.13)$$

8 Conclusion

The examples discussed in the present paper do not exhaust the possibilities of our approach to the problem of integrable discretizations. Other examples may be found in [1]–[6] and references therein, as well as, hopefully, in the subsequent publications.

The research of the author is financially supported by the DFG (Deutsche Forschungsgemeinschaft).

References

[1] Yu.B.Suris. Bi–Hamiltonian structure of the qd algorithm and new discretizations of the Toda lattice. – *Physics Letters A*, 1995, **206**, 153–161.

[2] Yu.B.Suris. A discrete–time relativistic Toda lattice. – *J.Physics A: Math. & Gen.*, 1996, **29**, 451–465.

[3] Yu.B.Suris. Integrable discretizations of the Bogoyavlensky lattices. *J. Math. Phys.*, 1996, **37**, 3982–3996.

[4] Yu.B.Suris. A discrete time peakons lattice. *Physics Letters A*, 1996, **217**, 321–329.

[5] Yu.B.Suris. New integrable systems related to the relativistic Toda lattice. *J. Phys. A: Math. and Gen.* (submitted for publication); solv-int/9605006.

[6] Yu.B.Suris. On some integrable systems related to the Toda lattice. *J. Phys. A: Math. and Gen.* (submitted for publication); solv-int/96050010.

[7] A.G.Reyman, M.A.Semenov-Tian-Shansky. Group–theoretical method in the theory of finite–dimensional integrable systems. – *In: Encyclopaedia of Math.Sciences, V.16, Dynamical systems VII. Springer, 1993.*

[8] W.W.Symes. The QR algorithm and scattering for the finite nonperiodic Toda lattice. *Physica D*, 1982, **4**, 275–280.

[9] M.Adler, P. van Moerbecke. Completely integrable systems, Euclidean Lie algebras, and curves. *Adv. Math.*, 1980, **38**, 267–317.

[10] M.Toda. *Theory of nonlinear lattices*, Springer, 1981.

[11] Yu.B.Suris. On the bi–Hamiltonian structure of Toda and relativistic Toda lattices. – *Physics Letters A*, 1993, **180**, 419–429.

[12] M.Bruschi, O.Ragnisco. On a new integrable Hamiltonian system with nearest neighbour interaction. – *Inverse Problems*, 1989, **5**, 983–998.

[13] Yu.B.Suris. On the algebraic structure of the Bruschi–Ragnisco lattice. – *Physics Letters A*, 1993, **179**, 403–406.

Chapter 3

Discrete Geometry

Discrete Conformal Maps and Surfaces

Alexander I. Bobenko *

FB Mathematik, Technische Universität Berlin,
Strasse des 17. Juni 136, Berlin, Germany[†]
and
Department of Mathematics, University of Massachusetts,
Amherst, MA 01003
Email: bobenko@sfb288.math.tu-berlin.de

1 Definition of a Discrete Conformal Map

Definition 1.1. A map $f \colon \mathbb{Z} \to \mathbb{C}$ is called *discrete conformal (discrete holomorphic)* if the cross-ratios of all its elementary quadrilaterals are equal to -1:

$$q_{n,m} := q(f_{n,m},\, f_{n+1,m},\, f_{n+1,m+1},\, f_{n,m+1}) :=$$
$$\frac{(f_{n,m} - f_{n+1,m})(f_{n+1,m+1} - f_{n,m+1})}{(f_{n+1,m} - f_{n+1,m+1})(f_{n,m+1} - f_{n,m})} = -1. \tag{1}$$

This definition appeared [1] in 1991 and is motivated by the following properties:

- $f \colon D \subset \mathbb{C} \to \mathbb{C}$ is a (smooth) conformal (holomorphic or antiholomorphic) map if and only if $\forall (x,\, y) \in D$

$$\lim_{\varepsilon \to 0} q\bigl(f(x,\, y),\, f(x + \epsilon,\, y),\, f(x + \epsilon,\, y + \epsilon),\, f(x,\, y + \epsilon)\bigr) = -1. \tag{2}$$

- Definition 1.1 is Möbius invariant:

$$f \qquad \text{and} \qquad \tilde{f} = \frac{af + b}{cf + d} \tag{3}$$

are discrete conformal simultaneously.

---

*Partially supported by the SFB288 and by NSF grant DMS93–12087
†permanent address

- The dual, $f^*\colon \mathbb{Z}^2 \to \mathbb{C}$, to a discrete conformal map f, is defined in [2] by

$$f^*_{n+1,m} - f^*_{n,m} = \frac{1}{f_{n+1,m} - f_{n,m}}, \quad f^*_{n,m+1} - f^*_{n,m} = -\frac{1}{f_{n,m+1} - f_{n,m}}.$$

$$(4)$$

The smooth limit of this duality is $(f^*)' = 1/\bar{f}'$ where f is holomorphic and f^* is antiholomorphic.

- Equation (1) is integrable. The Lax pair

$$\Psi_{n+1,m} = U_{n,m}\Psi_{n,m} \tag{5}$$
$$\Psi_{n,m+1} = V_{n,m}\Psi_{n,m}$$

found by Nijhoff and Capel in [3] is of the form

$$U_{n,m} = \begin{pmatrix} 1 & -u_{n,m} \\ \frac{\lambda}{u_{n,m}} & 1 \end{pmatrix}, \quad V_{n,m} = \begin{pmatrix} 1 & -v_{n,m} \\ -\frac{\lambda}{v_{n,m}} & 1 \end{pmatrix}, \tag{6}$$

where

$$u_{n,m} = f_{n+1,m} - f_{n,m}, \qquad v_{n,m} = f_{n,m+1} - f_{n,m} \tag{7}$$

Let us mention also that all the properties are preserved [2],[3] if $q = -1$ is replaced by

$$q_{n,m} = \frac{\alpha_n}{\beta_m}.$$

The discrete conformal maps defined above are quadrilateral patterns with the combinatorics of the square grid. Ramified coverings can be modelled by quadrilateral patterns with more complicated combinatorics when N edges [1] may meet at a vertex. In this case $\mathbb{Z}^2$ in the definition should be replaced by a quad–graph G [4].

2 Examples

- $Z :=$ discrete z

$$Z(n, m) := n + im.$$

---

[1] usually one assumes that N is even and $N \geq 4$.

- EXP := discrete e^z

$$\text{EXP}_\gamma(n, m) := \exp(2n \, \text{arcsinh} \gamma + 2im \arcsin \gamma), \ \gamma \in \mathbb{R}.$$

- Various discrete rational, trigonometric and hyperbolic functions (for example TANH := discrete $\tanh z$) can be obtained from the first two examples by various combinations of the transformations (3), (4) (the Bäcklund–Darboux transformations).

- Z^γ: = discrete z^γ.

Equation (1) can be supplemented with the following nonautonomous constraint:

$$\gamma(f_{n,m} + \delta) = 2(n - \alpha) \frac{(f_{n+1,m} - f_{n,m})(f_{n,m} - f_{n-1,m})}{f_{n+1,m} - f_{n-1,m}} \tag{8}$$

$$+ 2(m - \beta) \frac{(f_{n,m+1} - f_{n,m})(f_{n,m} - f_{n,m-1})}{f_{n,m+1} - f_{n,m-1}}.$$

Theorem 2.1. $f\colon \mathbb{Z}^2 \to \mathbb{C}$ *is a solution to the system* (1), (8) *if and only if there exists a solution to* (5), (6), *which satisfies the following differential equation in* λ:

$$\frac{d}{d\lambda}\Psi_{n,m} = A\Psi_{n,m}, \quad A = \frac{1}{\lambda}A_0 + \frac{1}{\lambda-1}A_1 + \frac{1}{\lambda+1}A_{-1}, \tag{9}$$

where the matrices A_0, A_1, A_{-1} *are* λ-*independent. The constraint* (8) *is compatible with* (1).

In the case $\gamma = 1$ the constraint (8) and the corresponding monodrony problem (9) were obtained in [5]. The calculation of the coefficients of A is rather tedious. Correcting misprints in the monodrony problem presented in [5] and generalizing it to the case $\gamma \neq 1$ important for us, we get

$$A_0 = \begin{pmatrix} -\dfrac{\gamma}{4} & (\alpha - n)\dfrac{u_{n,m}u_{n-1,m}}{u_{n,m} + u_{n-1,m}} + (\beta - m)\dfrac{v_{n,m}v_{n,m-1}}{v_{n,m} + v_{n,m-1}} \\ 0 & \dfrac{\gamma}{4} \end{pmatrix},$$

$$A_1 = \frac{m - \beta}{v_{n,m} + v_{n,m-1}}\begin{pmatrix} v_{n,m} & v_{n,m}\,v_{n,m-1} \\ 1 & v_{n,m-1} \end{pmatrix} + \frac{\beta}{2}\begin{pmatrix} 1 & 0 \\ 0 & 1 \end{pmatrix} \tag{10}$$

$$A_{-1} = \frac{n - \alpha}{u_{n,m} + u_{n-1,m}}\begin{pmatrix} u_{n,m} & u_{n,m}\,u_{n-1,m} \\ 1 & u_{n-1,m} \end{pmatrix} + \frac{\alpha}{2}\begin{pmatrix} 1 & 0 \\ 0 & 1 \end{pmatrix}.$$

Remark 1. The constraint (8) is not Möbius invariant and can be easily generalized by applying a general Möbius transformation to $f_{n,m}$. The generalized constraint is similarly to (8), but with a quadratic polynomial of $f_{n,m}$ in the left hand side. So defined generalized class is invariant with respect to the Möbius and dual transformations.

Remark 2. The monodromy problem (9) coincides with the one of the Painlevé VI equation [6], which shows that the system (1) and (8) can be solved in terms of the Painlevé transcendents.

Let us assume $\gamma < 2$ and denote $\mathbb{Z}_+^2 = \{(n, m) \in \mathbb{Z}^2 : n, m \geq 0\}$. Motivated by the asymtotics of the constraint (8) at $n, m \to \infty$ and the properties

$$z^\gamma(\mathbb{R}_+) \in \mathbb{R}_+, z^\gamma(i\mathbb{R}_+) \in e^{\gamma \pi i/2}\mathbb{R}_+,$$

of the holomorphic z^γ it is natural to give the following definition of the "discrete z^γ" which we denote by Z^γ.

Definition 2.1. $Z^\gamma \colon \mathbb{Z}^2 \to \mathbb{C}$ is the solution of (1), (8) with $\alpha = \beta = \delta = 0$ and with initial conditions

$$Z^\gamma(0, 0) = 0, \quad Z^\gamma(1, 0) = 1, \quad Z^\gamma(0, 1) = e^{\gamma \pi i/2}.$$

It is easy to see that $Z^\gamma(n, 0) \in \mathbb{R}_+, Z^\gamma(0, m) \in e^{\gamma \pi i/2}\mathbb{R}_+, \quad \forall n, m \in \mathbb{N}$.

Conjecture 2.1. $Z^\gamma : \mathbb{Z}_+^2 \to \mathbb{C}$ is an embedding, i.e. different open elementary quadrilaterals of the pattern $Z^\gamma(\mathbb{Z}_+^2)$ do not intersect.

Computer experiments made by Tim Hoffmann confirm this conjecture.

Conjecture 2.2. Z^γ is the only embedded discrete conformal map $f : \mathbb{Z}_+^2 \to \mathbb{C}$ with

$$f(0, 0) = 0, \quad f(n, 0) \in \mathbb{R}_+, \quad f(0, m) \in e^{\gamma \pi i/2}\mathbb{R}_+ \quad \forall n, m \in \mathbb{N}.$$

We hope to prove these conjectures by combining geometrical methods with the modern theory of the Painlevé equations [7], [6].

In the discrete as well as in the smooth case (up to constant factor) one has

$$(Z^\gamma)^* = \overline{Z^{2-\gamma}}.$$

3　Discrete Surfaces and Coordinate Systems

Almost all notions of this section belong to the conformal (Möbius) geometry. One can easily extend the notion of the cross-ratio (1) to points in $\mathbb{R}^3$ identifying a sphere S passing through $X_1, X_2, X_3, X_4 \in \mathbb{R}^3$ with the Riemann sphere $\mathbb{CP}^1$. The cross-ratio is real when the four points are concircular. A direct generalization of the definitions of Section 1 to $\mathbb{R}^3$ yields the following definition [2].

Definition 3.1. A *discrete I-surface* (discrete isothermic surface) is a map $F: \mathbb{Z}^2 \to \mathbb{R}^3$ for which

$$q(F_{n,m}, F_{n+1,m}, F_{n+1,m+1}, F_{n,m+1}) = \frac{\alpha_n}{\beta_m}, \qquad \alpha, \beta: \mathbb{Z} \to \mathbb{R}. \qquad (11)$$

All the properties of the discrete conformal maps listed in Section 1 hold (for simplicity we set $q_{n,m} = -1$):

- Infintesimal quadrilaterals of the smooth isothermic surfaces satisfy (2).

- Definition 3.1 is Möbius invariant (now with respect to the Möbius transformations in $\mathbb{R}^3 \cup \{\infty\}$).

- The dual discrete I-surface is defined by

$$F^*_{n+1,m} - F^*_{n,m} = \frac{F_{n+1,m} - F_{n,m}}{\|F_{n+1,m} - F_{n,m}\|^2},$$

$$F^*_{n,m+1} - F^*_{n,m} = -\frac{F_{n,m+1} - F_{n,m}}{\|F_{n,m+1} - F_{n,m}\|^2}.$$

There exists a Lax pair [2] for (11). Special classes of the discrete I-surfaces can be characterized as follows:

- Discrete M-surfaces (minimal) [2] : The dual surface F^* lies on a sphere. The Gauss map of F is F^*.

- Discrete H-surfaces (constant mean curvature) [1],[8]: A dual surface F^* is "parallel" to F, i.e. it lies in constant distance of F

$$\|F_{n,m} - F^*_{n,m}\| = \frac{1}{H} = \text{const.}$$

Then H is the mean curvature for both F and F^*.

Discrete integrable systems are closely related to the Bäcklund–Darboux (BD) transformations of their smooth analogues. The loop group interpretation (see for example [9]) of the BD-transformation naturally yields the permutability theorem: given two BD-transformations D_1, D_2, there exist transformations D'_1, D'_2 such that

$$D'_1 D_2 = D'_2 D_1 \qquad (12)$$

holds. Here the D's lie in the corresponding loop group. Equation (12) becomes the Lax representation of the discrete 2-dimensional net. Moreover the commuting diagram can be generalized to the N-dimensional case for arbitrary $D_1, ..., D_N$: the edges of an N–dimensional cube can be completed with the BD transformations so that the diagram commutes.

The case $N = 3$ is geometrically interesting. Motivated by the interpretation of (11) as the permutability theorem for the Bäcklund–Darboux transformations, we suggest

Definition 3.2. A *discrete I-system* is a map $F : \mathbb{Z}^3 \to \mathbb{R}^3$ for which:

$$q(F_{n,m,l}, F_{n+1,m,l}, F_{n+1,m+1,l}, F_{n,m+1,l}) = \frac{\alpha_n}{\beta_m},$$

$$q(F_{n,m,l}, F_{n+1,m,l}, F_{n+1,m,l+1}, F_{n,m,l+1}) = \frac{\alpha_n}{\gamma_l},$$

$$q(F_{n,m,l}, F_{n,m+1,l}, F_{n,m+1,l+1}, F_{n,m,l+1}) = \frac{\beta_n}{\gamma_l}$$

hold for some $\alpha, \beta, \gamma \colon \mathbb{Z} \to \mathbb{R}$.

All the coordinate surfaces of a discrete I-system are discrete I-surfaces. An analytical description of the discrete I-systems via the BD transformations for isothermic surfaces is given in [10].

A discrete I-system is uniquely determined by its Cauchy data

$$F(\bullet, 0, 0,), \quad F(0, 0, \bullet), \quad F(0, \bullet, 0,) \colon \mathbb{Z} \to \mathbb{R}^3, \qquad \alpha, \beta, \gamma \colon \mathbb{Z} \to \mathbb{R}.$$

A direct geometrical proof of this is presented in [8].

The following generalization is motivated by the smooth limit and the Möbius invariance of the curvature line parametrization.

Definition 3.3. A *discrete C-surface* (discrete curvature line parametrized surface) is a map $F \colon \mathbb{Z}^2 \to \mathbb{R}^3$ such that all elementary quadrilaterals have negative cross-ratios (i.e. they are concircular and embedded).

A 2–parametric family of spheres is called a Ribeaucour sphere congruence if the curvature lines of the two enveloping surfaces do correspond [11].

Definition 3.4. Two discrete C–surfaces $F, \tilde{F} \colon \mathbb{Z}^2 \to \mathbb{R}^3$ envelope a *discrete R-sphere congruence* if $\forall n, m \in \mathbb{Z}$ the vertices of the elementary hexahedron $(F_{n,m}, F_{n+1,m}, F_{n+1,m+1}, F_{n,m+1}, \tilde{F}_{n,m}, \tilde{F}_{n+1,m}, \tilde{F}_{n+1,m+1}, \tilde{F}_{n,m+1})$ lie on a sphere.

The discrete R–sphere congruences allow a natural quaternionic description, which in the special case of the discrete I–surfaces yields their Lax representation.

Definition 3.5. A *discrete O-system* (discrete triply–orthogonal coordinate system) is a map $F \colon \mathbb{Z}^3 \to \mathbb{R}^3$ for which all elementary quadrilaterals have negative cross-ratios.

This definition is motivated by the Dupin theorem [11], which claims that the coordinate surfaces of a smooth triply–orthogonal coordinate system intersect along their curvature lines.

Theorem 3.1. $F\colon \mathbb{Z}^3 \to \mathbb{R}^3$ *is a non–degenerate discrete O–system if and only if all its elementary hexahedra*

$$H_{n,m} = (F_{n,m,l},\ F_{n+1,m,l},\ F_{n+1,m+1,l},\ F_{n,m+1,l},$$
$$F_{n,m,l+1},\ F_{n+1,m,l+1},\ F_{n+1,m+1,l+1},\ F_{n,m+1,l+1})$$

lie on spheres and are embedded.

Remark 3. A less restrictive version of Definitions 3.3, 3.4, 3.5 and Theorem 3.1 includes the condition $q \in \mathbb{R}$ only (i.e. does not assume the embeddedness).

The discrete O–systems are sphere packings with the combinatorics of the cube grid. Since the spheres comprise the "dual lattice" it is natural to label them by half–integer numbers: the vertices of the hexahedreon $H_{n,m}$ lie on the sphere $S_{n+\frac{1}{2},m+\frac{1}{2},l+\frac{1}{2}}$, the vertex $F_{n,m}$ is the intersection of 8 spheres $S_{n\pm\frac{1}{2},m\pm\frac{1}{2},l\pm\frac{1}{2}}$.

The cross-ratios of the faces (the index labels the "center" of the corresponding face)

$$R_{n,m+\frac{1}{2},l+\frac{1}{2}} = q(F_{n,m,l},\ F_{n,m+1,l},\ F_{n,m+1,l+1},\ F_{n,m,l+1})$$
$$R_{n+\frac{1}{2},m,l+\frac{1}{2}} = q(F_{n,m,l},\ F_{n+1,m,l},\ F_{n+1,m,l+1},\ F_{n,m,l+1})$$
$$R_{n+\frac{1}{2},m+\frac{1}{2},l} = q(F_{n,m,l},\ F_{n+1,m,l},\ F_{n+1,m+1,l},\ F_{n,m+1,l})$$

satisfy

$$R_{n+\frac{1}{2},m,l+\frac{1}{2}}\, R_{n+\frac{1}{2},m+1,l+\frac{1}{2}} =$$
$$R_{n,m+\frac{1}{2},l+\frac{1}{2}}\, R_{n+1,m+\frac{1}{2},l+\frac{1}{2}}\, R_{n+\frac{1}{2},m+\frac{1}{2},l}\, R_{n+\frac{1}{2},m+\frac{1}{2},l+1}.$$

The last equation holds for any 8 points on a sphere and by modular transformations of the cross-ratios of the n and l faces

$$T_{n+\frac{1}{2},m,l+\frac{1}{2}} := R_{n+\frac{1}{2},m,l+\frac{1}{2}}$$
$$T_{n,m+\frac{1}{2},l+\frac{1}{2}} := q(F_{n,m,l}, F_{n,m+1,l+1}, F_{n,m+1,l}, F_{n,m,l+1}) = 1 - R_{n,m+\frac{1}{2},l+\frac{1}{2}}$$
$$T_{n+\frac{1}{2},m+\frac{1}{2},l} := q(F_{n,m,l}, F_{n,m+1,l+1}, F_{n,m+1,l}, F_{n,m,l+1}) = (1 - R_{n+\frac{1}{2},m+\frac{1}{2},l}^{-1})^{-1}.$$

is transformed to a gauge–invariant form of the 3D Hirota bilinear difference equation (see, for example, [12]) on a sublattice. Directing the n– m– and l–axes to right, front and up respectively one can write this equation as

$$T_f T_b = \frac{(1 - T_l)(1 - T_r)}{(1 - T_u^{-1})(1 - T_d^{-1})}, \tag{13}$$

where the labels denote the f(ront), b(ack), l(eft), r(ight), u(p), and d(own)
faces of a hexahedron.

4   Circle Patterns of Schramm

Recently, coming from questions in approximation theory, Schramm in his
fundamental paper [13] proposed a more restrictive definition of discrete conformal maps than Definition 1.1. He considers circle patterns with the combinatorics of $\mathbb{Z}^2$ in the plane with the following characteristic properties:

1. On each circle there are 4 vertices.

2. Each vertex has 2 pairs of touching circles in common, the pairs intersecting orthogonally.

3. Let C be a circle of the pattern and C_1, C_2, C_3, C_4 its neighbors,
   intersecting C orthogonally, then $C_i \cap C_j \in C \quad \forall i \neq j$.

This definition is obviously Möbius invariant. A pattern is embedded,
provided the open discs of the circles, which are not neighbors, are disjoint.
If this holds for the half–neighbors (touching circles) we say that the pattern
is immersed (or planar).

Adding temporarily the midpoints of the circles one obtains a refinement
of the lattice with the following properties:

- There are two kinds of vertices which alternate: at the original vertices
  of the pattern there are 2 perpendicular outgoing pairs of edges, whereas
  the outgoing edges at the added vertices have the same length. This
  nicely relates the properties of the smooth case: $f_x \perp f_y$ and $|f_x| = |f_y|$
  respectively.

- The quadrilaterals of the refined lattice are of the "kite" shape, in particular, they have cross-ratio -1, thus providing a discrete conformal
  map in the sense of Definition 1.1.

- Take an immersed Schramm circle pattern, construct its refinement as
  above, build the dual (4) of it, delete the added points. This recipe
  provides us with another Schramm immersion, which we call *dual*.

We return to the lattice formed by the intersections of circles, which we
denote by $f \colon \mathbb{Z}^2 \to \mathbb{C}$. Define a function on faces by

$$T_{n+\frac{1}{2}, m+\frac{1}{2}} := q(f_{n+1,m}, f_{n+1,m+1}, f_{n,m+1}, f_{n,m})$$

and another function S on vertices by

$$S_{n,m} := q(f_{n,m-1}, f_{n+1,m}, f_{n,m+1}, f_{n-1,m}).$$

It is easy to see that both these functions are negative–valued. A more elaborated calculation [13] proves that they satisfy the discrete Cauchy–Riemann equations

$$\frac{S_u}{S_d} = \left(\frac{1 - T_r}{1 - T_l}\right)^2, \qquad \frac{S_r}{S_l} = \left(\frac{1 - T_u^{-1}}{1 - T_d^{-1}}\right)^2. \tag{14}$$

Here we use the notation of Section 3, where

$$S_u = S_{n,m+1}, \ S_d = S_{n,m}, \ T_r = T_{n+\frac{1}{2},m+\frac{1}{2}}, \ T_l = T_{n-\frac{1}{2},m+\frac{1}{2}},$$
$$S_r = S_{n+1,m}, \ S_l = S_{n,m}, \ T_u = T_{n+\frac{1}{2},m+\frac{1}{2}}, \ T_d = T_{n+\frac{1}{2},m-\frac{1}{2}}.$$

Taking a quadrilateral formed by the 4 vertices on a circle, the cross-ratios of the 4 neighboring quadrilaterals satisfy the compatibility condition of (14)

$$T^2 = \frac{(1 - T_r)(1 - T_l)}{(1 - T_u^{-1})(1 - T_d^{-1})}, \tag{15}$$

where T is the cross-ratio of the center quadrilateral. This is exactly the $3D$ Hirota equation with a translational symmetry in the front–back direction (cf. (13)).

The negative solutions of the discrete Cauchy–Riemann equations (14) are in one to one correspondence with the Möbius equivalence classes of the Schramm circle patterns [13]. Moreover, for a negative solution of (15) there exists a one–parametric family of negative solutions of (14). (S is unique up to a multiplication by a positive constant $S \mapsto \lambda S, \ \lambda \in \mathbb{R}_+$) and consequently a one–parametric family (associated family) of circle patterns. Obviously, λ in this construction plays the role of a spectral parameter. Equation (15) possesses a maximum principle, which allows proof of global results. In particular it was proven in [13], that the only embedding of the whole $\mathbb{Z}^2$ is the standard circle pattern (where all circles have constant radius).

Let us mention also an alternative description of this geometry. The equation for the radii of the neighboring circles is

$$r^2(r_u + r_d + r_l + r_r) = r_u r_d r_l r_r (r_u^{-1} + r_d^{-1} + r_l^{-1} + r_r^{-1}), \tag{16}$$

which is probably also integrable.

It is possible to generalize Schramm's patterns replacing $\mathbb{Z}^2$ by a quad-graph G (see Section 1). Instead of having 4 vertices on every circle, one allows various numbers N of vertices (and as a consequence N neighboring and N half–neighboring circles). Such a singular point is natural to call a *branch point* of order $N/4 - 1$. In case of even $N = 2M$ cross-ratios can be prescribed to faces and there is a generalization of (15), which takes various cross-ratios of the central circle into account.

Let $F_1, \dots F_{2M}$ be the consequtive vertices on a circle C labeled counterclockwise. Denote by $C_1, \dots, C_{2M}$ the neighboring circles of C with the intersection points F_i, $F_{i+1} = C \cap C_i$. Denote by H_i, and G_i the vertices on C_i neighboring F_i and F_{i+1} such that the vertices G_i, F_i, F_{i+1}, H_i are consequtive on C_i. The cross-ratios [2] around C

$$R_i: = q(F_i, F_{i+2}, F_{i+1}, F_{i-1}), \qquad \tilde{R}_i: = q(G_i, F_{i+1}, H_i, F_i)$$

satisfy the following equation

$$\frac{R_1 R_3 \dots R_{2M-1}}{R_2 R_4 \dots R_{2M}} = \frac{\tilde{R}_1 \tilde{R}_3 \dots \tilde{R}_{2M-1}}{\tilde{R}_2 \tilde{R}_4 \dots \tilde{R}_{2M}}. \tag{17}$$

In addition R's are subject to constraints

$$R_i > 1,$$
$$R_i m(R_{i+1} m(R_{i+2} \dots R_{i+k-1} m(R_{i+k}) \dots)) > 1, \qquad k < 2M - 3$$
$$R_i m(R_{i+1} m(R_{i+2} \dots R_{i+2M-4} m(R_{i+2M-3}) \dots)) = 1, \quad i = 1, \dots 2M, \tag{18}$$

where $m(R) = 1 - \frac{1}{R}$ and the indices are taken $mod(2M)$. Note that only three constraints of (18) are independent. Given a solution to the equation and constraints above one can define the field S on vertices by using (14) and so get an associated family (S is defined up to a multiplication by a positive constant λ) of the generalized Schramm circle patterns. For $M = 2$ the system (17),(18) is equivalent to (15).

The discrete conformal map Z^γ of Section 2 with $\gamma = 4/N$, $N \in \mathbb{N}$, $N > 4$ is an example of such a generalized Schramm circle pattern. (Recall that the central points of the circles are also included). In this case the only branch point is at the origin. We call the combinatorics of this pattern combinatorics of the plane with one branch point of order $N/4 - 1$.

Conjecture 4.1. Up to a similarity $Z^{4/N}$ is the only embedded Schramm circle pattern with the combinatorics of the plane with one branch point of order $N/4 - 1$.

---

[2] which differ from the choice in (15) by the modular transformation $R = 1 - T$. Thus $R > 1$.

5 Concluding remarks: discrete conformal surfaces and coverings

One can try to globalize the above ideas. Take a topological surface and pack it with simply closed loops, which model the topology of the generalized Schramm's circle patterns. If, to the combinations of four vertices described above, one can assign negative numbers R, which satisfy (17) and (18) one can define local coordinate charts, which are local Schramm's circle patterns and then talk about a *discrete conformal covering*.

It is tempting to suggest also the following generalization of Schramm's circle patterns for surfaces. Let $\mathbf{S}$ be the space of spheres in $\mathbb{R}^3$ and $S : \mathbb{Z}^2 \to \mathbf{S}$ a discrete sphere congruence such that: the neighbouring spheres intersect orthogonally $S_{n,m} \perp S_{n+1,m}$, $S_{n,m} \perp S_{n,m+1}$ $\forall n, m$ and the half-neighbouring spheres are tangent $S_{n,m} \| S_{n+1,m+1}$, $S_{n,m} \| S_{n+1,m-1}$ $\forall n, m$. Then the touching points of half-neighbours build a net $F : \mathbb{Z}^2 \to \mathbb{R}^3$, which is natural to call a *discrete conformal surface*.

Acknowledgment

I would like to thank Tim Hoffmann and Ulrich Pinkall for many fruitful discussions.

References

[1] Pinkall U.: Discrete minimal surfaces, Conference, Granada 1991, Bobenko A.: Discrete constant mean curvature surfaces, Conference, Oberwolfach 1991

[2] Bobenko A., Pinkall U.: Discrete isothermic surfaces, J. reine angew. Mathematik 475, 187-208 (1996)

[3] Nijhoff F.W., Capel H.W.: The discrete Korteweg-de Vries equation, Acta Applicandae Mathematicae 39 (1995), 133

[4] Bobenko A., Pinkall U.: Discretization of surfaces and integrable systems, (in preraration)

[5] Nijhoff F.W.: On some "Schwarzian" equations and their discrete analogues, Preprint 1996

[6] Jimbo, M.T., Miwa, T.: Monodromy Perserving Deformation of Linear Ordinary Differential Equations with Rational Coefficients, II. Physica **2 D**, 407-448 (1981)

[7] Its, A.R.: The Painlevé Transcendents as Nonlinear Special Functions. In: Levi D., Winternitz P. (eds.) The Painlevé Transcendents, Their Asymptotics and Physical Applications. (pp. 40-60), New York: Plenum 1992

[8] Hertrich-Jeromin U., Hoffmann T., Pinkall U.: A discrete version of the Darboux transform for isothermic surfaces, Preprint 1996

[9] Its A.R.: Liouville theorem and inverse scattering problem. In: Zapiski Nauchn. Semin. LOMI 133 (1984), 133-125, [Engl. transl.] Journal of Soviet Mathematics, 31, No 6 (1985), 3330-3338

[10] Cieśliński J.: The Bäcklund transformation for discrete isothermic surfaces, this volume.

[11] Blaschke W.: Vorlesungen über Differentialgeometrie, III, Berlin, Springer, 1929

[12] Krichever I., Lipan O., Wiegmann P., Zabrodin A.: Quantum integrable systems and elliptic solutions of classical discrete nonlinear equations, hep-th/9604080 preprint 1996

[13] Schramm O.: Circle patterns with the combinatorics of the square grid, Preprint 1996

The Bäcklund transformation for discrete isothermic surfaces[*]

Jan Cieśliński

Filia Uniwersytetu Warszawskiego w Białymstoku

Instytut Fizyki, ul. Lipowa 41, 15-424 Białystkok, Poland

E-mail: janek@beta.uw.bialystok.pl, janek@fuw.edu.pl

Abstract

Discrete isothermic surfaces has been recently introduced by Bobenko and Pinkall. We construct the Bäcklund transformation which
is an analogue of the classical Darboux–Bianchi transformation for
smooth isothermic surfaces. We use the Sym formula expressing explicitly the immersion in terms of the wave function of the associated
4×4 Lax pair. The Lax pair can be conveniently rewritten in terms of
the Clifford algebra $\mathcal{C}(4,1)$. The 3-dimensional net generated by subsequent Bäcklund transformations forms a triply orthogonal system of
discrete isothermic surfaces.

1 Introduction

The discrete geometry has quite a long history, see [21]. Recently one can observe the growing interest in the subject [3, 4, 14, 15]. Discrete surfaces can be
defined as maps $\mathbf{Z}^2 \longrightarrow \mathbf{R}^n$. Usually one is interested in maps characterized
by some nice geometric properties.

In the present paper we consider discrete isothermic surfaces [3]. We
apply the soliton surfaces approach ([22, 23], see also [2, 8, 9, 18]) based
on the so called Sym formula. The same approach turned out to be very
useful in the continuous case [3, 6, 12]. Using the Clifford algebra $\mathcal{C}(4,1)$
we construct the Bäcklund (or Darboux-Bianchi) transformation generating
discrete isothermic surfaces. In this way we obtain 3 families of mutually
orthogonal discrete isothermic surfaces. Two families consist of standard
discrete isothermic surfaces (cross ratio for any elementary quadrilateral is
negative) and the third one consists of "twisted isothermic surfaces" (cross
ratio is positive).

[*]The research supported partially by the KBN grant 2 P03B 185 09.

2 Smooth isothermic surfaces

Surfaces admitting conformal parameterization of curvature lines are called isothermic surfaces. In other words, an isothermic immersion $F : \mathbf{R}^2 \ni \Omega \mapsto \mathbf{E}^3$ can be equipped with local coordinates u, v in which fundamental forms read as follows:

$$I \equiv dF \cdot dF = e^{2\vartheta}(du^2 + dv^2) , \quad II \equiv -dN \cdot dF = e^{2\vartheta}(k_2 du^2 + k_1 dv^2) ,$$

where $k_1 = k_1(u, v)$, $k_2 = k_2(u, v)$ are principal curvatures and $\vartheta = \vartheta(u, v)$. They satisfy the following system of nonlinear partial differential equations (Gauss-Mainardi-Codazzi equations):

$$\vartheta_{,uu} + \vartheta_{,vv} + k_1 k_2 e^{2\vartheta} = 0, \quad k_{1,u} = (k_2 - k_1)\vartheta_{,u} , \quad k_{2,v} = (k_1 - k_2)\vartheta_{,v} , \quad (1)$$

where $\vartheta_{,u} := \partial\vartheta/\partial u$ etc. The immersion F^*, defined by $F^*_{,u} = e^{-\vartheta}F_{,u}$ and $F^*_{,v} = -e^{-\vartheta}F_{,v}$, is also isothermic. The surface F^* is called dual surface or Christoffel transform of F. In the sequel we usually denote $F \equiv F_+, F^* \equiv F_-$. The nonlinear system (1) is integrable [12] and its Lax pair is given by

$$\Psi_{,u} = \tfrac{1}{2} \mathbf{e}_1 \left(-\vartheta_{,v} \, \mathbf{e}_2 - k_2 e^{\vartheta} \mathbf{e}_3 + \zeta \sinh \vartheta \, \mathbf{e}_4 + \zeta \cosh \vartheta \, \mathbf{e}_5 \right) \Psi ,$$
$$\Psi_{,v} = \tfrac{1}{2} \mathbf{e}_2 \left(-\vartheta_{,u} \, \mathbf{e}_1 - k_1 e^{\vartheta} \mathbf{e}_3 + \zeta \cosh \vartheta \, \mathbf{e}_4 + \zeta \sinh \vartheta \, \mathbf{e}_5 \right) \Psi . \quad (2)$$

where $\mathbf{e}_j$ satisfy the following relations ($\mathbf{I}$ is the identity matrix)

$$\mathbf{e}_1^2 = \mathbf{e}_2^2 = \mathbf{e}_3^2 = \mathbf{e}_4^2 = -\mathbf{e}_5^2 = \mathbf{I} , \quad \mathbf{e}_j \mathbf{e}_k + \mathbf{e}_k \mathbf{e}_j = 0 \quad (j \neq k) ,$$
$$i\mathbf{e}_1 \mathbf{e}_2 \mathbf{e}_3 \mathbf{e}_4 \mathbf{e}_5 = \mathbf{I} . \quad (3)$$

It means that $\mathbf{e}_k$ generate the Clifford algebra $\mathcal{C}(4, 1)$ with an additional constraint on the product of all generators.

Matrix representations of the Clifford algebra generators

We mention here two matrix representations of relations (3). The first one [6, 12] is given by:

$$\mathbf{e}_1 = \begin{pmatrix} 0 & i\sigma_2 \\ -i\sigma_2 & 0 \end{pmatrix}, \quad \mathbf{e}_2 = \begin{pmatrix} -\sigma_1 & 0 \\ 0 & -\sigma_1 \end{pmatrix}, \quad \mathbf{e}_3 = \begin{pmatrix} -\sigma_2 & 0 \\ 0 & \sigma_2 \end{pmatrix},$$

$$\mathbf{e}_4 = \begin{pmatrix} 0 & \sigma_2 \\ \sigma_2 & 0 \end{pmatrix}, \quad \mathbf{e}_5 = \begin{pmatrix} i\sigma_3 & 0 \\ 0 & i\sigma_3 \end{pmatrix}, \quad (4)$$

where σ_k $(k = 1, 2, 3)$ are the so called Pauli matrices:

$$\sigma_1 = \begin{pmatrix} 0 & 1 \\ 1 & 0 \end{pmatrix}, \quad \sigma_2 = \begin{pmatrix} 0 & -i \\ i & 0 \end{pmatrix}, \quad \sigma_3 = \begin{pmatrix} 1 & 0 \\ 0 & -1 \end{pmatrix} .$$

Another representation [3] reads as follows:

$$ie_1 = \begin{pmatrix} -\mathbf{i} & 0 \\ 0 & \mathbf{i} \end{pmatrix}, \quad ie_2 = \begin{pmatrix} -\mathbf{j} & 0 \\ 0 & \mathbf{j} \end{pmatrix}, \quad ie_3 = \begin{pmatrix} -\mathbf{k} & 0 \\ 0 & \mathbf{k} \end{pmatrix},$$

$$ie_4 = \begin{pmatrix} 0 & \mathbf{I} \\ -\mathbf{I} & 0 \end{pmatrix}, \quad ie_5 = \begin{pmatrix} 0 & \mathbf{I} \\ \mathbf{I} & 0 \end{pmatrix}, \tag{5}$$

where $\mathbf{I}, \mathbf{i}, \mathbf{j}, \mathbf{k}$ are standard quaternions (i.e., $\mathbf{ij} = \mathbf{k}$, $\mathbf{jk} = \mathbf{i}$, $\mathbf{ki} = \mathbf{j}$, $\mathbf{i}^2 = \mathbf{j}^2 = \mathbf{k}^2 = -\mathbf{I}$, for example: $\mathbf{i} = -i\sigma_1$, $\mathbf{j} = -i\sigma_2$, $\mathbf{k} = -i\sigma_3$). Of course, there are many other (equivalent) matrix representations: if e_j satisfy (3) then, for any non-degenerate A, matrices Ae_jA^{-1} satisfy (3) as well. By the way, the representation (5) can be obtained from (4) using the matrix

$$A = \begin{pmatrix} 1 & i & i & 1 \\ 1 & -i & -i & 1 \\ -1 & i & -i & 1 \\ -1 & -i & i & 1 \end{pmatrix}.$$

The Sym formula

The so called Sym formula $F = \Psi^{-1}\Psi_{,\zeta}$ proved to be very useful in the effective application of the soliton theory to geometry [8, 22, 23]. After an appropriate modification the formula can be applied to isothermic surfaces [12].

Theorem 1 ([6]) *Let $\Psi = \Psi(u, v; \zeta)$ be a solution of (2) such that $\Psi(u, v; 0)$ commutes with e_4 and e_5. Then the isothermic surface F_+ implicitly defined by ϑ, k_1, k_2 and the corresponding dual surface F_- are explicitly given by*

$$F_\pm := \frac{1}{2}(\mathbf{I} \mp e_{45})\Psi^{-1}\Psi_{,\zeta}\,|_{\zeta=0}, \tag{6}$$

where we denote $e_{ij} := e_i e_j$.

Isothermic surfaces produced by the formula (6) are immersed in the linear spaces spanned by $\frac{1}{4}e_k(e_4 + e_5)$ and $\frac{1}{4}e_k(e_4 - e_5)$ $(k = 1, 2, 3)$, respectively. In both cases we assume that these elements form orthonormal bases.

The Darboux matrix

One of the standard methods to construct special immersions associated with integrable systems is the Darboux–Bäcklund transformation [7, 9, 23]. The new function $\tilde{\Psi} := D\Psi$ is constructed in such a way that the corresponding new Lax pair is of the same form (the most important restriction on D is implied by the fact that the dependence of the Lax pair on ζ should not change). The Darboux matrix for the Lax pair (2) is given by ([6]):

$$D = \kappa e_2(p_1 e_1 + p_2 e_2 + p_3 e_3) + \zeta e_2(\cosh\chi\, e_4 + \sinh\chi\, e_5) , \qquad (7)$$

where κ is a real constant and p_1, p_2, p_3, χ are real functions depending on $\Psi(u, v; -i\kappa)$ in the standard way. Namely:

$$e^{\chi}\begin{pmatrix} p_1 \\ p_2 \\ p_3 \end{pmatrix} := \frac{1}{|a|^2 + (b-1)^2}\begin{pmatrix} 1 - |a|^2 - b^2 \\ 2\mathrm{Re}a \\ 2\mathrm{Im}a \end{pmatrix} , \quad p_1^2 + p_2^2 + p_3^2 = 1 ,$$

$$\begin{pmatrix} \bar{a} & b \\ b & -a \end{pmatrix} := AB^{-1} , \qquad \begin{pmatrix} A \\ B \end{pmatrix} := \Psi(x^1, x^2; -i\kappa)\begin{pmatrix} \bar{a}_0 & b_0 \\ b_0 & -a_0 \\ 1 & 0 \\ 0 & 1 \end{pmatrix} ,$$

and, finally, $a_0 \in \mathbf{C}$, $b_0 \in \mathbf{R}$ are constant. To apply the Sym formula it is necessary to use unimodular Darboux matrix dividing D by the constant $\sqrt{\zeta^2 + \kappa^2}$. The Darboux-Bianchi transformation for the surface (6), generated by the unimodular Darboux matrix,

$$\tilde{F}_{\pm} = F_{\pm} + \frac{2}{\kappa}e^{\pm\chi}\left(\pm p_1 e^{\mp\vartheta}F_{\pm,1} + p_2 e^{\mp\vartheta}F_{\pm,2} - p_3\mathbf{n}_{\pm}\right) ,$$

($\mathbf{n}_{\pm}$ is the normal vector to $F_{\pm}$), is identical with the classical Darboux-Bianchi transformation for isothermic surfaces (compare [6] with [1]).

3   Discrete isotermic surfaces

Discrete isothermic surfaces have been defined by Bobenko and Pinkall [3]. The important notion necessary for this definition is the cross ratio.

The cross ratio

Let identify $\mathbf{E}^3$ with the space spanned by the generators e_1, e_2, e_3 of the Clifford algebra, or, equivalently by quaternions $\mathbf{i}, \mathbf{j}, \mathbf{k}$. Cross ratio of (ordered) four points X_1, X_2, X_3, X_4 of $\mathbf{E}^3$ is defined as unordered pair of (complex conjugate) eigenvalues of Q where (see [3]):

$$Q(X_1, X_2, X_3, X_4) := (X_1 - X_2)(X_2 - X_3)^{-1}(X_3 - X_4)(X_4 - X_1)^{-1} . \qquad (8)$$

The eigenvalues coincide iff they are real. Then $Q = q\mathbf{I}$ ($q \in \mathbf{R}$) and we may identify $Q = q$. The cross ratio (8) is invariant under conformal transformations of $\mathbf{E}^3$ [3].

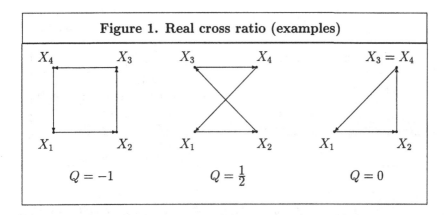

Figure 1. Real cross ratio (examples)

X_4 X_3 X_3 X_4 $X_3 = X_4$

X_1 X_2 X_1 X_2 X_1 X_2

$Q = -1$ $Q = \frac{1}{2}$ $Q = 0$

Lemma 1 ([3]) $Q(X_1, X_2, X_3, X_4)$ *is real iff* X_1, X_2, X_3, X_4 *lie on a circle.*

Given a triangle $X_1 X_2 X_3$ and the circle S^1 enscribed on the triangle we have a bijective map: $S^1 \ni X_4 \mapsto Q(X_1, X_2, X_3, X_4) \in \mathbf{R} \cup \{\infty\}$ (see also Figure 1).

Definition 2 ([3]) *A discrete surface* $F = F_{nm} \equiv F(n,m) \in \mathbf{E}^3$ *is isothermic if the cross ratio of any elementary quadrilateral,*

$$Q_{nm} := Q(F_{nm}, F_{n+1,m}, F_{n+1,m+1}, F_{n,m+1}) \, ,$$

is real and satisfies the relation $Q_{n+1,m+1} Q_{nm} = Q_{n+1,m} Q_{n,m+1}$ *(which is equivalent to* $Q_{nm} = f(n) g(m)$ *where* f *and* g *are some real functions).*

The definition corresponds, in a sense, to curvature coordinates. One can consider another definition, $Q_{nm} = -1$, corresponding to conformal parameterization of curvature lines (compare Table 2) [3].

The Lax pair and the Sym formula

The following Lax pair (compare [3]) is a natural discretization of (2)

$$\begin{aligned}
\Psi_{n+1,m} &= \mathbf{e}_1(\mathbf{u} + \zeta \sinh \varphi \, \mathbf{e}_4 + \zeta \cosh \varphi \, \mathbf{e}_5)\Psi_{nm} \equiv U_{nm}\Psi_{nm} \, , \\
\Psi_{n,m+1} &= \mathbf{e}_2(\mathbf{v} + \zeta \cosh \psi \, \mathbf{e}_4 + \zeta \sinh \psi \, \mathbf{e}_5)\Psi_{nm} \equiv V_{nm}\Psi_{nm} \, ,
\end{aligned} \tag{9}$$

where $\mathbf{u} := u_1\mathbf{e}_1 + u_2\mathbf{e}_2 + u_3\mathbf{e}_3$, $\mathbf{v} := v_1\mathbf{e}_1 + v_2\mathbf{e}_2 + v_3\mathbf{e}_3$ and all dependent variables $(u_k, v_k, \varphi, \psi)$ depend on n, m. From now on we will omit the explicit dependence on n, m using the notation $\Psi_{nm} \equiv \Psi$, $U_{nm} \equiv U$, $V_{nm} \equiv V$,

Table 2. Discrete analogues of some smooth geometric properties	
planar quadrilaterals	conjugate nets (II fundamental form diagonal)
cross ratio Q_{nm} real	curvature nets (both fundamental forms diagonal)
cross ratio real and $Q_{n+1,m+1}Q_{nm} = Q_{n+1,m}Q_{n,m+1}$	curvature nets on isothermic surfaces
$Q_{nm} = -1$	conformal curvature nets on isothermic surfaces

$\Psi_{n+1,m} = \Psi'$, $\Psi_{n,m+1} = \dot{\Psi}$, etc. The compatibility conditions for the Lax pair (9) are given by the following nonlinear system (a discretization of (1)):

$$\psi' + \dot{\psi} = \dot{\varphi} + \varphi \,,$$

$$-\mathbf{e}_2 \mathbf{u}\,\dot{}\,\mathbf{e}_2 \cosh(\psi - \varphi) = \mathbf{u}\cosh(\psi' - \varphi) + \mathbf{v}\sinh(\dot{\varphi} - \varphi) \,,$$

$$-\mathbf{e}_1 \mathbf{v}'\mathbf{e}_1 \cosh(\psi - \varphi) = \mathbf{u}\sinh(\psi' - \psi) + \mathbf{v}\cosh(\dot{\varphi} - \psi) \,, \tag{10}$$

$$2\langle \mathbf{u}\,|\,\mathbf{v}\rangle \cosh(\psi' - \varphi) + \mathbf{u}^2\sinh(\psi' - \psi) + \mathbf{v}^2\sinh(\dot{\varphi} - \varphi) = 0 \,,$$

where $\langle \mathbf{u}\,|\,\mathbf{v}\rangle := \frac{1}{2}(\mathbf{u}\mathbf{v} + \mathbf{v}\mathbf{u}) = u_1 v_1 + u_2 v_2 + u_3 v_3$. From (10) it follows

$$(\mathbf{u}\,\dot{}\,)^2 = \mathbf{u}^2 \,, \qquad (\mathbf{v}')^2 = \mathbf{v}^2 \,, \tag{11}$$

(which means that $\mathbf{u}^2$ depends only on n and $\mathbf{v}^2$ depends only on m) and also

$$\mathbf{e}_2 \mathbf{u}\,\dot{}\,\mathbf{e}_2 \mathbf{v} = -\mathbf{e}_1 \mathbf{v}'\mathbf{e}_1 \mathbf{u} \,. \tag{12}$$

Remark 3 *If* $\mathbf{e}^2 = \mathbf{I}$ *then the Clifford product* $-\mathbf{eve}$ *is the reflection of* $\mathbf{v}$ *with respect to the plane orthogonal to* $\mathbf{e}$ *(e.g.,* $-\mathbf{e}_1\mathbf{v}'\mathbf{e}_1 = -v_1'\mathbf{e}_1 + v_2'\mathbf{e}_2 + v_3'\mathbf{e}_3$*).*

Proposition 4 ([3]) *Applying the Sym formula (6) to the Lax pair (9) one obtains discrete isothermic surfaces with* $Q_{nm} = -\mathbf{v}^2\mathbf{u}^{-2}$. *What is more, any discrete isothermic surface can be obtained in this way.*

4   Algebraic representation of the Lax pair

The reduction group [20] for the discrete Lax pair (9) is described in Table 3 (V satisfies the same conditions as U). In the left part of Table 3 the

Table 3. Group properties of the Lax pair (9)	
$\beta(U)U = c\,\mathbf{I}\ (0 \neq c \in \mathbf{R})$	$U^T(\zeta)\,\mathbf{e}_{34}U(\zeta) = c\,\mathbf{e}_{34}$
$\alpha(U(\overline{\zeta})) = U(\zeta)$	$U^\dagger(\overline{\zeta})\,\mathbf{e}_5 U(\zeta) = c\,\mathbf{e}_5$
$U(-\zeta) = \mathbf{e}_{45}U(\zeta)\mathbf{e}_{45}^{-1}$	

group conditions are expressed in terms of α and β. The "main automorphism" α of $\mathcal{C}$ is the anti-linear map $\alpha : \mathcal{C} \to \mathcal{C}$ defined by

$$\alpha(\mathbf{e}_1\mathbf{e}_2 \ldots \mathbf{e}_k) = (-1)^k \mathbf{e}_1\mathbf{e}_2 \ldots \mathbf{e}_k\ , \qquad \alpha(aX + bY) = \bar{a}\,\alpha(X) + \bar{b}\,\alpha(Y)\ ,$$

where $a, b \in \mathbf{C}$ and $X, Y \in \mathcal{C}$. The "main anti-automorphism" β of $\mathcal{C}$ is the linear map $\beta : \mathcal{C} \mapsto \mathcal{C}$ defined for Clifford products of k vectors by

$$\beta(w_1 w_2 \ldots w_k) = w_k \ldots w_2 w_1\ .$$

Elements $Y \in \mathcal{C}$ such that $\beta(Y)Y = cI$ ($c \in \mathbf{R}$) form the Clifford group Γ.

The properties given in the left part of Table 3 can be checked immediately. U and V are products of two Clifford vectors so they belong to Γ. For any vector $w = \sum_{\mu=1}^5 w_\mu \mathbf{e}_\mu$ we have $w^2 = w_1^2 + w_2^2 + w_3^2 + w_4^2 - w_5^2$. Therefore,

$$U^{-1}(\zeta) = \frac{(\mathbf{u} + \zeta \sinh \varphi\ \mathbf{e}_4 + \zeta \cosh \varphi\ \mathbf{e}_5)\mathbf{e}_1}{\mathbf{u}^2 - \zeta^2} = \frac{\beta(U)}{\mathbf{u}^2 - \zeta^2}\ , \qquad (13)$$

and similarly for V. Thus we have shown that c (appearing in Table 3) is equal to $\mathbf{u}^2 - \zeta^2$. The condition of the second row of Table 3 means that the coefficients of U, V by $\mathbf{I}$ and $\mathbf{e}_{ij}$ are real.

The standard methods to construct the Darboux matrix have been formulated for the matrix Lax pairs. Therefore it is useful to derive the matrix form of the discussed constraints (the right part of Table 3).

4×4 matrices and the Clifford algebra $\mathcal{C}(4,1)$

Lemma 5 ([6]) *Any complex 4×4 matrix is a linear combination of the matrices $\mathbf{I}, \mathbf{e}_k$ and $\mathbf{e}_j\mathbf{e}_k$ ($k, j = 1, \ldots, 5$; $j < k$) (the basis will be denoted shortly by $\mathbf{f}_1, \mathbf{f}_2, \ldots, \mathbf{f}_{16}$), where $\mathbf{e}_k$ are traceless and satisfy relations (3).*

Lemma 6 *Let $\mathbf{e}_k$ be represented by (4) and X be any 4×4 matrix. Then:*

$$X^T = \mathbf{e}_{34}\beta(X)\mathbf{e}_{34}^{-1}\ , \qquad \overline{X} = \mathbf{e}_{12}\alpha(X)\mathbf{e}_{12}^{-1}\ , \qquad X^\dagger = \mathbf{e}_5\alpha\beta(X)\mathbf{e}_5^{-1}\ .$$

Table 4. Constraints on the Darboux matrix

$\beta(N)N \in \Gamma$		$P^T = e_{34}(I - P)e_{34}^{-1}$	$V_{im}^\perp = \overline{e_{34} V_{im}}$
			$V_{ker}^\perp = \overline{e_{34} V_{ker}}$
$N^\dagger = e_5 \beta(N) e_5^{-1}$	$\mu_1 = \overline{\lambda_1}$	$P^\dagger = e_5 P e_5^{-1}$	$V_{im}^\perp = e_5 V_{ker}$
	$\lambda_1, \mu_1 \in \mathbf{R}$	$P^\dagger = e_5(I - P)e_5^{-1}$	$V_{im}^\perp = e_5 V_{im}$
			$V_{ker}^\perp = e_5 V_{ker}$
$N = -e_{45} N e_{45}^{-1}$	$\mu_1 = -\lambda_1$	$e_{45} P e_{45}^{-1} = I - P$	$V_{ker} = e_{45} V_{im}$

Proof: By Lemma 5 we can verify Lemma 6 decomposing X in the basis $\mathbf{f}_\nu$. Note that $e_k^T = e_{34} e_k e_{34}^{-1}$, $\overline{e}_k = -e_{12} e_k e_{12}^{-1}$ and $e_k^\dagger = -e_5 e_k e_5^{-1}$.   □

Lemma 6 allows us to translate the automorphisms α, β into the matrix form. The matrix form of the reduction group in the discrete case is identical as the reduction group in the continuous case (compare [6]). However, in this paper we do not demand U, V, Ψ to be unimodular (in the continuous case the additional constraint $c = 1$ was imposed). Note that if $Y \in \Gamma$ is represented by a 4×4 matrix then $|\beta(Y)Y| = \sqrt{\det Y}$.

Remark 7 Ψ *satisfies the group conditions of Table 3. It is enough to assume that they are satisfied by $\Psi_{00}(\zeta)$. The group conditions have to be satisifed also by $D := \tilde{\Psi}\Psi^{-1}$. We can also impose another constraint: $\Psi_{nm}(0)$ commutes with e_4 and e_5.*

5 The construction of the Darboux matrix and the Clifford algebra

The general construction of the Darboux matrix in Clifford algebras is not available yet and we are compelled to use a matrix representation. However, the results of this paper suggest that such construction should be possible soon. Assume the Darboux matrix in the following form:

$$D = N(\zeta - \lambda_1 + (\lambda_1 - \mu_1)P), \tag{14}$$

where λ_1, μ_1 are complex constants and N, P are n, m-dependent matrices ($\det N \neq 0$, $P^2 = P$). This ansatz is very well known (compare [19, 24]), for motivation (in the continuous case) see [7]. Note that λ_1 and μ_1 are the only zeros of D (i.e. $\det D(\lambda_1) = \det D(\mu_1) = 0$). The 4×4 matrix D, being linear in ζ, should have 4 zeros. Thus the sum of multiplicities of λ_1 and μ_1 is 4.

The necessary condition for the transformed Lax pair to have the same form with respect to ζ (in our case: linear in ζ) is P given by ([7, 24])

$$\ker P = \Psi(\lambda_1) V_{ker} \,, \qquad \mathrm{im} P = \Psi(\mu_1) V_{im} \,,$$

where V_{ker} and V_{im} are constant vector spaces such that $V_{ker} \oplus V_{im} = \mathbf{C}^4$. From (14) we derive

$$\mathcal{D}^{-1} = \frac{(\zeta - \mu_1 + (\mu_1 - \lambda_1)P)N^{-1}}{(\zeta - \lambda_1)(\zeta - \mu_1)} \,,$$

and $\beta(D) = (\zeta - \lambda_1 + (\lambda_1 - \mu_1)\beta(P))\beta(N)$. Hence:

Corollary 8 *The condition $D \in \Gamma$ is equivalent to $\beta(P) = \mathbf{I} - P$ and $N \in \Gamma$.*

The remaining group properties of Table 3 imply (in a quite standard way, similar as in the continuous case [7]) the corresponding constraints on λ_1, μ_1, N and P (or, what is equivalent, on V_{ker} and V_{im}) collected in Table 4.

The unitary reduction (the second row of Table 3) gives two possibilities: $\mu_1 = \overline{\lambda}_1$ or $\lambda_1, \mu_1 \in \mathbf{R}$. From now on we will consider only the first case denoting $\lambda_1 = i\kappa$, $\kappa \in \mathbf{R}$. Using the representation (4) we can formulate the constraints on P as follows:

$$\mathrm{im} P = \begin{pmatrix} \overline{a} & b \\ b & -a \\ 1 & 0 \\ 0 & 1 \end{pmatrix} , \qquad \ker P = \begin{pmatrix} 1 & 0 \\ 0 & 1 \\ -a & b \\ b & \overline{a} \end{pmatrix} .$$

where $a \in \mathbf{C}$ and $b \in \mathbf{R}$ are some functions of n, m. Assuming that D commutes with $\mathbf{e}_4, \mathbf{e}_5$ the final conclusion reads as follows.

Corollary 9 *The matrix given by (7) is the Darboux matrix also in the discrete case.*

In fact we have a little bit more freedom (both in the continuous and discrete case, compare [6]) but in this paper we confine ourselves to Darboux matrices of the form (7).

6 The Bäcklund transformation and a triple system of discrete isothermic surfaces

The Lax pair (9) and the Darboux matrix (7) are similar in a striking way. Let us collect all these 3 equations together.

$$\Psi' = U\Psi \,, \qquad U = e_1(u + \zeta e_5 \exp(\varphi e_{45})) \,,$$
$$\Psi^\cdot = V\Psi \,, \qquad V = e_2(v + \zeta e_4 \exp(\psi e_{45})) \,, \qquad (15)$$
$$\tilde{\Psi} = D\Psi \,, \qquad D = e_2(\kappa p + \zeta e_4 \exp(\gamma e_{45})) \,,$$

where $p := p_1 e_1 + p_2 e_2 + p_3 e_3$. Note that κ (the length of the vector κp) depends only on the index k numbering subsequent Bäcklund transforms (similarly as u^2 depends only on n and v^2 depends only on m). The compatibility conditions for (15) yield the nonlinear system (10) and the following 8 equations which define the Bäcklund transformation for the Lax pair (9):

$$e_2 \tilde{u} e_2 \kappa p = -e_1 \kappa' p' e_1 u \,, \qquad e_2 \tilde{v} e_2 \kappa p = e_2 \kappa^\cdot p^\cdot e_2 v \,, \qquad (16a)$$

$$\tilde{\psi} + \psi = \gamma^\cdot + \gamma \,, \qquad \tilde{\varphi} + \varphi = \gamma' + \gamma \,, \qquad (16b)$$

$$-e_2 \tilde{v} e_2 \sinh(\gamma - \psi) = \kappa p \sinh(\tilde{\psi} - \psi) + v \sinh(\psi - \gamma^\cdot) \,,$$
$$-e_2 \tilde{u} e_2 \cosh(\gamma - \varphi) = \kappa p \sinh(\tilde{\varphi} - \varphi) + u \cosh(\varphi - \gamma') \,,$$
$$-e_2 \kappa^\cdot p^\cdot e_2 \sinh(\gamma - \psi) = \kappa p \sinh(\tilde{\psi} - \gamma) + v \sinh(\gamma - \gamma^\cdot) \,, \qquad (16c)$$
$$-e_1 \kappa' p' e_1 \cosh(\gamma - \varphi) = \kappa p \cosh(\tilde{\varphi} - \gamma) + u \sinh(\gamma' - \gamma) \,.$$

As a consequence of (16) we obtain $\kappa^\cdot = \kappa' = \kappa$ (which is evident anyway). Using equations (16c) we can rewrite equations (16a) as:

$$2\kappa \langle v \,|\, p \rangle \sinh(\psi - \gamma^\cdot) + \kappa^2 \sinh(\tilde{\psi} - \psi) + v^2 \sinh(\gamma^\cdot - \gamma) = 0 \,,$$
$$2\kappa \langle u \,|\, p \rangle \cosh(\gamma' - \varphi) + \kappa^2 \sinh(\tilde{\varphi} - \varphi) + u^2 \sinh(\gamma' - \gamma) = 0 \,.$$

A triply orthogonal system of discrete isothermic surfaces

We proceed to the last but very interesting observation. Given an isothermic surface $F : \mathbf{Z}^2 \to \mathbf{E}^3$, let us consider the map $\Phi : \mathbf{Z}^3 \to \mathbf{E}^3$ given by $\Phi(n, m, k) := F_k(n, m)$, where $F_0 \equiv F$, $F_1 := \tilde{F}$ is the Bäcklund transform of F, and F_{k+1} is the Bäcklund transform of F_k $(k = 1, 2, \ldots)$.

Proposition 10 *Let* $Q(F, F', F'^\cdot, F^\cdot) = -v^2 u^{-2}$ *(where* $u = u(n) \in \mathbf{R}$ *and* $v = v(m) \in \mathbf{R}$*) and* $\tilde{F}$ *be the Bäcklund transform of* F *with the parameter* κ. *Then* $Q(F, F', \tilde{F}', \tilde{F}) = -\kappa^2 u^{-2}$ *and* $Q(F, F^\cdot, \tilde{F}^\cdot, \tilde{F}) = \kappa^2 v^{-2}$.

Proof: To apply the formula (8) we have to identify elements of the form $X = \frac{1}{2}(\mathbf{I} - \mathbf{e}_{45})\mathbf{x}(a\mathbf{e}_4 + b\mathbf{e}_5)$ (where $\mathbf{x} = x_1\mathbf{e}_1 + x_2\mathbf{e}_2 + x_3\mathbf{e}_3$) with $(a+b)\mathbf{x}$. We denote shortly: $X \approx (a+b)\mathbf{x}$. Considering dual surfaces we identify $\frac{1}{2}(\mathbf{I} + \mathbf{e}_{45})\mathbf{x}(a\mathbf{e}_4 + b\mathbf{e}_5)$ with $(a-b)\mathbf{x}$ and all computations are almost identical. We denote also $\Psi_0 := \Psi(n, m; 0)$ (the formula (6) is evaluated at $\zeta = 0$). From (6) we compute $F' - F = \frac{1}{2}(\mathbf{I} - \mathbf{e}_{45})\Psi^{-1}U^{-1}U_{,\zeta}\Psi|_{\zeta=0}$ etc. Therefore, taking into account that Ψ_0 commutes with $\mathbf{e}_{45}$, we have:

$$\Psi_0(F' - F)\Psi_0^{-1} = \tfrac{1}{2}(\mathbf{I} - \mathbf{e}_{45})U^{-1}U_{,\zeta} \approx \mathbf{u}^{-1}e^{\varphi},$$

$$\Psi_0(F^{\cdot} - F)\Psi_0^{-1} = \tfrac{1}{2}(\mathbf{I} - \mathbf{e}_{45})V^{-1}V_{,\zeta} \approx \mathbf{v}^{-1}e^{\psi},$$

$$\Psi_0(\tilde{F} - F)\Psi_0^{-1} = \tfrac{1}{2}(\mathbf{I} - \mathbf{e}_{45})D^{-1}D_{,\zeta} \approx (\kappa\mathbf{p})^{-1}e^{\gamma},$$

$$\Psi_0(F'^{\cdot} - F')\Psi_0^{-1} = \tfrac{1}{2}(\mathbf{I} - \mathbf{e}_{45})U^{-1}(V')^{-1}V'_{,\zeta}U \approx \mathbf{u}^{-1}\mathbf{e}_1(\mathbf{v}')^{-1}\mathbf{e}_1\mathbf{u}e^{\psi'},$$

$$\Psi_0(F'^{\cdot} - F^{\cdot})\Psi_0^{-1} = \tfrac{1}{2}(\mathbf{I} - \mathbf{e}_{45})V^{-1}(U^{\cdot})^{-1}U^{\cdot}_{,\zeta}V \approx \mathbf{v}^{-1}\mathbf{e}_2(\mathbf{u}^{\cdot})^{-1}\mathbf{e}_2\mathbf{v}e^{\varphi^{\cdot}},$$

$$\Psi_0(\tilde{F}' - \tilde{F})\Psi_0^{-1} = \tfrac{1}{2}(\mathbf{I} - \mathbf{e}_{45})D^{-1}\tilde{U}^{-1}\tilde{U}_{,\zeta}D \approx (\kappa\mathbf{p})^{-1}\mathbf{e}_2\tilde{\mathbf{u}}^{-1}\mathbf{e}_2\kappa\mathbf{p}e^{\tilde{\varphi}},$$

$$\Psi_0(\tilde{F}' - F')\Psi_0^{-1} = \tfrac{1}{2}(\mathbf{I} - \mathbf{e}_{45})U^{-1}(D')^{-1}D'U \approx \mathbf{u}^{-1}\mathbf{e}_1(\kappa'\mathbf{p}')^{-1}\mathbf{e}_1\mathbf{u}e^{\gamma'},$$

$$\Psi_0(\tilde{F}^{\cdot} - F^{\cdot})\Psi_0^{-1} = \tfrac{1}{2}(\mathbf{I} - \mathbf{e}_{45})V^{-1}(D^{\cdot})^{-1}D_{,\zeta}V \approx \mathbf{v}^{-1}\mathbf{e}_2(\kappa^{\cdot}\mathbf{p}^{\cdot})^{-1}\mathbf{e}_2\mathbf{v}e^{\gamma^{\cdot}},$$

$$\Psi_0(\tilde{F}^{\cdot} - \tilde{F})\Psi_0^{-1} = \tfrac{1}{2}(\mathbf{I} - \mathbf{e}_{45})D^{-1}\tilde{V}^{-1}\tilde{V}_{,\zeta}D \approx (\kappa\mathbf{p})^{-1}\mathbf{e}_2\tilde{\mathbf{v}}^{-1}\mathbf{e}_2\kappa\mathbf{p}e^{\tilde{\psi}}.$$

Using the above results, the formula (8), equations (12) and (16ab), the first equation of (10), and $\mathbf{e}_1^2 = \mathbf{e}_2^2 = \mathbf{p}^2 = 1$, we check in the straightforward way

$$Q(F, F', F'^{\cdot}, F^{\cdot}) = \Psi_0^{-1}\mathbf{u}^{-2}\mathbf{e}_1\mathbf{v}'\mathbf{e}_1\mathbf{u}\mathbf{v}^{-1}\mathbf{e}_2(\mathbf{u}^{\cdot})^{-1}\mathbf{e}_2\mathbf{v}^2\Psi_0 e^{\varphi - \psi' + \varphi^{\cdot} - \psi} = -v^2u^{-2},$$

$$Q(F, F', \tilde{F}', \tilde{F}) = \Psi_0^{-1}\mathbf{u}^{-2}\mathbf{e}_1\kappa'\mathbf{p}'\mathbf{e}_1\mathbf{u}(\kappa\mathbf{p})^{-1}\mathbf{e}_2\tilde{\mathbf{u}}^{-1}\mathbf{e}_2(\kappa\mathbf{p})^2\Psi_0 e^{\varphi - \gamma' + \tilde{\varphi} - \gamma} = -\kappa^2u^{-2},$$

$$Q(F, F^{\cdot}, \tilde{F}^{\cdot}, \tilde{F}) = \Psi_0^{-1}\mathbf{v}^{-2}\mathbf{e}_2\kappa^{\cdot}\mathbf{p}^{\cdot}\mathbf{e}_2\mathbf{v}(\kappa\mathbf{p})^{-1}\mathbf{e}_2\tilde{\mathbf{v}}^{-1}\mathbf{e}_2(\kappa\mathbf{p})^2\Psi_0 e^{\psi - \gamma^{\cdot} + \tilde{\psi} - \gamma} = \kappa^2v^{-2}.$$

By the way, we also proved the first part of Proposition 4. □

Therefore, starting from a given isothermic surface and performing subsequent Bäcklund transformations we obtain a triple family of discrete isothermic surfaces. They can be considered to be mutually orthogonal (in the discrete sense, compare table 2).

7 Recent developments

The 3-dimensional lattice consisting of subsequent Bäcklund transforms of a given discrete isothermic surface seems to be closely related to the construction of quadrilateral lattices recently given by Doliwa and Santini [16]. The orthogonal reduction of quadrilateral lattices, leading to the discretization of Lamé equations, is already constructed [11]. The most natural deeper reduction is the isothermic reduction.

Preparing the final version of my paper I got an interesting recent preprint on the Darboux transformations for isothermic surfaces [17]. The results of this paper seems to be closely related to my research and certainly can be helpful in the construction of isothermic 3-dimensional lattices.

Quaternions turned out to be useful to compute the cross ratio in $\mathbf{E}^3$. Using Clifford algebras one can easily compute the cross ratio of 4 points in n-dimensional space. Indeed, we can use exactly the same formula (8). The real part of the corresponding usual complex cross ratio is given by $T(Q)$, the coefficient of Q by the unit $\mathbf{I}$. The square of the imaginary part is given by $T^2(Q) - \beta(Q)Q$ (for more details see [10]).

Acknowledgements

I am grateful to Adam Doliwa and Alexander Bobenko for stimulating discussions. Thanks are due to KBN (Polish Committe for Scientific Researches) and to the Organizers of SIDE II for financial support.

References

[1] L.Bianchi: "Ricerche sulle superficie isoterme and sulla deformazione delle quadriche", *Annali di Matematica* 1905, serie III, tomo XI, pp. 93–157 [in Italian].

[2] A.I.Bobenko: "Surfaces in Terms of 2 by 2 Matrices. Old and New Integrable Cases", [in:] *Harmonic maps and integrable systems* (Aspects of Mathematics, vol. 23), eds. A.P.Fordy, J.C.Wood; Vieweg, Brunswick 1994.

[3] A. Bobenko, U. Pinkall: "Discrete Isothermic Surfaces", *J. reine angew. Math.* **475** (1996) 187–208.

[4] A.Bobenko, U.Pinkall: "Discrete Surfaces with constant Negative Gaussian Curvature and the Hirota Equation", to be published in *J. Diff. Geom.*

[5] F. Burstall, U. Hertrich–Jeromin, F. Pedit, U. Pinkall: "Curved Flats and Isothermic Surfaces", to be published in *Math. Zeit.*

[6] J.Cieśliński: "The Darboux–Bianchi transformation for isothermic surfaces. Classical results versus the soliton approach", *Diff. Geom. Appl.* **7** (1997), 1–28.

[7] J.Cieśliński: "An algebraic method to construct the Darboux matrix", *J. Math. Phys.* **36** (1995) 5670–5706.

[8] J.Cieśliński: "A generalized formula for integrable classes of surfaces in Lie algebras", *preprint IFT/11/96*, Warsaw 1996. Submitted to *J. Math. Phys.*

[9] J.Cieśliński: "The Darboux–Bianchi–Bäcklund transformations and soliton surfaces", to be published in the proceedings of the *1-st Non-orthodox School on Nonlinearity and Geometry* (Warsaw, September 1995).

[10] J.Cieśliński: "The cross ratio and Clifford algebras", in preparation.

[11] J.Cieśliński, A.Doliwa, P.M.Santini: "The Integrable Discrete Analogues of Orthogonal Coordinate Systems are Multidimensional Circular Lattices", in preparation.

[12] J. Cieśliński, P. Goldstein, A. Sym: "Isothermic surfaces in E^3 as soliton surfaces", *Phys. Lett.* **A 205** (1995) 37–43.

[13] J.F.Cornwell: *Group Theory in Physics*, vol. II, Academic Press, London 1984.

[14] A.Doliwa, P.M.Santini: "Integrable dynamics of a discrete curve and the Ablowitz–Ladik hierarchy", J. Math. Phys. **36** (1995) 1259–1273.

[15] A.Doliwa, P.M.Santini: "The integrable dynamics of a discrete curve", [in:] *Symmetries and Integrability of Difference Equations* (CRM Proceedings and Lecture Notes, Vol. 9), pp. 91–102, eds. D.Levi, L.Vinet, P.Winternitz; AMS, Providence 1996.

[16] A.Doliwa, P.M.Santini: "Multidimensional quadrilateral lattices are integrable", *preprint ROME1-1162/96*, Dip. di Fisica, Univ. di Roma "La Sapienza", Rome 1996.

[17] U.Hertrich–Jeromin, T.Hoffmann, U.Pinkall: "A discrete version of the Darboux transforms for isothermic surfaces", *SfB288 preprint* No. 239, Berlin 1996.

[18] H.S.Hu: "Solitons and Differential Geometry", [in:] *Soliton Theory and Its Applications*, pp. 297–336, Springer, Berlin 1995.

[19] D. Levi, O. Ragnisco, M. Bruschi: "Extension of the Zakharov–Shabat Generalized Inverse Method to Solve Differential–Difference and Difference–Difference Equations", *Nuovo Cim.* **A 58** (1980) 56–66.

[20] A.V.Mikhailov: "The reduction problem and the inverse scattering method", *Physica* **D 3** (1981) 73–117.

[21] R.Sauer: *Differenzengeometrie*, Springer, Berlin 1970 [in German].

[22] A.Sym: "Soliton Surfaces II. Geometric Unification of Solvable Nonlinearities", *Lett. Nuovo Cim.* **36** (1983) 307–312.

[23] A.Sym: "Soliton surfaces and their application. Soliton geometry from spectral problems.", [in:] *Geometric Aspects of the Einstein Equations and Integrable Systems* (Lecture Notes in Physics **239**), pp. 154–231, ed. R.Martini, Springer, Berlin 1985.

[24] V.E.Zakharov, S.V.Manakov, S.P.Novikov, L.P.Pitaievsky: *Theory of solitons*, Nauka, Moscow 1980.

Integrable Discrete Geometry with Ruler and Compass

Adam Doliwa*

Istituto Nazionale di Fisica Nucleare, Sezione di Roma [†]
P-le Aldo Moro 2, I–00185 Roma, Italy
Instytut Fizyki Teoretycznej, Uniwersytet Warszawski
ul. Hoża 69, 00-681 Warszawa, Poland
Email: doliwa@roma1.infn.it, doliwa@fuw.edu.pl

Abstract

Recent results of the geometry of integrable lattices are reviewed in a unified setting. The basic idea is to apply simple geometric constructions in a process of building the lattices from initial-boundary data.

Keywords: Integrable systems, discrete geometry

1 Introduction

Many of the research papers of the XIX-th century geometers is devoted to studies of special classes of surfaces, like for example:

i) surfaces admitting the Chebyshev net parametrization by asymptotic lines (pseudospherical surfaces)

ii) surfaces admitting the conjugate net parametrization by geodesic lines (surfaces of Voss)

iii) surfaces admitting the orthogonal net parametrization by asymptotic lines (minimal surfaces)

---

*The work is supported partially by the Polish Committee of Scientific Research (KBN) under the Grant Number 2P03 B 18509.

[†]Current address

iv) surfaces admitting the isothermic net parametrization by curvature lines (isothermic surfaces).

To investigate such surfaces the statements about their geometric properties had been usually expressed in the language of differential equations. This way in the old books of differential geometry one can find, for example, the sine-Gordon equation (pseudospherical surfaces and surfaces of Voss) or the Liouville equation (minimal surfaces) and many others which are now included in the list of the integrable systems. One can find there not only the equations but broad classes of their exact solutions, including N-soliton solutions and even some solutions in terms of the theta functions as well. Also the transformations (of Bianchi, Bäcklund, Darboux, Moutard, Ribaucour, Combescure) between the solutions of these equations or between corresponding surfaces enjoyed in that period a particular attention. These classical results have been summarized in monographs [2, 11]; as a good introduction to them can serve [17].

Before we leave the *differential* geometry and proceed to the *difference* (or discrete) geometry let us add more advanced examples to the list above, together with the corresponding integrable partial differntial equations

v) the Laplace–Darboux sequence of conjugate nets – generalized Toda system [10, 28]

vi) orthogonal systems of coordinates – the Lamé equations [25]

vii) multi-conjugate systems of coordinates – the Darboux equations [11, 35].

It is worth to mention that near all the (integrable) systems studied in that period can be put in a general scheme as special reductions of the multi-conjugate systems.

Is this strong coincidence of results of the classical differential geometry and the modern theory of integrable systems (see also [34, 3, 14, 21] and references cited therein) only an accident or it is something more? A good place to face this question seems to be the rapidly deverloping field of the integrable discrete equations. Although they were studied since many years, nowadays this is the area which focuses the recent results of integrable systems theory, statistical physics, quantum field theory, special function theory, group theory and algebra [18].

To construct an integrable discretization of the given soliton equation one can follow several different approaches. The standard ones are:

i) a discrete version of the Lax pair [1]

ii) the Hirota method via a bilinear form [20, 12]

iii) extensions of the Zakharov – Shabat dressing method [27, 6]

iv) direct linearization using linear integral equations [30, 29, 7]

v) Bäcklund transformation approach [26].

We would like to add a new item to this list, i.e.

vi) integrable discretization of the corresponding geometric model [4, 5, 15, 13, 16, 9].

In fact, in the geometric literature [33] there were known nice discrete versions of the pseudospherical surfaces and Voss surfaces. Recently, the geometric properties of the discrete pseudospherical surfaces have been formulated in terms of a partial difference equation [4] which appears to be the Hirota integrable discretization of the sine-Gordon equation [19]. Also the isothermic surfaces have been discretized and the corresponding system of integrable difference equations has been found [5].

By a discrete surface we mean a mapping

$$\mathbf{x} : \mathbb{Z}^2 \to \mathbb{E}^3 \ , \tag{1}$$

from 2-dimensional integer lattice (one can consider also different graphs which describe discrete surfaces of more complicated topology) into 3-dimensional Euclidean space. To discretize given class of surfaces we must find then restrictions on the mapping (1) which in the limit of small lattice parameter reproduce the known properties of the continuous model. But how to select *integrable* discretization among all possible ones? From the known examples one can extract the following simple principle:

The integrable restrictions on the mapping (1) allow to construct the discrete surface from the initial data using ruler and compass only.

The initial data for discrete surfaces are usually two intersecting curves. Going into multidimensions both the above principle and the initial data must be generalized, but the main idea remains the same.

The aim of the paper is to review some resent results concerning integrable geometry of discrete surfaces and multidimensional lattices. The exposition follows the unified point of view of the "ruler and compass" constructions.

In Section 2 we will be concerned with discrete pseudospherical surfaces, discrete surfaces of Voss and discrete isothermic surfaces. Section 3 is devoted to the study of the Laplace sequence of discrete quadrilateral surfaces (discrete conjugate nets). Section 4 deals with multidimensional quadrilateral lattices (discrete analogues of the multi-conjugate systems of Darboux) and with their reduction to multidimensional circular lattices (discrete analogue of the orthogonal systems of Lamé). In Section 5 we conclude the paper presenting some remarks, open problems and research directions.

I am greatly indebted to J. Cieśliński and P. M. Santini for numerous discussions in the course of which many results presented in this paper came into being. I wish to express my gratitude to A. Sym for encourage to study works of Bianchi and Darboux and for bringing reference [33] to my attention. I would like to thank the organizers of the SIDE II Workshop for the invitation and kind hospitality.

2 Discrete integrable surfaces

2.1 Pseudospherical surfaces

Discrete pseudospherical surfaces are defined by imposing two restrictions on the general mapping (1) being discrete analogues of the asymptotic and Chebyshev parametrization (for details, see [33, 4]):

i) For each point $\mathbf{x}$ of the lattice there is a plane P which contains the point $\mathbf{x}$ and its four nearest neighbours $T_1\mathbf{x}$, $T_2\mathbf{x}$, $T_1^{-1}\mathbf{x}$ and $T_2^{-1}\mathbf{x}$, i.e. the tangent planes to parametric curves coincide with the tangent plane do the surface.
ii) The lengths $\Delta_i = |D_i\mathbf{x}|$ and $T_j\Delta_i$ $(i \neq j)$ of the opposite sides of an elementary quadrilateral are equal.

Here T_i denotes the shift operator in the i-th direction of the lattice, and $D_i = T_i - 1$ is the corresponding partial difference.

Denoting by θ_i $(i = 1, 2)$ the angle between the plane P in the point $\mathbf{x}$ and the similar plane T_iP in the point $T_i\mathbf{x}$ one can observe that the properties i) and ii) imply

$$\frac{\sin\theta_1}{\Delta_1} = \frac{\sin\theta_2}{\Delta_2} = \text{const.} \quad ; \qquad (2)$$

in particular the angle θ_i does not vary in the second direction of the lattice. This means that the Gauss map

$$\mathbf{N} : \mathbb{Z}^2 \to S^2 \qquad (3)$$

given by the vector $\mathbf{N}$ normal to the plane P forms on the unit sphere S^2 the Chebyshev net.

Let ω denote the angle between the edges $D_1\mathbf{x}$ and $D_2\mathbf{x}$ of the lattice, then ω satisfies the following integrable difference equation [4]

$$D_1 D_2 \omega = 2\arg\left(1 - T_1\left(ke^{-i\omega}\right)\right) + 2\arg\left(1 - T_2\left(ke^{-i\omega}\right)\right) , \qquad (4)$$

where $k = \tan(\frac{1}{2}\theta_1)\tan(\frac{1}{2}\theta_2)$.

Note that the four nearest neighbours $T_i^{\pm 1}\mathbf{x}$ ($i = 1, 2$) of the point $\mathbf{x}$ form a planar quadrilateral. Let us consider such new *quadrilateral lattice* (i.e. lattice with planar quadrilaterals) with vertices in points of the original lattice labelled by integers with the sum equal to an even number. By ω_u, ω_l, ω_d and $\omega_r = \omega$ we denote the angles made by the edges of the original lattice meeting in the "central" point $\mathbf{x}$ whose coordinates sum up to the odd number. Apart from the obvious relation

$$\omega + \omega_u + \omega_l + \omega_d = 2\pi \; , \tag{5}$$

the Chebyshev condition implies

$$T_1 T_2 \omega_l = \omega \quad , \quad T_1 T_2^{-1} \omega_d = \omega_u \; ; \tag{6}$$

observe that $T_1 T_2$ and $T_1 T_2^{-1}$ are the shift operators along the new quadrilateral lattice.

The above constraints can be resolved introducing new field Ω such that

$$\begin{aligned}
\omega_l &= T_1 T_2^{-1}\Omega + \Omega \quad , \\
\omega_d &= \pi - (T_1 T_2 \Omega + \Omega) \quad , \\
\omega &= T_1 T_2 (T_1 T_2^{-1}\Omega + \Omega) \quad , \\
\omega_u &= \pi - T_1 T_2^{-1}(T_1 T_2 \Omega + \Omega) \quad .
\end{aligned}$$

Similarly, we can define Ω for the second quadrilateral lattice with vertices labelled by integers with the sum equal to an odd number. The integrable nonlinear difference equation which characterizes the discrete pseudospherical surfaces in terms of the field Ω is the Hirota discrete sine-Gordon equation [19]

$$\sin\left(\frac{D_1 D_2 \Omega}{2}\right) = k \sin\left(\frac{T_1 T_2 \Omega + T_1 \Omega + T_2 \Omega + \Omega}{2}\right) \; . \tag{7}$$

The initial boundary problem for discrete pseudospherical surfaces can be formulated as follows. Given two discrete space curves $\mathbf{x}_i \in \mathbb{E}^3$ ($i = 1, 2$) meeting in a point $\mathbf{x}_1(0) = \mathbf{x}_2(0)$ in such a way that their tangent planes coincide. Let, in addition, the (varying) lengths of their segments Δ_i and their torsion angles θ_i satisfy the constraint (2). Then the conditions i) and ii) determine completely the discrete pseudospherical surface. In particular, the point $T_1 T_2 \mathbf{x}$ belongs to the line of intersection of the two planes $T_1 P$ and $T_2 P$; it belongs also to the circle made by rotating the triangle with sides of lengths Δ_1 and Δ_2 around the base $T_1 \mathbf{x} T_2 \mathbf{x}$. To make the intersection possible the constraint (2) is necessary.

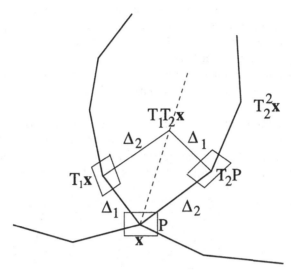

FIGURE 1.

2.2   Surfaces of Voss

Discrete surfaces of Voss are defined by two conditions [33] being discrete
analogues of the conjugate and geodesic parametrization:

i) For each point $\mathbf{x}$ *of the lattice there is a plane* P *which contains all the
vertices* $\mathbf{x}$, $T_1\mathbf{x}$, $T_2\mathbf{x}$ *and* $T_1T_2\mathbf{x}$ *of the elementary quadrilateral* Q
*ii) The opposite angles formed by the edges meeting in a vertex of the lattice
are equal.*

Like in the previus case one can show that the angle θ_i between the planes P
and T_iP does not vary in the second direction of the lattice, i.e. the normal
vector to the planes P forms on the sphere the Chebyshev net again. The
planarity condition i) implies for the four angles of elementary quadrilaterals
an analogue of the formula (5) and the condition ii) states that pairs of
opposite angles in a vertex are connected by the analogue of the formula (6).

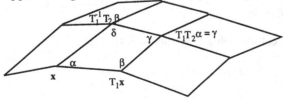

FIGURE 2.

In fact, like in the continuous case, the discrete pseudospherical surfaces and surfaces of Voss are in a sense dual to each other [17, 33] and are characterized by the same integrable equations (4) or (7). Note that the quadrilateral surface constructed from the discrete pseudospherical surface in the previous paragraph is **not** a surface of Voss.

It is worth pointing out that both discrete pseudospherical surfaces and surfaces of Voss allow continous deformations which do not change their properties [33, 4].

The initial boundary problem for discrete surfaces of Voss is given by two discrete space curves $\mathbf{x}_i \in \mathbb{E}^3$ $(i = 1, 2)$ meeting in a point $\mathbf{x}_1(0) = \mathbf{x}_2(0)$ in such a way that the condition ii) is fulfilled. In addition, in points of the initial curves we prescribe "initial directions" which must be compatible with the condition i).

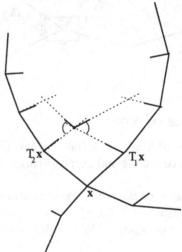

FIGURE 3.

These data define the "initial curves" of normal vectors $\mathbf{N}$ on the unit sphere S^2. The Chebyshev property of the Gauss map (3) allows to find, using only the compass, the normal vectors to the discrete surface of Voss in other points of the lattice, and then the new points of the surface.

2.3   Isothermic surfaces

Discrete isothermic surfaces can be defined by the condition [5]:

All the elementary quadrilaterals of the lattice are harmonic.

Let us recall the construction of harmonic quadrilaterals [32]. Given three points $\mathbf{x}$, $T_1\mathbf{x}$ and $T_2\mathbf{x}$, they define uniquely the circle S. Tangent lines to the circle in points $T_1\mathbf{x}$ and $T_2\mathbf{x}$ meet at $\mathbf{y}$. The point $T_1T_2\mathbf{x}$ is the second intersection of the line $\mathbf{xy}$ with the circle S.

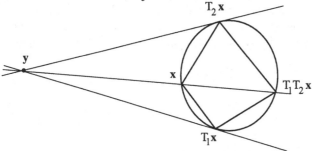

FIGURE 4.

The initial boundary problem for discrete isothermic surfaces can be then stated by giving two curves intersecting in a point. The ambient space may be of arbitrary dimension $M \geq 2$.

Other interesting examples of discrete surfaces like discrete constant mean curvature surfaces and discrete minimal surfaces can be obtained from isothermic surfaces by imposing additional restrictions, see [5].

3 Laplace–Darboux sequence of quadrilateral surfaces

The discrete analogue of the generalized Toda system [27, 20, 12] playes a particular role among all the integrable systems. Its suitable limits with respect to the lattice parameters give rise to the various types of soliton equations including the Kadomtsev–Petviashvili equation, Korteweg–de Vries equation, Benjamin–Ono equation, sine–Gordon equation or their differential-difference versions. Moreover, the discrete Toda system provides unification of many particular examples of solvable models of statistical physics and quantum field theory [22, 24, 23].

To obtain the discrete Toda system we construct the Laplace–Darboux sequence for discrete quadrilateral surfaces [13] which are the discrete analogues of surfaces in conjugate net parametrization. The construction is as follows. We start from an arbitrary discrete surface $\mathbf{x} : \mathbb{Z}^2 \to \mathbb{E}^M$ $(M \geq 2)$ with all the elementary quadrilaterals planar, i. e. the position vector $\mathbf{x}$ of the points of the initial lattice satisfies the discrete Laplace equation

$$D_1 D_2 \mathbf{x} = A_1^{(0)} D_1 \mathbf{x} + A_2^{(0)} D_2 \mathbf{x} \ , \tag{8}$$

where $A_i^{(0)} : \mathbb{Z}^2 \to \mathbb{R}$ $(i = 1, 2)$ are real "initial" functions which allow to construct the initial quadrilateral surface when the initial curves are given.

The intersections of the directions of the opposite sides of the quadrilaterals of the lattice define new quadrilateral surfaces, called their Laplace–Darboux transforms. This procedure can be applied recursively giving the Laplace–Darboux sequence $\mathbf{x}^{(k)}$ $(k \in \mathbb{Z})$ of quadrilateral lattices and functions $A_i^{(k)}$.

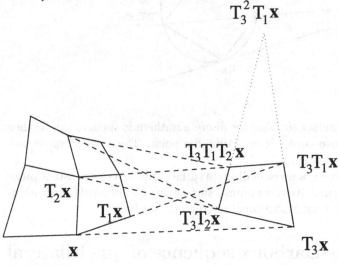

FIGURE 5.

Considering the parameter $k \in \mathbb{Z}$ along the sequence as a new discrete variable and denoting by T_3 the shift operator along it we obtain the following equations for $A_i : \mathbb{Z}^3 \to \mathbb{R}$ $(i = 1, 2)$

$$T_3(A_1 + 1) = A_2 T_2 \left(\frac{A_1 + 1}{A_2} \right) , \tag{9}$$

$$T_3^{-1}(A_2 + 1) = A_1 T_1 \left(\frac{A_2 + 1}{A_1} \right) .$$

In the construction of the Laplace–Darboux sequence we use only ruler, no compass is needed. This means that the proper place to consider the Laplace–Darboux sequence of quadrilateral surfaces is the projective space $\mathbb{P}^M$.

The "projective" representation of the equations (9) involves the cross-ratio K of four collinear points $T_1^{-1}\mathbf{x}$, $\mathbf{x}$, $T_3\mathbf{x}$ and $T_2^{-1}T_3\mathbf{x}$, and reads

$$\frac{((T_1 T_3 K) + 1)(T_2 T_3^{-1} K + 1)}{(K + 1)(T_1 T_2 K + 1)} = \frac{(T_1 K)(T_2 K)}{K\, T_1 T_2 K} , \tag{10}$$

which is the so called gauge invariant form [24] of Hirota's bilinear difference equation (discrete Toda system) [20].

4 Multidimensional quadrilateral lattices

In this Section we present the geometric discretization of the multi-conjugate coordinate systems. The associated differential equations were derived by Darboux [10] and rediscovered (and solved), in the matrix case, about ten years ago by Zakharov and Manakov [35], using a $\bar{\partial}$ approach.

The N-dimensional quadrilateral lattices are defined as mappings

$$\mathbf{x} : \mathbb{Z}^N \to \mathbb{R}^M \ , \ M \geq N \tag{11}$$

which satisfy the following constraint [16]:

All the elementary quadrilaterals Q_{ij} with vertices $\mathbf{x}$, $T_i\mathbf{x}$, $T_j\mathbf{x}$ and $T_iT_j\mathbf{x}$ $(i \neq j)$ are planar.

These *linear* costraints result in a system of $\frac{N(N-1)}{2}$ discrete Laplace equations

$$D_iD_j\mathbf{x} = (T_iA_{ij})D_i\mathbf{x} + (T_jA_{ji})D_j\mathbf{x} \ , \ i \neq j, \ i,j = 1,\ldots,N \ , \tag{12}$$

where $A_{ij} : \mathbb{Z}^N \to \mathbb{R}$ are $N(N-1)$ real functions of the lattice variable (we use the shifted coefficients of the Laplace equation (12) for convenience).

In order to make the construction of the quadrilateral lattice possible the functions A_{ij} need to satisfy the compatibility conditions

$$D_kA_{ij} = A_{ij}T_jA_{jk} + A_{ik}T_kA_{kj} - A_{ik}T_kA_{ij} \ , \ i \neq j \neq k \neq i \ , \tag{13}$$

between the Laplace equations (12).

The above equations have been recently obtained by Bogdanov and Konopelchenko [6] as a natural discrete integrable analogue of the Darboux–Zakharov–Manakov equations, but their geometric meaning was unknown.

4.1 Construction of multidimensional quadrilateral lattices

In order to construct uniquely the N-dimensional quadrilateral lattice, one has to give $\frac{N(N-1)}{2}$ arbitrary intersecting quadrilateral surfaces or, equivalently, N arbitrary intersecting curves (in general position) in $\mathbb{R}^M$ plus $N(N-1)$ arbitrary functions of two discrete variables [16]

$$A_{ij}^{(0)}(n_i, n_j) \ , \ A_{ji}^{(0)}(n_i, n_j) \ , \ 1 \leq i < j \leq N \ . \tag{14}$$

The construction is genuine 3-dimensional and is based on the following simple fact from *linear* geometry:

Three planes in a 3-dimensional space intersect, in general, in a single point.

In particular, the point $T_1T_2T_3\mathbf{x}$ is the point of intersection of the three planes

$$\langle T_1\mathbf{x}, T_1T_2\mathbf{x}, T_1T_3\mathbf{x}\rangle \ , \quad \langle T_2\mathbf{x}, T_1T_2\mathbf{x}, T_2T_3\mathbf{x}\rangle \ , \quad \langle T_3\mathbf{x}, T_1T_3\mathbf{x}, T_2T_3\mathbf{x}\rangle \ ,$$

in the three dimensional subspace $\langle \mathbf{x}, T_1\mathbf{x}, T_2\mathbf{x}, T_3\mathbf{x}\rangle$ of $\mathbb{R}^M$.

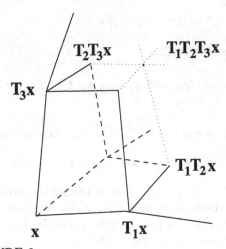

FIGURE 6.

The above construction is compatible for arbitrary $N \geq 3$ (see, for example the discussion in [16]) and allows to build uniquely the whole N-dimensional lattice or, equivalently, the functions A_{ij} which solve the equation (13) with the values on the initial surfaces given by (14).

Let us note that the Laplace–Darboux sequence of quadrilateral surfaces can be viewed as a degenerate 3-dimensional quadrilateral lattice with elementary "hexahedra" of the form presented below.

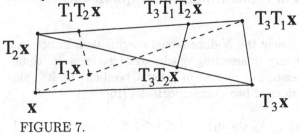

FIGURE 7.

4.2 Reduction to circular lattices

One of the most important integrable reductions of the multi-conjugate co-ordinate systems are the orthogonal systems of Lamé [25, 11]. On the level of discrete geometry the integrable discretization of the orthogonal systems is given by multidimensional circular lattices, which are quadrilateral lattices with additional restriction:

All the elementary quadrilaterals can be inscribed in circles.

This additional property in the limit of the small lattice parameter gives the usuall orthogonality of intersecting lines. Moreover it is integrable in a sense that the circularity constraint "propagates" in the linear construction of the quadrilateral lattice when it is satisfied by the initial data [9].

The circularity condition implies similar restriction for the elementary cells of arbitrary possible dimension. In particular the following fact holds.

Multidimensional circular lattice is a multidimensional quadrilateral lattice such that its every elementary cell of maximal dimension N is inscribed in an $(N-1)$ dimensional sphere.

As it was mentioned, in order to construct uniquely the N-dimensional circular lattice, one has to give $\frac{N(N-1)}{2}$ arbitrary intersecting circular surfaces. The compass is neccesary in drawing the initial surfaces only.

5 Final remarks

We have presented a geometric picture which unifies many of the known integrable discrete (and in consequence also differential) systems. Their integrability scheme is governed by simple geometric constructions.

It is important to note that to remove singularities [31] in the building of the lattice (and in the corresponding solutions of the equations) the projective picture appears to be helpful.

On the level of differential geometry many of the integrable systems can be obtained through additional restrictions imposed on the Darboux multi-conjugate systems [11]. It seems that similar feature happens on the discrete level as well, the discrete Toda system and the circular lattices are good examples, see also [8].

In general, the geometric restrictions come as additional structures in the projective geometry scheme (e. g. the notion of a circle) what is in agreement with Cayley and Klein approach to various geometries.

It may be interesting to consider also projective geometries over other fields, like complex or matrix fields. In particular, one may expect that the multidimensional quadrilateral lattices in the scheme of projective geometry over finite fields will give the integrable cellular automata.

References

[1] M. J. Ablowitz and J. F. Ladik, *Nonlinear differential-difference equations*, J. Math. Phys. **16** (1975) 598.

[2] L. Bianchi, *Lezioni di Geometria Differenziale*, 3-a ed., Zanichelli, Bologna, 1924.

[3] A. Bobenko, *Surfaces in terms of 2 by 2 matrices. Old and new integrable cases*, in *Harmonic Maps and Integrable Systems*, eds. A. Fordy and J. Wood, Vieweg 1994.

[4] A. Bobenko and U. Pinkall, *Discrete Surfaces with Constant Negative Gaussian Curvature and the Hirota equation*, Sfb 288 Preprint 127 (1994), to appear in J. Diff. Geom.

[5] A. Bobenko and U. Pinkall, *Discrete isothermic surfaces*, J. reine angew. Math. **475** (1996), 187–208.

[6] L. V. Bogdanov and B. G. Konopelchenko, *Lattice and q-difference Darboux–Zakharov–Manakov systems via $\bar{\partial}$ method*, J. Phys. A: Math. Gen. **28** (1995) L173-L178.

[7] H. W. Capel and F. W. Nijhof, *Integrable Lattice Equations*, in *Important Developments in Soliton Theory*, p. 38–57, eds. A. S. Fokas and V. E. Zakharov, Springer, 1993.

[8] J. Cieśliński, *The Bäcklund transformation for discrete isothermic surfaces*, this volume.

[9] J. Cieśliński, A. Doliwa and P. M. Santini, *The Integrable Discrete Analogues of Orthogonal Coordinate Systems are Multidimensional Circular Lattices*, Preprint ROME1-1167/97, Dipartimento di Fisica, Università di Roma "La Sapienza", 1997.

[10] G. Darboux, *Leçons sur la théorie générale des surfaces I – IV*, Gauthier – Villars, Paris, 1887 – 1896.

[11] G. Darboux, *Leçons sur les systémes orthogonaux et les coordonnées curvilignes*, 2-éme éd., complétée, Gauthier – Villars, Paris, 1910.

[12] E. Date, M. Jimbo and T. Miwa, *Method for Generating Discrete Soliton Equations I, II*, J. Phys. Soc. Jpn. **51** (1982), 4116–4131.

[13] A. Doliwa, *Geometric discretisation of the Toda system*, Preprint ROME1–1160/96, Dipartimento di Fisica, Università di Roma "La Sapienza", 1996, solv-int/9612006.

[14] A. Doliwa and P. M. Santini, *An Elementary Geometric Characterisation of the Integrable Motions of a Curve*, Phys. Lett. A **185** (1994) 373

[15] A. Doliwa and P. M. Santini, *Integrable dynamics of a discrete curve and the Ablowitz-Ladik hierarchy*, J. Math. Phys. **36** (1995), 1259–1273.

[16] A. Doliwa and P. M. Santini, *Multidimensional Quadrilateral Lattices are Integrable*, Preprint ROME1–1162/96, Dipartimento di Fisica, Università di Roma "La Sapienza", 1996, solv-int/9612007.

[17] L. P. Eisenhart, *A Treatise on the Differential Geometry of Curves and Surfaces*, Ginn and Company, Boston, 1909.

[18] *Symmetries and Integrability of Difference Equations*, eds. D. Levi, L. Vinet and P. Winternitz, CRM Proceedings and Lecture Notes, Vol. 9, AMS, Providence, 1996.

[19] R. Hirota, *Nonlinear partial difference equations. III. Discrete sine-Gordon equation*, J. Phys. Soc. Jpn. **43** (1977), 2079–2186.

[20] R. Hirota, *Discrete Analogue of a Generalized Toda Equation*, J. Phys. Soc. Jpn. **50** (1981), 3785–3791.

[21] B. G. Konopelchenko, *Induced Surfaces and Their Integrable Dynamics*, Studies Appl. Math. **96** (1996) 9-51.

[22] I. G. Korepanov, *Algebraic integrable dynamical systems, 2+1 - dimensional models in wholly discrete space-time, and the inhomogeneous models in 2-dimensional statistical physics*, solv-int/9506003.

[23] I. Krichever, O. Lipan, P. Wiegmann and A. Zabrodin, *Quantum Integrable Systems and Elliptic Solutions of Classical Discrete Nonlinear Equations*, preprint ESI 330 (1996), hep-th/9604080.

[24] A. Kuniba, T. Nakanishi and J. Suzuki, *Functional relations in solvable lattice models, I: Functional relations and representation theory, II: Applications*, Int. Journ. Mod. Phys. A **9** (1994) 5215–5312.

[25] G. Lamé, *Leçons sur les coordonnées curvilignes et leurs diverses applications*, Mallet–Bachalier, Paris, 1859.

[26] D. Levi, *Nonlinear differential-difference equations as Bäcklund transformations*, J. Phys. A: Math. Gen. **14** (1981) 1083–1098.

[27] D. Levi, L. Pilloni and P. M. Santini, *Integrable three-dimensional lattices*, J. Phys. A **14** (1981), 1567–1575.

[28] A. V. Mikhailov, *Integrability of a two-dimensional generalization of the Toda chain*, JETP Lett. **30** (1979) 414–418.

[29] F. W. Nijhoff and H. W. Capel, *The direct linearisation approach to hierarchies of integrable PDEs in 2+1 dimensions: I. Lattice equations and the differential-difference hierarchies*, Inverse Problems **6** (1990) 567–590.

[30] G. R. W. Quispel, F. W. Nijhoff, H. W. Capel and J. van der Linden, *Linear integral equations and nonlinear difference-difference equations*, Physica A **125**, (1984) 344–380

[31] A. Ramani, B. Grammaticos and V. Papageorgiou, *Singularity Confinement*, in [18], p. 303–318.

[32] P. Samuel, *Projective Geometry*, Springer, 1988.

[33] R. Sauer, *Differenzengeometrie*, Springer, Berlin, 1970.

[34] A. Sym, *Soliton Surfaces and Their Applications (Soliton Geometry from Spectral Problems)*, in *Geometric Aspects of the Einstein Equations and Integrable Systems*, Lecture Notes in Physics 239, Springer, Berlin, 1985, p. 154–231.

[35] V. E. Zakharov and S. E. Manakov, *Construction of Multidimensional Nonlinear Integrable Systems and Their Solutions*, Funk. Anal. Ego Prilozh. **19** (1985) 11–25.

Self-dual Einstein spaces and a discrete Tzitzeica equation. A permutability theorem link

W.K. Schief

School of Mathematics,

The University of New South Wales,

Sydney, NSW 2052,

Australia

Abstract

The generation of integrable (differential-)difference equations by means of Bäcklund transformations and associated permutability theorems has become a standard technique in soliton theory. Here, it is shown that a permutability theorem for an integrable discrete version of the classical Tzitzeica equation leads, in the natural continuum limit, to a novel fully symmetric form of the equation governing self-dual Einstein spaces in four dimensions. As a by-product of this construction, the associated linear representation is found which turns out to be the Lax pair for the self-dual Yang-Mills equations with four translational symmetries and the gauge group of volume preserving diffeomorphisms.

1   Introduction

The generation of integrable differential-difference and fully discrete equations has become a well-developed branch of soliton theory. Typically, these (differential-)difference equations translate to integrable continuous counterparts by taking appropriate continuum limits. The limiting processes involved are usually quite subtle, with continuous variables entering via a multiple scales process (see [1] for a recent review).

There exists, however, a distinct method of constructing integrable discrete systems which reduce to continuous ones by means of a natural limiting procedure in which the discrete independent variables may be regarded as truly discretized versions of the continuous variables. This technique has its origin in work by Levi and Benguria [2] and has been refined and generalized by several authors [3]-[5]. The main idea is to reinterpret Bäcklund transformations and associated nonlinear superposition principles as integrable differential-difference and pure difference versions of their continuous counterparts respectively. Indeed, the prototypical example for this is the classical permutability theorem [6]

$$\tan\left(\frac{\omega_{12} - \omega}{4}\right) = \frac{\beta_2 + \beta_1}{\beta_2 - \beta_1} \tan\left(\frac{\omega_2 - \omega_1}{4}\right) \tag{1}$$

which may be regarded as an integrable discrete sine-Gordon equation [1] if indices on ω are interpreted as increments in the respective discrete variables.

The permutability theorem (1) is based on the classical Bäcklund transformation for the sine-Gordon equation. Interestingly, this Bäcklund transformation may be derived from a constrained Moutard transformation. In fact, iteration of the classical Moutard transformation [7] has been shown to generate the permutability theorem [8]

$$\tau\tau_{123} = \tau_1\tau_{23} - \tau_2\tau_{13} + \tau_3\tau_{12} \tag{2}$$

which represents a discrete counterpart of the 2+1-dimensional sine-Gordon system introduced by Konopelchenko and Rogers in [9]. Here, τ is assumed to be a complex function. As in the continuous case, however, the discrete sine-Gordon system (2) may be specialized to the real single equation

$$\frac{\sin(\omega_{123} - \omega_3)\sin(\omega_{13} - \omega)\sin(\omega_{12} - \omega_{23})\sin(\omega_1 - \omega_2)}{\sin(\omega_{123} - \omega_2)\sin(\omega_{12} - \omega)\sin(\omega_{13} - \omega_{23})\sin(\omega_1 - \omega_3)} = 1 \tag{3}$$

representing a discrete version of the 2+1-dimensional integrable sine-Gordon equation set down by Darboux [10] in connection with triply orthogonal systems of surfaces. This is achieved by imposing bilinear constraints on (2) analogous to those employed in the continuous case [8, 11].

The permutability theorems displayed above share the feature that they are purely algebraic. In general, this is not the case. For instance, the permutability theorem for the Boussinesq equation reads

$$u_{12} = u_1 + u_2 - u + \ln(u_{1x} - u_{2x}). \tag{4}$$

Nevertheless, this is now to be interpreted as a differential-difference equation and, in fact, it may be obtained from a single (natural) continuum limit performed on the standard discrete Kadomtsev-Petviashvili equation [12].

In a recent development [13], it has been shown that the classical Tzitzeica equation [14]

$$\omega_{xy} = e^\omega - e^{-2\omega} \tag{5}$$

governing affine spheres possesses a non-algebraic permutability theorem. It contains derivatives with respect to the original continuous variables x and y and, in the usual reinterpretation, increments with respect to the discrete variables n_1 and n_2. In the natural continuum limit, the permutability theorem becomes a quadratic partial differential equation in four dimensions where all independent variables appear on an equal footing. The corresponding Lax pair has been constructed algorithmically and has been identified with the Lax pair for the self-dual Yang-Mills equations with four translational symmetries and the gauge group of volume preserving diffeomorphisms [15].

Here, an integrable discrete version of the Tzitzeica equation is presented along with a Bäcklund transformation based on a discrete Moutard transformation. The corresponding permutability theorem is derived which may be interpreted as a fully symmetric, discrete version of the above-mentioned differential-difference equation. This implies, in turn, that this discrete equation may be regarded as a discretisation of the equation governing self-dual Einstein spaces [16]. It is also recorded that the four-dimensional discrete equation discussed here contains Hirota's discrete 'master' equation, namely the discrete Toda lattice [12], in a three-dimensional reduction and an integrable generalized discrete Tzitzeica equation [17] in a two-dimensional reduction.

2   A discrete Tzitzeica equation

The differential equation

$$(\ln h)_{xy} = h - h^{-2} \tag{6}$$

is well-known in affine geometry. It was set down as long ago as 1910 by Tzitzeica [14] in connection with a particular class of hyperbolic surfaces now known as *affine spheres*. In this section, an integrable discrete version of the Tzitzeica equation is proposed. A geometric interpretation of the discrete Tzitzeica equation in terms of discrete surfaces is the topic of a forthcoming paper [17]. Thus, it may be directly verified that the triad

$$\phi_{11} - \phi_1 = \frac{H_1 - 1}{H_1(H - 1)}(\phi_1 - \phi) + \frac{\lambda A}{H - 1}(\phi_{12} - \phi_1)$$

$$\phi_{12} + \phi = H(\phi_1 + \phi_2) \tag{7}$$

$$\phi_{22} - \phi_2 = \frac{H_2 - 1}{H_2(H - 1)}(\phi_2 - \phi) + \frac{\lambda^{-1} B}{H - 1}(\phi_{12} - \phi_2)$$

is compatible if and only if A, B and H satisfy the nonlinear system

$$A_2 = \frac{H_1}{H}A, \quad B_1 = \frac{H_2}{H}B$$

$$H_{12} = \frac{H(H-1)}{H^2(H_1 + H_2 - H_1 H_2) - H + ABH_1 H_2}. \tag{8}$$

Here, indices denote increments in the discrete variables n_1 and n_2, for instance $H_1 = H(n_1 + 1, n_2)$, and λ is a constant parameter. In the following, the above system is referred to as the *discrete Tzitzeica system*. Remarkably, the linear equations $(8)_{1,2}$ imply the existence of a potential τ defined according to

$$\tau_{11} = c\frac{\tau_1^2}{\tau A}, \quad \tau_{12} = \frac{\tau_1 \tau_2}{\tau H}, \quad \tau_{22} = \hat{c}\frac{\tau_2^2}{\tau B}, \tag{9}$$

where c and $\hat{c}$ are arbitrary constants. This may be exploited to express A, B and H in terms of τ only. Hence, the discrete Tzitzeica system simplifies to the *discrete Tzitzeica equation*

$$\begin{vmatrix} \tau & \tau_1 & \tau_{11} \\ \tau_2 & \tau_{12} & \tau_{112} \\ \tau_{22} & \tau_{122} & \tau_{1122} \end{vmatrix} + c\hat{c}\tau_{12}^3 = 0. \tag{10}$$

The justification of the terms discrete Tzitzeica system/equation is seen as follows. If one regards the variables n_1 and n_2 as discretizations of some continuous variables x and y, that is

$$x = \epsilon_1 n_1, \quad y = \epsilon_2 n_2 \tag{11}$$

for small ϵ_i, then the Taylor expansions

$$f_1 = f + \epsilon_1 f_x + O(\epsilon_1^2), \quad f_2 = f + \epsilon_2 f_y + O(\epsilon_2^2) \tag{12}$$

apply. Now, the form of the linear triad (7) suggests the natural expansion

$$H = 1 + \tfrac{1}{2}\epsilon_1 \epsilon_2 h, \quad A = \tfrac{1}{2}\epsilon_1^3 a, \quad B = \tfrac{1}{2}\epsilon_2^3 b \tag{13}$$

so that the discrete Tzitzeica system reduces to

$$(\ln h)_{xy} = h - abh^{-2}, \quad a = a(x), \quad b = b(y) \tag{14}$$

in the limit $\epsilon_i \to 0$. This is a gauge-invariant form of the Tzitzeica equation (6) in the sense that the change of variables $x \to X(x), y \to Y(y)$, $h \to (X_x Y_y)^{-1}h, a \to X_x^{-3}a, b \to Y_y^{-3}b$ preserves the form of (14). Hence, for non-vanishing a and b one may set $a = b = 1$ without loss of generality. In this case, introduction of a 'τ-function' defined by

$$h = -2(\ln \tau)_{xy} \tag{15}$$

tranforms the Tzitzeica equation into the cubic equation

$$\begin{vmatrix} \tau & \tau_x & \tau_{xx} \\ \tau_y & \tau_{xy} & \tau_{xxy} \\ \tau_{yy} & \tau_{xyy} & \tau_{xxyy} \end{vmatrix} + \tfrac{1}{4}\tau^3 = 0 \tag{16}$$

which is but the continuum limit of the discrete Tzitzeica equation (10) for an appropriate choice of the parameters c and $\hat{c}$.

3 A Bäcklund transformation

In his classical paper [14], Tzitzeica showed that the Tzitzeica equation is form-invariant under a variant of the Moutard transformation [7]. In the discrete case, this transformation reads as follows.

Theorem (The discrete Tzitzeica transformation). *The linear triad (7) and the discrete Tzitzeica system (8) are form-invariant under*

$$\phi \to \phi' \sim \frac{S}{\psi}, \quad H \to H' = \frac{\psi_1 \psi_2}{\psi \psi_{12}} H$$

$$A \to A' = \frac{\psi_1^2}{\psi \psi_{11}} A, \quad B \to B' = \frac{\psi_2^2}{\psi \psi_{22}} B \tag{17}$$

where

$$S = \frac{\lambda - \mu}{\lambda + \mu}\left[\psi\phi - \frac{H}{(\lambda-\mu)(H-1)}(\lambda\Delta_1\psi\Delta_2\phi - \mu\Delta_2\psi\Delta_1\phi)\right] \tag{18}$$

and ψ is another solution of the triad (7) with parameter μ. Here, Δ_i denotes the usual difference operator, i.e. $\Delta_i f = f_i - f$.

It is noted that the skew-symmetric bilinear quantity S obeys the relations

$$\Delta_1 S = \psi\phi_1 - \psi_1\phi, \quad \Delta_2 S = \psi_2\phi - \psi\phi_2. \tag{19}$$

In fact, the discrete analogue of the Moutard transformation consists of the invariance $(17)_{1,2}$ for the linear equation

$$\phi_{12} + \phi = H(\phi_1 + \phi_2) \tag{20}$$

where S is now an *arbitrary* solution of (19). However, in order that the form of the remaining linear equations $(7)_{1,3}$ be preserved, the integration constant in S has to be so chosen that S takes the form (18). Finally, on taking into account the invariance

$$\tau \to f(n_1)g(n_2)\tau, \quad ff_{11} = f_1^2, \quad gg_{22} = g_2^2 \tag{21}$$

of the discrete Tzitzeica equation, the new τ-function turns out to be

$$\tau' = \psi\tau \tag{22}$$

without loss of generality. This relation is the basis of the considerations in the subsequent section.

The Bäcklund equations for the discrete Tzitzeica equation are now obtained by replacing the eigenfunction ψ in the corresponding linear triad of the form (7) by means of (22). Thus, in terms of the τ-functions, the *Bäcklund transformation* reads

$$\tau\tau'_{12} + \tau_{12}\tau' = \tau_1\tau'_2 + \tau_2\tau'_1$$

$$\begin{vmatrix} \tau & \tau_1 & \tau_{11} \\ \tau' & \tau'_1 & \tau'_{11} \\ \tau_2 & \tau_{12} & \tau_{112} \end{vmatrix} = c\mu\tau_1 \begin{vmatrix} \tau_1 & \tau_{12} \\ \tau'_1 & \tau'_{12} \end{vmatrix}, \quad \begin{vmatrix} \tau & \tau_2 & \tau_{22} \\ \tau' & \tau'_2 & \tau'_{22} \\ \tau_1 & \tau_{12} & \tau_{122} \end{vmatrix} = \hat{c}\mu^{-1}\tau_2 \begin{vmatrix} \tau_2 & \tau_{12} \\ \tau'_2 & \tau'_{12} \end{vmatrix} \tag{23}$$

which may be shown to be symmetric in τ and τ' by virtue of $(23)_1$. Incidentally, the latter may be identified as the 2+1-dimensional discrete sine-Gordon system (2) if one interprets the relation between τ and τ' as a shift on a one-dimensional lattice labelled by a third discrete variable, n_3, say and permutes the indices $1, 2$ and 3 appropriately.

In conclusion, it is noted that the Bäcklund equations for the classical Tzitzeica equation in the form (16) may be derived by performing the continuum limit discussed in the previous section on the discrete Bäcklund equations (23). They become

$$\tau\tau'_{xy} + \tau_{xy}\tau' = \tau_x\tau'_y + \tau_y\tau'_x$$

$$\begin{vmatrix} \tau & \tau_x & \tau_{xx} \\ \tau' & \tau'_x & \tau'_{xx} \\ \tau_y & \tau_{xy} & \tau_{xxy} \end{vmatrix} = \tfrac{1}{2}\mu\tau \begin{vmatrix} \tau & \tau_y \\ \tau' & \tau'_y \end{vmatrix}, \quad \begin{vmatrix} \tau & \tau_y & \tau_{yy} \\ \tau' & \tau'_y & \tau'_{yy} \\ \tau_x & \tau_{xy} & \tau_{xyy} \end{vmatrix} = \tfrac{1}{2}\mu^{-1}\tau \begin{vmatrix} \tau & \tau_x \\ \tau' & \tau'_x \end{vmatrix}. \tag{24}$$

4   A permutability theorem

As in the continous case, the discrete Tzitzeica transformation may be iterated in a purely algebraic manner. Thus, let ϕ^3, ϕ^4 and ϕ be three solutions of the linear triad (7) with parameters λ_3, λ_4 and λ respectively. Then, the new eigenfunctions generated by ϕ^3 and ϕ^4 may be taken to be

$$\phi_3 = \phi - \frac{H}{(\lambda - \lambda_3)(H-1)}\left[\lambda\frac{\Delta_1\phi^3}{\phi^3}\Delta_2\phi - \lambda_3\frac{\Delta_2\phi^3}{\phi^3}\Delta_1\phi\right]$$

$$\phi_4 = \phi - \frac{H}{(\lambda - \lambda_4)(H-1)}\left[\lambda\frac{\Delta_1\phi^4}{\phi^4}\Delta_2\phi - \lambda_4\frac{\Delta_2\phi^4}{\phi^4}\Delta_1\phi\right] \tag{25}$$

while the new solutions of the discrete Tzitzeica equation read

$$\tau_3 = \phi^3\tau, \quad \tau_4 = \phi^4\tau. \tag{26}$$

It is emphasized that, at this stage, the indices 3 and 4 are merely labels and do not denote increments of discrete variables. Moreover, since the quantity $\phi_4^3 = \phi_4[\phi = \phi^3, \lambda = \lambda_3]$ is an eigenfunction associated with τ_4, another application of the discrete Tzitzeica transformation produces

$$\tau_{34} = \phi_4^3\tau_4 = \phi_4^3\phi^4\tau \tag{27}$$

or, explicitly,

$$\tau_{34} = \left[\phi^3\phi^4\tau - \frac{H}{(\lambda_3 - \lambda_4)(H-1)}(\lambda_3\Delta_1\phi^4\Delta_2\phi^3 - \lambda_4\Delta_2\phi^4\Delta_1\phi^3)\right]\tau. \tag{28}$$

Finally, by means of (26), the eigenfunctions ϕ^3 and ϕ^4 may be expressed in terms of τ-functions so that (28) becomes the *permutability theorem*

$$(\lambda_3 - \lambda_4)(\tau\tau_{12} - \tau_1\tau_2)(\tau\tau_{34} - \tau_3\tau_4)$$
$$= \lambda_3(\tau\tau_{14} - \tau_1\tau_4)(\tau\tau_{23} - \tau_2\tau_3) - \lambda_4(\tau\tau_{13} - \tau_1\tau_3)(\tau\tau_{24} - \tau_2\tau_4). \tag{29}$$

The above relation may be regarded as a nonlinear superposition principle linking four solutions of the discrete Tzitzeica equation generated by the discrete Tzitzeica transformation. It is readily seen that it is symmetric in the indices 3 and 4 which implies that the corresponding *Bianchi diagram* [18] is closed, that is $\tau_{34} = \tau_{43}$.

The permutability theorem (29) may now be reinterpreted if one assumes that τ represents a point on a discrete lattice labelled by n_3 and n_4, that is

$$\tau = \tau(n_1, n_2; n_3, n_4) \tag{30}$$

and regards the transformed objects τ_3 and τ_4 as its neighbours on this lattice, namely

$$\tau_3 = \tau(n_1, n_2; n_3 + 1, n_4), \quad \tau_4 = \tau(n_1, n_2; n_3, n_4 + 1). \tag{31}$$

Analogously, the identification

$$\tau_{34} = \tau(n_1, n_2; n_3 + 1, n_4 + 1) \tag{32}$$

is made. In this way, the permutability theorem (29) becomes a fully symmetric discrete equation in four dimensions, viz

$$
\begin{aligned}
&c_1(\tau\tau_{12} - \tau_1\tau_2)(\tau\tau_{34} - \tau_3\tau_4) \\
&+ c_2(\tau\tau_{14} - \tau_1\tau_4)(\tau\tau_{23} - \tau_2\tau_3) \\
&+ c_3(\tau\tau_{13} - \tau_1\tau_3)(\tau\tau_{24} - \tau_2\tau_4) = 0, \qquad \sum_i c_i = 0
\end{aligned}
\tag{33}
$$

and the transformation laws (25) in the form

$$
\begin{aligned}
\Delta_3\phi &= \frac{1}{(\lambda - \lambda_3)(\tau\tau_{12} - \tau_1\tau_2)}\left[\lambda\frac{\tau_2}{\tau_3}(\tau\tau_{13} - \tau_1\tau_3)\Delta_2\phi - \lambda_3\frac{\tau_1}{\tau_3}(\tau\tau_{23} - \tau_2\tau_3)\Delta_1\phi\right] \\
\Delta_4\phi &= \frac{1}{(\lambda - \lambda_4)(\tau\tau_{12} - \tau_1\tau_2)}\left[\lambda\frac{\tau_2}{\tau_4}(\tau\tau_{14} - \tau_1\tau_4)\Delta_2\phi - \lambda_4\frac{\tau_1}{\tau_4}(\tau\tau_{24} - \tau_2\tau_4)\Delta_1\phi\right]
\end{aligned}
\tag{34}
$$

constitute a corresponding 'Lax pair'. It turns out, however, that the latter is compatible modulo (29) *and* the original triad (7). At first glance, this seems to be an indication that the discrete equation (29) is not integrable on its own but only if constrained by the discrete Tzitzeica equation (10). This issue is touched upon in the following section.

5 Integrable reductions

The conjecture that all lower-dimensional (Lie point symmetry) reductions of integrable systems should be integrable has become widely accepted. A systematic study of the reductions of the discrete equation (29) is under way. Here, two canonical reductions are presented, leading to important integrable systems. Firstly, the ansatz

$$\tau(n_1, n_2, n_3, n_4) = \kappa_1^{n_1 n_4}\kappa_2^{n_2 n_4}\kappa_3^{n_3 n_4}\tau(n_1, n_2, n_3) \tag{35}$$

leads to a three-dimensional discrete equation of the form

$$\hat{c}_1\tau_1\tau_{23} + \hat{c}_2\tau_2\tau_{13} + \hat{c}_3\tau_3\tau_{12} = 0, \qquad \sum_i \hat{c}_i = 0 \tag{36}$$

provided that the constants κ_i obey the relation

$$(\lambda_3 - \lambda_4)(\kappa_3 - 1) = \lambda_3(\kappa_1 - 1) - \lambda_4(\kappa_2 - 1). \tag{37}$$

Even though the constants $\hat{c}_i$ are constrained by $(36)_2$, they may be regarded as arbitrary if one takes into account the transformation

$$\tau \to \tilde{c}_1^{n_2 n_3} \tilde{c}_2^{n_1 n_3} \tilde{c}_3^{n_1 n_2} \tau. \tag{38}$$

The equation $(36)_1$ which has come to be known as the discrete Kadomtsev-Petviashvili (KP) equation may be linked to Hirota's discrete Toda lattice equation [12]

$$\hat{c}_1 \tau_{\bar{1}'} \tau_{1'} + \hat{c}_2 \tau_{\bar{2}'} \tau_{2'} + \hat{c}_3 \tau_{\bar{3}'} \tau_{3'} = 0 \tag{39}$$

through the change of independent variables

$$n_1' = n_2 + n_3, \quad n_2' = n_1 + n_3, \quad n_3' = n_1 + n_2. \tag{40}$$

Here, barred indices designate backwards shifts on a lattice, for instance $\tau_{\bar{1}} = \tau(n_1 - 1, n_2, n_3)$. Hirota's equation is occasionally considered a 'master integrable equation' since it unifies important continuous integrable equations together with their hierarchies via multiple scales-type continuum limits.

Secondly, it is observed that the discrete equation (33) may be brought into the form

$$\begin{vmatrix} \tau_{12} & \tau_1 & \tau_{13} \\ \tau_2 & \tau & \tau_3 \\ \tau_{24} & \tau_4 & \tau_{34} \end{vmatrix} = \tilde{c} \begin{vmatrix} \tau_{13} & \tau_1 & \tau_{14} \\ \tau_3 & \tau & \tau_4 \\ \tau_{23} & \tau_2 & \tau_{24} \end{vmatrix}, \qquad \tilde{c} = \frac{c_2}{c_1} \tag{41}$$

so that the symmetry reduction

$$\tau(n_1, n_2, n_3, n_4) = \tau(n_1 - n_4, n_2 - n_3) \tag{42}$$

yields the two-dimensional discrete equation

$$\begin{vmatrix} \tau_{12} & \tau_1 & \tau_{1\bar{2}} \\ \tau_2 & \tau & \tau_{\bar{2}} \\ \tau_{\bar{1}2} & \tau_{\bar{1}} & \tau_{\bar{1}\bar{2}} \end{vmatrix} = \tilde{c} \begin{vmatrix} \tau_{1\bar{2}} & \tau_1 & \tau \\ \tau_{\bar{2}} & \tau & \tau_{\bar{1}} \\ \tau & \tau_2 & \tau_{\bar{1}2} \end{vmatrix} \tag{43}$$

or, equivalently,

$$\begin{vmatrix} \tau & \tau_1 & \tau_{11} \\ \tau_2 & \tau_{12} & \tau_{112} \\ \tau_{22} & \tau_{122} & \tau_{1122} \end{vmatrix} = \tilde{c} \begin{vmatrix} \tau_1 & \tau_{11} & \tau_{12} \\ \tau_2 & \tau_{12} & \tau_{22} \\ \tau_{12} & \tau_{112} & \tau_{122} \end{vmatrix}. \tag{44}$$

The latter constitutes an integrable *generalized* discrete Tzitzeica equation since it contains the discrete Tzitzeica equation as a degenerate case. Thus, let

$$\tau \to \alpha^{\frac{n_1^2}{2}} \beta^{\frac{n_2^2}{2}} \tau, \qquad \alpha, \beta = \text{const.} \tag{45}$$

Then, the discrete Tzitzeica equation (10) is retrieved in the limit $\alpha, \beta \to 0$ provided that $\tilde{c}/\alpha\beta$ is finite. The above limiting process has its origin in a geometric connection between the discrete Tzitzeica equation and its generalized version (44). A derivation of the latter in a geometric context may be found in [17].

It is natural to enquire about the significance of the 'Lax pair' (34) in the above-mentioned reductions. Remarkably, in both cases, suitable ansätze for the eigenfunction ϕ lead to *compatible* linear representations of the reduced equations. Thus, it may be shown that a reduction of the form

$$\phi(n_1, n_2, n_3, n_4) = \mu(\lambda)^{n_4} \phi(n_1, n_2, n_3) \tag{46}$$

produces a proper Lax pair for the discrete KP equation (36). On the other hand, if one assumes that

$$\phi(n_1, n_2, n_3, n_4) = \phi(n_1 - n_4, n_2 - n_3) \tag{47}$$

then the two equations (34) turn out to be part of the linear triad for the generalized discrete Tzitzeica equation (44) as discussed in [17].

6   Self-dual Einstein spaces

As pointed out in Section 4, the linear system (34) is compatible only for a restricted class of solutions of (29). The situation is different in the natural continuum limit. Thus, continuous variables x, y, s and t are introduced according to

$$x = \epsilon_1 n_1, \quad y = \epsilon_2 n_2, \quad s = \epsilon_3 n_3, \quad t = \epsilon_4 n_4 \tag{48}$$

so that in the limit $\epsilon_i \to 0$, the nonlinear partial differential equation

$$(\lambda_3 - \lambda_4)\Theta_{xy}\Theta_{st} = \lambda_3 \Theta_{xt}\Theta_{ys} - \lambda_4 \Theta_{xs}\Theta_{yt} \tag{49}$$

is obtained, where

$$\tau = \exp\Theta. \tag{50}$$

In addition, the linear system (34) beomes

$$\phi_s = \frac{1}{\lambda - \lambda_3}\left[\lambda\frac{\Theta_{xs}}{\Theta_{xy}}\phi_y - \lambda_3\frac{\Theta_{ys}}{\Theta_{xy}}\phi_x\right]$$

$$\phi_t = \frac{1}{\lambda - \lambda_4}\left[\lambda\frac{\Theta_{xt}}{\Theta_{xy}}\phi_y - \lambda_4\frac{\Theta_{yt}}{\Theta_{xy}}\phi_x\right] \tag{51}$$

which now turns out to be a proper Lax pair in that it is compatible modulo (49). The reason for this is the following. The discrete 'Lax pair' (34) is compatible only modulo the Tzitzeica triad (7). However, the contribution of the latter is of higher order in ϵ_i. Thus, in the limit $\epsilon_i \to 0$, the linear representations (7) and (34) 'decouple' and (51) may be considered independently.

The nonlinear equation (49) is of considerable importance in self-dual Yang-Mills theory and complex relativity. To this end, it is noted that in terms of the vector fields

$$A_\alpha = \lambda_3(-\partial_s + \frac{\Theta_{ys}}{\Theta_{xy}}\partial_x), \quad A_{\bar\alpha} = -\partial_t + \frac{\Theta_{xt}}{\Theta_{xy}}\partial_y$$

$$A_\beta = \lambda_4(-\partial_t + \frac{\Theta_{yt}}{\Theta_{xy}}\partial_x), \quad A_{\bar\beta} = \partial_s - \frac{\Theta_{xs}}{\Theta_{xy}}\partial_y \tag{52}$$

the Lax pair (51) reads

$$(A_\alpha + \lambda A_{\bar\beta})\phi = 0, \quad (A_\beta - \lambda A_{\bar\alpha})\phi = 0. \tag{53}$$

The compatibility condition then assumes the form

$$[A_\alpha, A_\beta] = 0, \quad [A_{\bar\alpha}, A_{\bar\beta}] = 0, \quad [A_\alpha, A_{\bar\alpha}] + [A_\beta, A_{\bar\beta}] = 0 \tag{54}$$

which is precisely the self-dual Yang-Mills equations subject to the 'maximal symmetry condition', that is four translational symmetries [15]. Furthermore, the vector fields A_κ are divergence free with respect to the volume form

$$\varepsilon = \Theta_{xy} \, dx \wedge dy \wedge ds \wedge dt \tag{55}$$

or, explicitly,

$$(\Theta_{xy}A_\kappa^a)_{,a} = 0, \quad A_\kappa = A_\kappa^a \partial_a. \tag{56}$$

This implies, in turn, that the vector fields A_κ may be identified with the generators of the group of volume preserving diffeomorphism of the auxiliary manifold labelled by (x, y, s, t). The latter and the maximal symmetry condition have been shown to be necessary and sufficient to construct all four-dimensional, self-dual, Ricci flat, complex-valued spaces [19, 20]. In terms of the curvature two form $R^a{}_b$, these are governed by $*R^a{}_b = R^a{}_b$ where $*$ is the Hodge star operator. Metrics which satisfy this condition automatically have vanishing Ricci tensor and hence satisfy the (complex) vacuum Einstein equations. Accordingly, the continuum limit (49) of the discrete equation (29) is equivalent to Plebanski's first *heavenly equation* [16]

$$\Omega_{PR}\Omega_{QS} - \Omega_{PS}\Omega_{QR} = 1 \tag{57}$$

which is standard in the theory of self-dual Einstein spaces. The exact nature of this connection has been set down in [13].

References

[1] F. Nijhoff and H. Capel, in *Proceedings of the International Symposium KdV' 95*, ed. M. Hazewinkel, H.W. Capel and E.M. de Jaeger (Kluwer Academic Publishers, Dordrecht, 1995) 133.

[2] D. Levi and R. Benguria, *Proc. Natl. Acad. Sci. USA* **77** (1980) 5025.

[3] D. Levi, *J. Phys. A: Math. Gen.* **14** (1981) 1083.

[4] B.G. Konopelchenko, *Phys. Lett. A* **87** (1982) 445.

[5] F.W. Nijhoff, H.W. Capel, G.L. Wiersma and G.R.W. Quispel, *Phys. Lett. A* **105** (1984) 267.

[6] L. Bianchi, *Lezioni di Geometria Differenziale* **2** (Pisa, 1902) 418.

[7] Th.-F. Moutard, *J. de L'Ecole Polytechnique, Cahier* **45** (1878) 1.

[8] J.J.C. Nimmo and W.K. Schief, to appear in *Proc. R. Soc. London A* (1996).

[9] B.G. Konopelchenko and C. Rogers, *Phys. Lett. A* **158** (1991) 391.

[10] G. Darboux, *Leçons sur les systèmes orthogonaux et les coordonnées curvilignes* (Gauthier-Villars, Paris, 1910).

[11] J.J.C. Nimmo and W.K. Schief, in preparation (1996).

[12] R. Hirota, *J. Phys. Japan* **50** (1981) 3785.

[13] W.K. Schief, to appear in *Phys. Lett. A* (1996).

[14] G. Tzitzeica, *C. R. Acad. Sci. Paris* **150** (1910) 955.

[15] M.J. Ablowitz and P. Clarkson, *Solitons, Nonlinear Evolution Equations and Inverse Scattering* (Cambridge University Press, Cambridge, 1991).

[16] J.F. Plebański, *J. Math. Phys.* **16** (1975) 2395.

[17] A. Bobenko and W.K. Schief, to appear in *Discrete Integrable Geometry and Physics*, eds. A. Bobenko and R. Seiler (Oxford University Press, 1997).

[18] G.L. Lamb, Jr, in *Bäcklund Transformations, Lecture Notes in Mathematics* **515**, ed. R.M. Miura (Springer, Heidelberg, 1976) 69.

[19] L.J. Mason and E.T. Newman, *Commun. Math. Phys.* **121** (1989) 659.

[20] S. Chakravarty, L. Mason and E.T. Newman, *J. Math. Phys.* **32** (1991) 1458.

Chapter 4

Asymptotic Analysis

New Solutions of the Nonstationary Schrödinger and Kadomtsev-Petviashvili Equations

Mark J. Ablowitz[†] and Javier Villarroel[‡]
[†] Department of Applied Mathematics
University of Colorado
Boulder, CO 80309-0526, USA
[‡] Universidad de Salamanca
Dept. de Matematicos puras y aplicadas
37008 Salamanca, SPAIN

Abstract

A method to obtain new real, nonsingular, decaying potentials of the nonstationary Schrödinger equation and corresponding solutions of the Kadomtsev-Petviashvili equation is developed. The solutions are characterized by the order of the poles of an eigenfunction of the Schrödinger operator and an underlying topological quantity, an index or charge. The properties of some of these solutions are discussed.

1 Introduction

In this paper we describe a method to obtain a new class of decaying potentials and corresponding solutions to the nonstationary Schrödinger and Kadomtsev-Petviashvili-I equation (KPI; here the "I" stands for one of the two physically interesting choices of sign in the equation). These equations have significant applications in Physics. The Schrödinger operator is, of course, centrally important in quantum mechanics and the KP equation is a ubiquitous nonlinear wave equation governing weakly nonlinear long waves in two dimensions with slowly varying transverse modulation. The nonstationary Schrödinger operator can be used to linearize the KPI equation via the inverse scattering transform (IST; [see e.g. 1]). In [2] it was shown by IST that discrete states associated with complex conjugate pairs of simple eigenvalues of the Schrödinger operator yield lump type soliton solutions which decay as $O(1/r^2)$, $r^2 = x^2 + y^2$.

In [3] it was shown, for the first time, that there are real, nonsingular, decaying potentials of the Schrödinger operator corresponding to its discrete spectrum, whose corresponding eigenfunctions have multiple poles/eigenvalues. This class of potentials is related to decaying solutions in the KPI equation which we refer to as "multipole lumps". In this paper we discuss methods to derive these solutions, give a number of explicit formulae, and discuss their properties. As do the ordinary lumps these solutions decay as $O(1/r^2)$. Thus, as opposed to the time independent Schrödinger equation, we now see that there are real, nonsingular decaying, potentials of the nonstationary Schrödinger equation which are related to eigenvalues with multiplicity. We note, however, that these potentials are not absolutely integrable. This underlies the fact that the potentials are characterized by the order of the pole and a topological number, an index or winding number, which we refer to as the charge.

The properties of the multilump solutions are interesting. They can be thought of as a collection of a number of individual humps which interact in a nontrivial manner. The number of humps depend on the index Q, and their maxima move with different asymptotic velocities as $t \to \pm\infty$.

2 Methods, Solutions and Properties

In what follows, we outline the methods and give the main results. The nonstationary Schrödinger equation is taken in the form

$$i\mu_y + \mu_{xx} + 2ik\mu_x + u\mu = 0. \tag{1}$$

As is well-known, the compatibility of (1) and

$$(4\partial_{xxx} + 12ik\partial_{xx} - 12k^2\partial_x + 6u\partial_x + 3u_x + 6iku + \partial_t - 3i\partial_y\partial_x^{-1})\mu = 0, \tag{2}$$

where $\partial_x^{-1} = \int_{-\infty}^{x}$, yields the KPI equation:

$$(u_t + 6uu_x + u_{xxx})_x = 3u_{yy}. \tag{3}$$

The discrete states we are concerned with here are the eigenvalues of the Fredholm integral equation

$$\hat{G}\mu(x, y; k) = 1 \tag{4}$$

where the operator $\hat{G}$ is defined as follows:

$$\hat{G}f(x, y) = f(x, y, k) - \int_{-\infty}^{\infty} dx' \int_{-\infty}^{\infty} dy'\, G(x - x', y - y', k)(uf)(x', y', k). \tag{5}$$

(Hereafter the limits on all double integrals are from $-\infty, \infty$.) The kernel, or Green's function, is given by

$$G(x, y; k) = \frac{1}{4\pi^2} \int \int \frac{\exp(ipx + iqy)}{p^2 + 2pk + q} \, dp \, dq. \tag{6}$$

Here we will concern ourselves with pure pole solutions of (4) i.e. $\mu(x, y; k) = \mu(k)$ is written:

$$\mu(k) = 1 + \sum_{m=1}^{n} \frac{\Phi_m}{k - k_m}. \tag{7}$$

This corresponds to pure discrete spectrum. Continuous spectrum can be added, but we shall not do so here.

Substitution of μ given by (7) into the integral equation (4), noting that the eigenfunctions Φ_m are homogeneous solutions of the integral equation, and using properties of the Green's function, yields the following equation (see e.g [1,2]):

$$-i(x - 2k_j y)\hat{G}_{k_j}\Phi_j + \hat{G}_{k_j}(\nu + i(x - 2k_j y)\Phi_j) = 1 - Q \tag{9a}$$

where in general we denote $\hat{G}_{k_j} \equiv \hat{G}(x, y; k = k_j), \nu(k_j) \equiv \lim_{k \to k_j} (\mu\text{–singular part of } \mu \text{ at } k_j)$; in this case, $\nu(k_j) = \lim_{k \to k_j} (\mu - \Phi_j/(k - k_j))$ and

$$Q(k_j, \Phi_j) \equiv \frac{i}{2\pi} \text{ sign (Im } k_j) \int \int u\Phi_j \, dx \, dy. \tag{9b}$$

Real, nonsingular potentials (the usual discrete states/lump solutions) are obtained when $n = 2N$, $k_{j+N} = \bar{k}_j$ and $\gamma_{j+N} = \bar{\gamma}_j, j = 1, \ldots N$ (in [4] a proof of this fact is given). In addition, Fredholm theory requires the constraint, $Q(k_j, \Phi_j) = 1$. We call $Q(k_j, \Phi_j)$ the charge or index. This is discussed more fully later. For simple poles which are related to usual lumps of KPI, $Q(k_j, \Phi_j) = 1$. From the fact that Φ_j are homogeneous solutions, using (9a) and the above definition of $\nu(k_j)$, it follows that

$$\nu(k_j) = -if(k_j)\Phi_j \tag{10}$$

where $f_j = f(k_j) = x - 2k_j y + \gamma_j + 12k_j^2 t$. Hence the following linear algebraic system of equations ensues:

$$1 + \sum_{\substack{m=1 \\ m \neq j}}^{2N} \frac{\Phi_m}{k_j - k_m} = -if_j\Phi_j, \qquad j = 1, \ldots 2N. \tag{11}$$

The corresponding "reflectionless" potentials are given by

$$u = -2i\frac{\partial}{\partial x} \sum_{m=1}^{2N} \Phi_m = 2\frac{\partial^2}{\partial x^2} \log F \tag{12}$$

where F is the determinant of the matrix

$$B_{\alpha\beta} = f_\alpha \delta_{\alpha\beta} - i(1 - \delta_{\alpha\beta})\frac{1}{k_\alpha - k_\beta} \quad , \quad \alpha, \beta = 1, 2 \ldots 2N. \tag{13}$$

The simplest case is a one lump solution, $N = 1$, $k_1 = a + ib$, where

$$F = z^2 + 4b^2(y')^2 + \frac{1}{4b^2}$$

with

$$z = x' - 2ay', \qquad x' = x - 12(a^2 + b^2)t - x_0, \qquad y' = y - 12at - y_0$$

and x_0, y_0 are constants. The foregoing discussion essentially summarizes what has been considered so far to be the discrete spectrum of the operator (1) and the corresponding lump solutions of the KPI equation. The simplest case is a one lump solution. A one lump solution, and the governing formulae, follow from an eigenfunction consisting of a pair of simple poles k_j, $\bar{k}_j$ with the condition Q=1 enforced.

A remarkable fact, discussed below, is that there exists a class of localized decaying solutions to KPI beyond the above class of "elementary" lump solutions. One such new solution is

$$u(x, y, t) = 2\frac{\partial^2}{\partial x^2}\log F \tag{14}$$

where

$$F = (z^2 - 4b^2(y')^2 - 24bt + \delta_R)^2 + \left(2y'(1 + 2bz) + \frac{\gamma_I}{b} - \delta_I\right)^2$$
$$+ \frac{1}{b^2}\left(\left(z - \frac{1}{2b}\right)^2 + 4b^2 y'^2 + \frac{1}{4b^2}\right), \tag{15}$$

and

$$z = x' - 2ay', \qquad x' = x - 12(a^2 + b^2)t - x_0, \qquad y' = y - 12at - y_0,$$

with

$$x_0 = \frac{\gamma_I a - \gamma_R b}{b}, \qquad y_0 = \frac{\gamma_I}{2b}. \tag{16}$$

This particular solution of KPI was constructed some time ago [5], but without reference to the underlying scattering problem. It was studied more recently in [6], again purely in terms of KP solutions and related dynamics. The above solution, although sharing some of the standard properties of the elementary lump type solutions, is not encompassed by the aforementioned class of lump solutions. Hence the earlier picture regarding the discrete spectrum of the Schrödinger operator is incomplete and needs to be modified. In recent work [3] we have found the following:

i) The associated Fredholm theory of (4) (Fredholm conditions) allows the quantity Q to take on any positive integer values besides Q=1. Values of Q different to 1 yield equations for the residues manifestly different from (10).

ii) The quantity Q is a topological invariant of the Schrödinger operator. Indeed, Q = winding number of η where η is connected with the the residue Φ of the pole at, say, k_1 by : $\Phi = (\text{Ln } \eta)' + O(1/r^2)$, $r \equiv \sqrt{x^2 + y^2}$, where the derivative is taken along a contour Γ_∞ that surrounds the origin once.

iii) Multiple poles may exist.

iv) A complete description of the discrete spectrum requires one to give the number of poles, their position k_j, their multiplicity r_j and the corresponding index Q_j. In addition, certain relevant constants $\gamma_j, \delta_j \ldots$ are required.

The approach discussed here (and in [3]) serves to incorporate a wide class of new localized solutions into the KPI equation, including the solution (15), via the scattering theory of the nonstationary Schrödinger operator.

More concretely, the spectral description via (1) that is associated with the solution (14-15) corresponds to an eigenfunction μ which has the following structure:

$$\mu = 1 + \frac{\Phi_1}{k - k_1} + \frac{\Psi_1}{(k - k_1)^2} + \frac{\Phi_{\bar{1}}}{k - \bar{k}_1} \qquad (17)$$

where both of the relevant indexes $Q(k_1, \Phi_1) = Q(\bar{k}_1, \Phi_{\bar{1}}) = 2$. As outlined below, in this case the following system of equations is obtained.

a) At $k = k_1$

$$\Phi_1 = -if(k_1)\Psi_1; \qquad \nu(k_1) = \tfrac{1}{2}((f + \gamma)^2 - 2iy + \delta)\Psi_1 \equiv -\tfrac{1}{2}g\Psi_1 \qquad (18a)$$

where $f(k_1)$ is defined below (10) and in addition there are two constraints:

$$Q(k_1,, \Phi_1) = 2 \quad \text{and} \quad \int \int u\Psi_1 \, dx \, dy = 0;$$

b) at $k = \bar{k}_1$

$$\left[\frac{d\nu}{dk} + if(k)\nu\right]_{k=\bar{k}_1} = \tfrac{1}{2}\bar{g}\Phi_{\bar{1}} \qquad (18b)$$

with $Q(\bar{k}_1, \Phi_{\bar{1}}) = 2$, where $\bar{g}$ is the complex conjugate of g which is given above, and we recall by the definition of ν,

$$\nu(k_1) = \lim_{k \to k_1} \left[\mu - \frac{\Psi_1}{(k - k_1)^2} - \frac{\Phi_1}{k - k_1}\right],$$

$$\nu(\bar{k}_1) = \lim_{k \to \bar{k}_1} \left(\mu - \frac{\Phi_{\bar{1}}}{k - \bar{k}_1}\right).$$

Indeed, substitution of (17) into the integral equation (4), where Ψ_1 is a homogeneous solution, yields the following at $k = k_1$:

$$-i(x - 2k_1 y)\hat{G}_{k_1}\Psi_1 + \hat{G}_{k_1}(\Phi_1 + if\Psi_1) = -Q_2 \tag{19a}$$

and also

$$\hat{G}_{k_1}\nu + i\hat{G}_{k_1}(f\Phi_1) - if\hat{G}_{k_1}\Phi_1 - \tfrac{1}{2}\hat{G}_{k_1}(f^2 + 2iy)\Psi_1$$
$$+f\hat{G}_{k_1}(f\Psi_1) + \tfrac{1}{2}Q = 1. \tag{19b}$$

If $Q_2 = 0$, and $Q = 2$, then since, $G_{k_1}\Psi_1 = 0$, we have that $\hat{G}_{k_1}(\Phi_1 + if\Psi_1) = 0$ and

$$\Phi_1 = -i(x - 2k_1 y + \gamma)\Psi_1$$

Hence (19b) reduces to

$$\hat{G}_{k_1}\left(\nu + if\Phi_1 - \tfrac{1}{2}(f^2 + 2iy)\Psi_1\right) = 0$$

This equation implies that

$$\nu(k_1) = \tfrac{1}{2}((f + \gamma)^2 + \delta - 2iy)\Psi_1$$

where δ is a constant that evolves in time (from (2)) as $\delta(t) = \delta(0) + 24ik_1 t$.

Similarly, substitution of (17) into (4) and evaluating at $k = \bar{k}_1$ yields the fact that $\Phi_{\bar{1}}$ is a homogeneous solution and the following equations:

$$\hat{G}_{\bar{k}_1}(\nu + if\Phi_{\bar{1}}) = 1 - Q_{\bar{1}}.$$

Also after using the above equation,

$$\hat{G}_{\bar{k}_1}\left\{\frac{\partial\nu}{\partial k} + if\nu - \tfrac{1}{2}(2iy + f^2)\Phi_{\bar{1}}\right\} - if(1 - \tfrac{1}{2}Q_{\bar{1}}) + (i\theta_2 - \tfrac{1}{2}\theta_3) = 0$$

where

$$\theta_2 \equiv \frac{i}{2\pi}\,\text{sign }(\text{Im }\bar{k}_1)\int u\nu(x, y)\,dx\,dy,$$
$$\theta_3 \equiv \frac{i}{2\pi}\,\text{sign }(\text{Im }\bar{k}_1)\int fu\Phi_{\bar{1}}(x, y)\,dx\,dy.$$

If $Q_{\bar{1}} = 2$ then the equations above reduce to

$$\hat{G}_{\bar{k}_1}(\nu + if\Phi_{\bar{1}}) = -1$$

and also

$$\hat{G}_{\bar{k}_1}\left\{\frac{\partial\nu}{\partial k} + if\nu - \tfrac{1}{2}(2iy + f^2)\Phi_{\bar{1}}\right\} + (i\theta_2 - \tfrac{1}{2}\theta_3) = 0$$

For these equations to have a solution, the Fredholm alternative condition requires $\int \int u \Psi_1 \, dx \, dy = 0$; In this case we get

$$\frac{d\nu}{dk} + i(f + \gamma)\nu = \tfrac{1}{2}(2iy + (f + \gamma)^2 + \delta)\Phi_{\bar{1}}$$

for certain constants γ, δ that we do not need to specify.

Equations (18a,b) yield the system of equations

$$\frac{g}{2if}\Phi_1 - \alpha\Phi_{\bar{1}} = 1 \tag{20a}$$

$$\left[\frac{2\bar{\alpha}^3}{f\bar{f}} + \frac{\bar{\alpha}^2}{i}\left(\frac{1}{f} + \frac{1}{\bar{f}}\right) - \bar{\alpha}\right]\Phi_1 + \frac{\bar{g}}{2i\bar{f}}\Phi_{\bar{1}} = 1 \tag{20b}$$

$k_1 = a + ib, \alpha \equiv \dfrac{1}{2ib}$. Solving this linear system and using eq (12) we obtain the KP solution (14,15).

A detailed study of this configuration shows that it describes two mutually interacting humps each of which moves with distinct asymptotic velocities. For large t the amplitude $u(x_\pm, y_\pm) = 16b^2$ as $t \to \pm\infty$. It follows that these particles are not distinguishable.

An important issue is to determine the dynamics of the humps and also the nature of the interaction that allows humps to scatter in this way. Assuming $b > 0$, as $t \to -\infty$ the two maxima ($+$ denotes fast, $-$ denotes slower hump) are located at

$$x_\pm \sim 12(a^2 + b^2)t \pm \sqrt{24a^2|t|/b} + x_{\bar{1}} + 2ay_0 - 1/(2b),$$

$$y_\pm \sim 12at \pm \sqrt{6|t|/b} + y_0$$

and as $t \to +\infty$,

$$x_\pm \sim 12(a^2 + b^2)t \pm \sqrt{24bt} + x_0 + 2ay_0,$$

$$y_\pm \sim 12at + y_0.$$

As $t \to \pm\infty$ the humps diverge from one another, proportional to $|t|^{1/2}$. This situation is different from the "pure" lump case where each lump has the same velocity as $t \to \pm\infty$.

For this two particle system the center of mass system is defined by

$$X = x_+ - x_-, \qquad Y = y_+ - y_-, \qquad r \equiv (X^2 + Y^2)^{1/2}$$

As $t \to -\infty$ the equations of motion read with $\alpha = \dfrac{\beta}{2a} = \sqrt{\dfrac{24a^2}{b}}$

$$\frac{d^2X}{dt^2} = -\frac{\alpha(a^2 + \beta)^{3/2}}{4r^3}$$

$$\frac{d^2Y}{dt^2} = -\frac{\beta(a^2 + \beta)^{3/2}}{4r^3}i$$

The interaction between lumps corresponds to an attractive interaction potential given by

$$V(r) = -\frac{(\alpha^2 + \beta^2)^2}{8r^2}.$$

Notice that although the lumps attract each other still they do not form a bound state. We see that the attractive interaction is not strong enough to bind them together and hence they eventually split apart horizontally as they travel parallel to the y axis

From a dynamical perspective an important quantity is the angle Ω the humps get deflected due to interaction. While for standard lumps this angle is zero (there is no phase shift) here one obtains that

$$\Omega = \operatorname{arc} \tan \frac{1}{2|\alpha|}$$

Hence by properly choosing the free parameter α we can obtain a solution that scatters off in any desired angle.

The method extends to higher order multipoles. Consider next an eigenfunction $\mu(k)$:

$$\mu(k) = 1 + \frac{\Psi_2}{(k - k_1)^3} + \frac{\Psi_1}{(k - k_1)^2} + \frac{\Phi_1}{k - k_1} + \frac{\Phi_{\bar{1}}}{k - \bar{k}_1} \tag{21}$$

Substituting (21) into (4) and noting that the functions $\Psi_2, \Phi_{\bar{1}}$ are homogeneous solutions yields the following equations and constraints.
a) At $k = k_1$,

$$\Psi_1 = 2i(f/g)\Phi_1 = -if\Psi_2, \qquad \nu(k_1) = -(h/3g)\Phi_1$$

with the constraints:

$$Q(k_1, \Phi_1) = 3, \qquad \int\int u\Psi_j \, dx \, dy = 0, \quad j = 1, 2.$$

b) at $k = \bar{k}_1$,

$$\left[\frac{d^2\nu}{dk^2} + 2if\frac{d\nu}{dk} - \bar{g}\nu + \frac{1}{3}\bar{h}\Phi_{\bar{1}} \right]_{k=\bar{k}_1} = 0$$

with $Q(\bar{k}_1, \Phi_{\bar{1}}) = 3$. In the above formulae, g was given earlier, and we define $h(k_1) \equiv h \equiv if^3 + (3i\delta(t) + 6y)f + \beta(t)$. The above equations form a closed system for $\Psi_1, \Phi_{\bar{1}}$; then from equation (12) we obtain the reflectionless potential/multipole lump order-3 of KPI,

$$u = 2\frac{\partial^2}{\partial x^2}(\log F)$$

where

$$F = \left[z^3 - 12b^2(y')^2 z + 3z(\delta_R - 24bt) + 6by'(\delta_I - 2y') + \beta_R - 24t\right]^2$$
$$+ \left[8b^3(y')^3 - 6b(z')^2 y' + 3z(\delta_I - 2y') - 6by'(\delta_R - 24bt) + \beta_I\right]^2$$
$$+ \frac{9}{4b^2}\left[\left(z^2 - 4b^2(y')^2 + \delta_R - 24bt - \frac{z}{b} + \frac{1}{2b^2}\right)^2 + (4y'bz - \delta_I)^2 \right. \tag{22}$$
$$\left. + \frac{1}{b^2}\left(z - \frac{1}{b}\right)^2 + 4(y')^2 + \frac{1}{4b^4}\right]$$

where we note that the time dependence of the KPI solution is obtained from (2); $\gamma(t)$, $\delta(t)$ is determined to be as given before, and $\beta(t) = 24it + \beta_0$. Since the charges $Q(k_1, \Phi_1) = Q(\bar{k}_1, \Phi_{\bar{1}}) = 3$, we say that this solution has charge=3.

This multipole order-3 configuration for the KPI equation is composed of three mutually interacting humps, which move with distinct asymptotic velocities. As $t \to \pm\infty$ the three maxima are located at (assuming $b > 0$) (x'_+, y'_+), (x'_0, y'_0), (x'_-, y'_-) where $x'_0 = 12(a^2+b^2)t+x_0$, $y'_0 = 12at+y_0$, and as $t \to -\infty$ $x'_\pm \sim x'_0 - 4/3b$, $y'_\pm \sim y'_0 \pm \sqrt{18|t|/b}$, and as $t \to +\infty$, $x'_\pm \sim x'_0 \pm \sqrt{72bt} + 1/(6b)$, $y'_\pm \sim y'_0$.

To determine the nature of the interaction between the humps, define

$$x_{i,j} \equiv x_i - x_j, \qquad y_{i,j} \equiv y_i - y_j,$$
$$\vec{r}_{i,j} = (x_{i,j}, y_{i,j}), \qquad \hat{r}_{i,j} = \frac{1}{|\vec{r}_{i,j}|}\vec{r}_{i,j}.$$

The above trajectories of the humps are solutions of the following system:

$$\frac{d^2 x_i}{dt^2} = \sum_{j,j\neq i} \hat{r}_{i,j} f(|\vec{r}_{i,j}|)$$

where

$$f(r) \equiv \frac{-2A}{9r^3},$$

with $A = (18/b)^2$ as $t \to -\infty$, and $A = (72b)^2$ as $t \to +\infty$. The interaction of the humps corresponds to a sum of two particle forces $f(r)$ which depend only on the relative distance between them. They attract each other although, again, the interaction is not strong enough to bind them together. The boundary conditions are such that initially $r_{1,2} = r_{2,3}$ and this condition is preserved under the time evolution.

Higher order multipole lumps can be constructed in the same manner as outlined here but we will not dwell on that. Moreover, other classes of solutions can also be obtained via similar techniques. For example, if the eigenfunction μ has the spectral structure

$$\mu = 1 + \frac{\Psi_1}{(k-k_1)^2} + \frac{\Phi_1}{k-k_1} + \frac{\Psi_{\bar{1}}}{(k-\bar{k}_1)^2} + \frac{\Phi_{\bar{1}}}{k-\bar{k}_1}, \tag{23}$$

then the methods described above can be applied to find a closed system of equations for Ψ_1, $\Phi_{\bar{1}}$ with $Q(k_1, \Phi_1) = Q(\bar{k}_1, \Phi_{\bar{1}}) = 3$ and with the additional constraints:

$$\int \int u \Psi_j \, dx \, dy = 0, \qquad j = 1, \bar{1}.$$

Solving the system and using (12) yields the following solution:

$$u = 2 \frac{\partial^2}{\partial x^2} (\log F)$$

where

$$F = \ (z^3 - 12b^2(y')^2 z + 12t + \beta_R)^2 + (8b^3(y')^3 - 6bz^2y' + \beta_I)^2$$
$$+ \frac{9}{4b^2} \left[\left(z - \frac{1}{2b} \right)^2 + 4b^2 y'^2 + \frac{1}{4b^2} \right] \left[\left(z + \frac{1}{2b} \right)^2 + 4b^2(y')^2 + \frac{1}{4b^2} \right],$$
$$\tag{24}$$

This solution corresponds to a three humped structure with humps located on a conic $(x' - 2ay')^2 + 4b^2(y')^2 = (12t)^{2/3}$ moving with uniform speed and whose semiaxes are first decreasing and then increasing at a rate $|t|^{1/3}$. The position of the humps is given asymptotically by

$$x_1 \sim 12(a^2 + b^2)t + (12|t|)^{1/3} + x_0 + 2ay_0$$
$$x_2 = x_3 \sim 12(a^2 + b^2)t + \tfrac{1}{2}(12|t|)^{1/3} + x_0 + 2ay_0$$
$$y_1 = 0, \qquad y_{2,3} \sim 12at \pm \tfrac{1}{2}\sqrt{3}(12|t|)^{1/3} + y_0$$

as $t \to \pm\infty$

Galilean invariance of this three particle system follows from the fact that the momentum is conserved:

$$\sum_i \frac{dx_i}{dt} = 12(a^2 + b^2), \qquad \sum_i \frac{dy_i}{dt} = 12a.$$

Here the force between particles is (for convenience, we take $a = 0$)

$$f(r) \equiv \frac{-288}{27} \left(\frac{9}{4} + \frac{3}{16b^2} \right)^3 \frac{1}{r^5}.$$

In this case we have three particles that interact in a nontrivial manner. However after collision the particles follow their original trajectories without any scattering or even a phase shift.

Note also the overall simplicity of this solution as compared to the standard three lump solution (12-13) where the function F involves a 6×6 determinant.

Furthermore, consideration of spectral functions formed by superposition of, say, terms like (17) with (double) poles at $k_1, ...k_r$ and single poles at $\bar{k}_1, ...\bar{k}_r$ with $Q_1, ...Q_r = Q_{\bar{1}}, ...Q_{\bar{r}} = 2$ implies a $2r \times 2r$ linear system whose solution yields a $2r$ hump interacting solution of KPI. Similar considerations apply to superposition of terms like, say, (21) or (23). One can also consider a mixture of multiple poles with different charges $Q_1, ..Q_j$ corresponding to different locations of the associated poles. The equations which define the solution and overall interaction of humps can be found via the above method.

Thus we have demonstrated that a broad class of new real, nonsingular decaying solutions of KPI with interesting physical properties can be obtained.

3 Index

Next we discuss the notion of the charge or index Q. It turns out that Q is a topological invariant for the Schrödinger operator. Indeed,

$$Q = \text{index } \eta$$

where η is related with the residue Φ of the pole at, say, k_1 by

$$\Phi = (\text{Ln } \eta)' + O(1/r^2), \qquad r \equiv \sqrt{x^2 + y^2},$$

where the derivative is taken along a contour Γ_∞ that surrounds the origin once.

For example, in the case Q= 1, equation (10) with the definitions $\Phi_j = \Phi$, $k_j = k$, $f_j = f$, implies that $\Phi = \dfrac{i}{f} + O\left(\dfrac{1}{r^2}\right)$. Using

$$\Phi_{xx} + 2ik\Phi_x + i\Phi_y + u\Phi = 0$$

and Green's theorem we find that

$$\begin{aligned}
Q(k, \Phi) &\equiv \frac{i}{2\pi} \text{ sgn (Im } k) \int\int u\Phi \, dx \, dy = \frac{-1}{2\pi} \int_{\Gamma_\infty} \frac{dx - 2k \, dy}{f} \\
&= \frac{1}{2\pi i} \int_{\Gamma_\infty} \frac{df}{f} = \frac{1}{2\pi i} \int_{f(\Gamma_\infty)} \frac{dz}{z} = 1,
\end{aligned} \tag{25}$$

and we see from (25) that Q = winding number of f. Similarly for higher charges. For example, in the double pole solution with charge 2, substituting (17) into (1), integrating over x, y, using Green's theorem and the properties of the solution yields

$$Q = \frac{1}{2\pi i} \int_{\Gamma_\infty} \frac{dg}{g} = \frac{1}{2\pi i} \int_{g(\Gamma_\infty)} \frac{dz}{z} = 2 \tag{26}$$

and hence from (26), Q is the winding number of g. The definition of charge and

$$u(x, y, t) = -2i\frac{\partial}{\partial x}(\Phi_1 + \Phi_{\bar{1}})$$

establishes that $Q(k_1, \Phi_1) = Q(\bar{k}_1, \Phi_{\bar{1}})$. Similar comments apply for higher charges. From the definition of charge and the examples discussed earlier, it is also clear that $Q = (\text{index } F)/2$ where F is defined by equations (12), (15), (24) etc.

It should also be noted that such winding numbers also appear in two dimensional gauge theories [cf. 7], although the origin and interpretation is clearly different.

4 Coalescence of simple poles

Here we briefly mention how, by appropriately coalescing the simple poles (eigenvalues) of the Schrödinger operator, we can find discrete states which correspond to eigenvalues of higher multiplicity. Consider the case $n = 2N = 4$ in formulae (7); i.e. two pairs of conjugate simple eigenvalues in the upper/lower half planes. In the limiting process, take $k_1 = k_2 + \varepsilon$, $k_3 = \bar{k}_1$, $k_4 = \bar{k}_2$, $k_1 = a + ib$, $|\varepsilon| \ll 1$, and expand as follows:

$$\Phi_m = \frac{\Phi_m^{(-1)}}{\varepsilon} + \Phi_m^{(0)} + \Phi_m^{(1)}\varepsilon + \dots,$$

$$\gamma_{m,0} = \gamma_{m,0}^{(-1)}/\varepsilon + \gamma_{m,0}^{(0)} + \gamma_{m,0}^{(1)}\varepsilon + \dots, \qquad m = 1, \dots 4.$$

Substitute these expansions into (11) and equate powers of ε to zero. At order $1/\varepsilon^2$ there are various possibilities. But reality forces a condition on $\gamma_{m,0}^{(-1)}$. We restrict to:

$$\gamma_{1,0}^{(-1)} = \gamma_{4,0}^{(-1)} = -\gamma_{2,0}^{(-1)} = -\gamma_{3,0}^{(-1)} = -i.$$

In order for μ to have a finite limit as ε tends to zero, we are forced to take:

$$\Phi_1^{(-1)} + \Phi_2^{(-1)} = \Phi_3^{(-1)} + \Phi_4^{(-1)} = 0.$$

Proceeding to subsequent orders is straightforward. It is convenient to define new variables:

$$\Phi_1^{(0)} + \Phi_2^{(0)} \equiv \Phi_1, \qquad \Phi_3^{(0)} + \Phi_4^{(0)} = \Phi_{\bar{1}}, \qquad \Phi_1^{(-1)} \equiv \Psi_2.$$

We find that

$$\Phi_3^{(-1)} = \Phi_4^{(-1)} = 0, \qquad \Phi_1^{(-1)} = i\Phi_1/(f + \gamma^{(0)}),$$

$$\gamma^{(0)} \equiv \gamma_{1,0}^{(0)}, \qquad f \equiv f(k_1), \qquad \Phi_3^{(0)} = \Phi_4^{(0)},$$

and in the limit $\varepsilon \to 0$, the eigenfunction μ then has the spectral structure (17) and the system of equations (18) is obtained where

$$\delta(t) = \delta^{(0)} + 24ik_1 t, \qquad \left(\delta^{(0)} = \gamma_{1,0}^{(1)} - \gamma_{2,0}^{(1)}\right).$$

5 Concluding remarks

We summarize the main results.

i) A method has been developed to obtain a new class of real, nonsingular, decaying ("reflectionless") potentials/solutions to the nonstationary Schrödinger problem and the Kadomtsev-Petviashvili equation.

ii) The solutions are characterized by the order of the pole of the associated eigenfunction and a topological quantity, an index. Thus unlike the time independent Schrödinger operator, there are multiple eigenvalues corresponding to real, nonsingular decaying potentials.

iii) The solutions are obtained by solving a linear algebraic system, i.e. purely by algebraic means.

iv) The solutions are multi-humped. The dynamics of the humps are nontrivial unlike the case of standard lumps (index=1).

It is important to remark that the relevant ideas and methods are extendable to other integrable equations, including discrete equations such as the two dimensional Toda and Volterra lattices [cf. 4]. Full details will be published separately.

Acknowledgments

This work was partially sponsored by the Air Force Office of Scientific Research, Air Force Materials Command, USAF, under grant number F49620-94-0120. The US Government is authorized to reproduce and distribute reprints for governmental purposes notwithstanding any copyright notation thereon. The views and conclusions contained herein are those of the authors and should not be interpreted as necessarily representing the official policies or endorsements, either expressed or implied, of the Air Force Office of Scientific Research or the US Government. This work was also partially supported by NSF grant DMS-9404265 and by CICYT 0028/95 and Junta de Castilla-Leon JO 127 in Spain.

References

1. M.J. Ablowitz and P.A. Clarkson, *Solitons, Nonlinear Evolution Equations and Inverse Scattering*, Cambridge University Press, 1991, 516pp.

2. A.S. Fokas and M.J. Ablowitz, On the inverse scattering of the time dependent Schrödinger equations and the associated KPI equation, *Stud. Appl. Math.*, **69** (1983) 211.

3. M.J. Ablowitz and J. Villarroel, Solutions to the time dependent Schrödinger and Kadomtsev-Petviashvili equations, APPM* #277, to be published *Physical Review Letters*.

4. J. Villarroel, S. Chakravarty and M.J. Ablowitz, On a 2+1 Volterra system, *Nonlinearity*, **9** (1996) 1113.

5. R.S. Johnson and S. Thompson, A Solution of the inverse scattering problem for the Kadomtsev-Petviashvili equation by the method of separation of variables, *Phys. Lett. A*, **66** (1978) 279.

6. K.A. Gorshkov, D.E. Pelinovskii and Yu.A. Stepanyants, Normal and anomalous scattering, formation and decay of bound states of two-dimensional solitons described by the Kadomtsev-Petviashvili equation, *JETP*, **77** (1993) 237.

7. R.K. Dodd, J. Eilbeck, J. Gibbon and H. Morris, *Solitons and Nonlinear Wave Equations*, Academic Press, 1982.

APPM*: Department of Applied Mathematics Preprint.

On Asymptotic Analysis of Orthogonal Polynomials via the Riemann-Hilbert Method

Pavel Bleher and Alexander Its
Department of Mathematical Sciences
Indiana University-Purdue University at Indianapolis

Abstract

Semiclassical asymptotics for the orthogonal polynomials on the line with the weight $\exp(-NV(z))$, where $V(z) = \frac{tz^2}{2} + \frac{gz^4}{4}$, $g > 0$, $t < 0$, is a double-well quartic polynomial, is presented alongside with the asymptotics for the relevant recursive coefficients. The derivation and the proof of the asymptotics is based on the analysis of the corresponding differential- difference Lax pair and associated to it matrix Riemann-Hilbert problem. As an application of the semiclassical asymptoitics, the Sine and Airy kernels universalities can be proven in the matrix models with the double-well quartic interaction in the presence of two cuts.

1   Main Result

Let

$$V(z) = \frac{tz^2}{2} + \frac{gz^4}{4}, \qquad g > 0, \qquad t < 0, \tag{1.1}$$

be a double-well quartic polynomial, and let

$$P_n(z) = z^n + \ldots, \qquad n = 0, 1, 2, \ldots, \tag{1.2}$$

be orthogonal polynomials on a line with the weight $e^{-NV(z)}$,

$$\int_{-\infty}^{\infty} P_n(z) P_m(z)\, e^{-NV(z)} dz = h_n \delta_{mn}. \tag{1.3}$$

The polynomials $P_n(z)$ satisfy the basic recursive equation

$$z P_n(z) = P_{n+1}(z) + R_n P_{n-1}(z), \tag{1.4}$$

where

$$R_n = \frac{h_n}{h_{n-1}}. \tag{1.5}$$

In addition, integration by parts gives

$$P'_n(z) = NR_n[t + g(R_{n-1} + R_n + R_{n+1})]P_{n-1}(z) + (NR_{n-2}R_{n-1}R_n)P_{n-3}(z), \tag{1.6}$$

Since $P'_n(z) = nz^{n-1} + \ldots$, this implies the Freud equation (cf. [Fre]):

$$n = NR_n[t + g(R_{n-1} + R_n + R_{n+1})]. \tag{1.7}$$

From (1.5) and (1.7) it follows that

$$0 < R_n < \frac{t + \sqrt{t^2 + 4\lambda g}}{2g}, \qquad \lambda = \frac{n}{N}. \tag{1.8}$$

Let

$$\psi_n(z) = \frac{1}{\sqrt{h_n}} P_n(z) e^{-NV(z)/2}. \tag{1.9}$$

Then

$$\int_{-\infty}^{\infty} \psi_n(z)\psi_m(z)\,dz = \delta_{nm}. \tag{1.10}$$

In this communication we present the semiclassical asymptotics for the functions $\psi_n(z)$ and for the coefficients R_n in the limit when $N, n \to \infty$ in such a way that there exists $\varepsilon > 0$ such that the ratio $\lambda = n/N$ satisfies the inequalities

$$\varepsilon < \lambda < \lambda_{\mathrm{cr}} - \varepsilon, \qquad \lambda = \frac{n}{N}, \tag{1.11}$$

where

$$\lambda_{\mathrm{cr}} = \frac{t^2}{4g}. \tag{1.12}$$

Denote

$$\lambda' = \frac{n + \frac{1}{2}}{N}.$$

In what follows the potential function

$$U_0(z) = \frac{z^2}{4} \left[(gz^2 + t)^2 - 4\lambda'g \right],$$

is important. Introduce the turning points z_1 and z_2 as zeros of $U_0(z)$,

$$z_{1,2} = \sqrt{\frac{-t \mp 2\sqrt{\lambda'g}}{g}}. \tag{1.13}$$

The condition (1.11) implies that z_1 and z_2 are real and $z_2 > z_1 > C\sqrt{\varepsilon}$. Our main result is formulated in the following theorem.

Theorem 1. *Assume that $N, n \to \infty$ in such a way that (1.11) holds. Then there exists $C = C(\varepsilon) > 0$ such that*

$$\left| R_n - \frac{-t - (-1)^n \sqrt{t^2 - 4\lambda g}}{2g} \right| \leq CN^{-1}, \qquad \lambda = \frac{n}{N}. \tag{1.14}$$

In addition, for every $\delta > 0$, in the interval $z_1 + \delta < z < z_2 - \delta$,

$$\psi_n(z) = \frac{2C_n\sqrt{z}}{\sqrt{\sin\phi}} \cos\left\{ \frac{(n + \frac{1}{2})}{2} \left[\frac{\sin(2\phi)}{2} - \phi \right] + \frac{\pi - (-1)^n \chi}{4} + O(N^{-1}) \right\}, \tag{1.15}$$

where

$$\phi = \arccos x, \qquad \chi = \arccos y,$$

and

$$x = \frac{gz^2 + t}{2\sqrt{\lambda' g}}, \qquad y = \frac{2\sqrt{\lambda' g} - tx}{2\sqrt{\lambda' g}\, x - t} = \frac{-tgz^2 - t^2 + 4\lambda' g}{2\sqrt{\lambda' g}\, gz^2}, \qquad \lambda' = \frac{n + \frac{1}{2}}{N}. \tag{1.16}$$

If $z > z_2 + \delta$ or $0 \leq z < z_1 - \delta$, then

$$\psi_n(z) = (-1)^\sigma \frac{C_n\sqrt{z}}{\sqrt{\sinh\phi}} \exp\left\{ \Phi_n + \frac{(-1)^n \chi}{4} + O\left(\frac{1}{N(1 + |z|)} \right) \right\}, \tag{1.17}$$

where

$$\Phi_n = -\frac{(n + \frac{1}{2})}{2} \left[\frac{\sinh(2\phi)}{2} - \phi \right],$$

$$\sigma = \frac{1 - \operatorname{sign}(z - z_1)}{2} \left[\frac{n}{2} \right], \qquad \left[\frac{n}{2} \right] = l, \quad \text{if} \quad n = 2l \quad \text{or} \quad n = 2l + 1,$$

$$\phi = \cosh^{-1}|x|, \qquad \chi = \cosh^{-1}|y|$$

and x, y are given by (1.16).

If $z_k - \delta \leq z \leq z_k + \delta$, $k = 1, 2$ then

$$\psi_n(z) = \frac{D_n z}{\sqrt{|\varphi_N'(z)|}} \left[\operatorname{Ai}\left(N^{2/3} \varphi_N(z) \right) + O(N^{-1}) \right], \tag{1.18}$$

where $\operatorname{Ai}(z)$ is the Airy function, $\varphi_N(z)$ is an analytic function on $[z_k - \delta, z_k + \delta]$ such that for $(z - z_k^{(N)})(-1)^k \geq 0$,

$$\varphi_N(z) = \left[\frac{3}{2} \left| \int_{z_k^{(N)}}^z \sqrt{U_N(v)}\, dv \right| \right]^{2/3}, \qquad k = 1, 2, \tag{1.19}$$

where $z_k^{(N)} = z_k + O(N^{-1})$ is the closest to z_k zero of the function

$$U_N(z) = U_0(z) + N^{-1} \left(\frac{t}{2} + gR_n \right) = z^2 \left[\frac{(gz^2 + t)^2}{4} - \lambda' g \right] + N^{-1} \left(\frac{t}{2} + gR_n \right).$$

(1.20)

The constant factor C_n in (1.15) and (1.17) is

$$C_n = \frac{1}{2\sqrt{\pi}} \left(\frac{g}{\lambda} \right)^{\frac{1}{4}} (1 + O(N^{-1})),$$

(1.21)

and D_n in (1.18) is

$$D_n = N^{1/6} \sqrt{g} \, (-1)^{\sigma_0} (1 + O(N^{-1})), \quad \sigma_0 = (2 - k) \left[\frac{n}{2} \right].$$

(1.22)

Finally, h_n in (1.3) is

$$h_n = 2\pi \sqrt{R_n} \, \exp \left[\frac{Nt^2}{4g} - \frac{N\lambda}{2} \left(1 + \ln \frac{g}{\lambda} \right) + O(N^{-1}) \right].$$

(1.23)

The asymptotic formulae (1.15), (1.17) and (1.18) is an extension of the classical Plancherel–Rotach asymptotics of the Hermite polynomials (see [PR] and [Sze]), to the orthogonal polynomials with respect to the weight $e^{-NV(z)}$ where $V(z)$ is the quartic polynomial (1.1). These formulae are extended into the complex plane in z as well. Asymptotics (1.14) of the coefficients R_n is a Freud's type asymptotics. For the homogeneous function $V(z) = |z|^\alpha$ and some its generalizations, the asymptotics of R_n is obtained in the papers of Freud [Fre], Nevai [Nev 1], Magnus [Mag1,Mag2], Lew and Quarles [LQ], Máté, Nevai, and Zaslavsky [MNZ]. Semiclassical asymptotics of the functions $\psi_n(z)$ is proved for $V(z) = z^4$ by Nevai [Nev 1] and for $V(z) = z^6$ by Sheen [She]. See also somewhat weaker asymptotic results for general homogeneous $V(z)$ in the works of Lubinsky and Saff [LuS], Lubinsky, Mhaskar, and Saff [LMS], Levin and Lubinsky [LL], Rahmanov [Rah], and others. Application of these asymptotics to random matrices is discussed in the work of Pastur [Pas]. The distribution of zeros and related problems for orthogonal polynomials corresponding to general homogeneous $V(z)$ is studied in the recent work [DKM] by Deift, Kriecherbauer, and McLaughlin. Many results and references on the asymptotics of orthogonal polynomials are given in the comprehensive review article [Nev2] by Nevai. The problem of finding asymptotics of R_n for a quartic nonconvex polynomial is discussed in [Nev2], and it is known as "Nevai's problem".

The equation (1.14) shows that if $0 < \lambda < \lambda_{\mathrm{cr}}$ then

$$\lim_{N \to \infty; \, (2m)/N \to \lambda} R_{2m} = L(\lambda) = \frac{-t - \sqrt{t^2 - 4\lambda g}}{2g},$$

$$\lim_{N \to \infty; \, (2m+1)/N \to \lambda} R_{2m+1} = R(\lambda) = \frac{-t + \sqrt{t^2 - 4\lambda g}}{2g}.$$

Both $L(\lambda)$ and $R(\lambda)$ satisfy the quadratic equation

$$gu^2 + tu + \lambda = 0,$$

so that, when n grows, R_n jumps back and forth from one sheet of this parabola to another. At $\lambda = \lambda_{\mathrm{cr}}$ the two sheets merge, i.e., $L(\lambda_{\mathrm{cr}}) = R(\lambda_{\mathrm{cr}})$. For $\lambda > \lambda_{\mathrm{cr}}$,

$$\lim_{N \to \infty;\, n/N \to \lambda} R_n = Q(\lambda),$$

where $u = Q(\lambda)$ satisfies the quadratic equation

$$3gu^2 + tu - \lambda = 0$$

(which follows from the Freud equation (1.7) if we put $u = R_{n-1} = R_n = R_{n+1}$). We consider semiclassical asymptotics for $\lambda > \lambda_{\mathrm{cr}}$ and in the vicinity of λ_{cr} (double scaling limit) in a separate work. The difference in the asymptotics between the cases $\lambda < \lambda_{\mathrm{cr}}$ and $\lambda > \lambda_{\mathrm{cr}}$ is that for $\lambda < \lambda_{\mathrm{cr}}$ the function $\psi_n(z)$ is concentrated on two intervals, or two cuts, $[-z_2, -z_1]$ and $[z_1, z_2]$, and it is exponentially small outside of these intervals, while for $\lambda > \lambda_{\mathrm{cr}}$, z_1 becomes pure imaginary, and $\psi_n(z)$ is concentrated on one cut $[-z_2, z_2]$. The transition from two-cut to one-cut regime is discussed in physical works by Cicuta, Molinari, and Montaldi [CMM], Crnković and Moore [CM], Douglas, Seiberg, Shenker [DSS], Periwal and Shevitz [PeS], and others. It is remarkable that this transition is governed by the Hastings-McLeod special Painlevé II transcendent.

A general ansatz on the structure of the semiclassical asymptotics of the functions $\psi_n(z)$ for a "generic" polynomial $V(z)$ is proposed in the work [BZ] of Brézin and Zee. They consider n close to N, $n = N + O(1)$, and they suggest that for these n's,

$$\psi_n(z) = \frac{1}{\sqrt{f(z)}} \cos(N\zeta(z) - (N - n)\varphi(z) + \chi(z)),$$

with some fuctions $f(z)$, $\zeta(z)$, $\varphi(z)$, and $\chi(z)$. This fits in well the asymptotics (1.15), except for the factor $(-1)^n$ at χ in (1.15), which is related to the two-cut structure of $\psi_n(z)$.

Equation (1.7) also appears in the planar Feynman diagram expansions of Hermitian matrix models, which were introduced and studied in the classical papers [BIPZ], [BIZ], [IZ] by Brézin, Bessis, Itzykson, Parisi, and Zuber and in the well- known recent works by Brézin, Kazakov [BK], Douglas, Shenker [DS], and Gross, Migdal [GM] devoted to the matrix models for 2D quantum gravity (see also [Dem] and [Wit]). In fact, it is the latter context that broaden the interest to the Freud equation (1,7) and brought in the area new powerful analytic methods from the theory of integrable systems. It turns out [FIK1,2] that equation (1.7) admits 2×2 matrix Lax pair representation (see equation (1.24) below), which allows one to identify Freud equation (1.7) as a discrete Painlevé I equation and imbeds it in the framework of the Isomonodromy Deformation Method suggested in 1980 by Flaschka and Newell [FN] and by Jimbo, Miwa, and Ueno [JMU] (about analytical aspects of the method see e.g. [IN], [FI]). The relevant Riemann-Hilbert formalism for (1.7) was developed in [FIK1,2] as well. It was used in [FIK1-3] together with the Isomonodromy Method for the asymptotic analysis of the solution of (1.7), which is related to the double-scaling limit in the 2D quantum gravity studied in [BK], [DS], [GM].

The solution of (1.7) which is analysed in [FIK1-3] is different from the one associated to the orthogonal polynomials (1.3). It corresponds to the system of orthogonal polynomials on the certain rays in complex domain. Nevertheless, the basic elements of the Riemann-Hilbert isomonodromy scheme suggested in [FIK1-3] can be easily extended (not the concrete analysis of course) to the other systems of semi-classical orthogonal polynomials (see e.g. [FIK4]).

2   Sketch of the Proof

The proof is based on the approach of [FIK] combined with the Nonlinear Steepest Descent Method proposed recently by Deift and Zhou [DZ] for analyzing the asymptotics of oscillatory matrix Riemann-Hilbert problems. We follow [FIK] in formulation of the Lax pair (see equation (1.24) below) and master Riemann-Hilbert problem associated to the orthogonal polynomials (1.3) (see problem (i-iii) below). We then solve asymptotically the copresponding direct monodromy problem *assuming* ansatz (1.14) for the recursive coefficients R_n. Using the asymptotic solution of the direct monodromy problem we define explicitly a piecewise analytic function $\Psi^0(z)$ which is be expected to approximate the exact solution $\Psi_n(z)$ of the Riemann-Hilbert problem on the whole z- plane. Finally, we appeal to the Deift-Zhou method and justify rigorously that $\Psi^0(z)$ is indeed the asymptotic solution of the master Riemann-Hilbert problem. This proves the asymptotics (1.15-19) for the polynomials $P_n(z)$ and, *simultaneously*, the asymptotic equation (1.14) for the recursive coefficients R_n. The use of the method of [DZ] rather than the original approach of [FIK] at this point of the proof simplifies it dramatically.

The details of the proof will be presented elsewhere. Here we only indicate the basic ingredients of the scheme, i.e. the Riemann-Hilbert problem and the Lax pair.

Let $\Psi_n(z)$ be the 2×2 matrix- valued function which solves the following matrix Riemann-Hilbert problem on a line:

(i) $\Psi_n(z)$ is analytic in $\mathbb{C} - \mathbb{R}$

(ii) $\Psi_n(z) \sim \left(\displaystyle\sum_{k=0}^{\infty} \dfrac{\Gamma_k}{z^k} \right) e^{-\left(\frac{NV(z)}{2} - n \ln z + \lambda_n \right)\sigma_3}, \qquad z \to \infty,$

$$\lambda_n = \frac{1}{2} \ln h_n, \quad \sigma_3 = \begin{pmatrix} 1 & 0 \\ 0 & -1 \end{pmatrix},$$

$$\Gamma_0 = \begin{pmatrix} 1 & 0 \\ 0 & R_n^{-1/2} \end{pmatrix}, \qquad \Gamma_1 = \begin{pmatrix} 0 & 1 \\ R_n^{1/2} & 0 \end{pmatrix}.$$

(iii) $\Psi_{n+}(z) = \Psi_{n-}(z)S, \qquad \operatorname{Im} z = 0, \qquad S = \begin{pmatrix} 1 & -2\pi i \\ 0 & 1 \end{pmatrix}.$

Then, the orthogonal polynomial scalar function $\psi_n(z)$ defined in (1.9) is given by the equation,

$$\psi_n(z) = (\Psi_n(z))_{11}.$$

It must be emphasized that in the setting of the Riemann-Hilbert problem (i-iii) the real quantities R_n and λ_n *are not the given data*. They are evaluated via the solution $\Psi_n(z)$, which is determined by conditions (i-iii) uniquely without any prior specification of R_n and λ_n.

Simultaneously, function $\Psi_n(z)$ satisfies the Lax pair (cf. (3.1-7) in [FIK2]):

$$\begin{cases} \Psi_{n+1}(z) = U_n(z)\Psi_n(z), \\ \dfrac{d\Psi_n(z)}{dz} = N A_n(z)\Psi_n(z), \end{cases} \tag{1.24}$$

where

$$U_n(z) = \begin{pmatrix} R_{n+1}^{-1/2} z & -R_{n+1}^{-1/2} R_n^{1/2} \\ 1 & 0 \end{pmatrix},$$

and

$$A_n(z) = \begin{pmatrix} -\left(\frac{tz}{2} + \frac{gz^3}{2} + gzR_n\right) & R_n^{1/2}[t + gz^2 + g(R_n + R_{n+1})] \\ -R_n^{1/2}[t + gz^2 + g(R_{n-1} + R_n)] & \frac{tz}{2} + \frac{gz^3}{2} + gzR_n \end{pmatrix}.$$

The jump matrix S in (iii) constitutes the only nontrivial Stokes matrix (for the basic definitions related to the general monodromy theory we refer the reader to the monograph [Sib]) corresponding to the second equation in (1.24) with R_n generated by the orthogonal polynomials (1.3).

The above facts reduce the problem of the asymptotic analysis of the quantities $P_n(z)$ and R_n to the asymptotic solution of the matrix Riemann-Hilbert problem **(i-iii)**, i.e. to the asymptotic solution of the corresponding inverse monodromy problem for the differential equation,

$$\frac{d\Psi_n(z)}{dz} = N A_n(z)\Psi_n(z).$$

Remark 1. As it has already been mentioned above, the Freud equation (1.7) has a meaning of the discrete Painlevé I equation. We refer the reader to the papers [FIZ], [NPCQ], [GRP], [Mag3,4], [Meh] for more on the subject. As it was first noticed by Kitaev, equation (1.7) can be also interpreted as the Backlund-Schlezinger transform of the classical Painlevé IV equation so that the coefficients R_n coincide in fact with the special PIV function (see [FIK1,3] for more details). This PIV function in turn can be expressed in terms of the certain $n \times n$ determinants involving the parabolic cylinder functions (see [Mag4]). In the proving of Theorem 1 however we do not use these algebraic by their nature connections to the modern Painlevé theory. We use its analytical methods.

Remark 2. Following [FIK4], one can reduce the analysis of an *arbitrary* system of the orthogonal polynomials $\{P_n(z)\}$ on some contour L with some weight $\omega(z)$ to the analysis of the relevant 2×2 matrix Riemann-Hilbert problem. The RH problem is formulated for 2×2 matrix function $Y(z)$ which is analytic outside the contour L, normalized by the asymptotic condition

$$Y(z)z^{-n\sigma_3} \to I \qquad z \to \infty, \qquad \sigma_3 = \begin{pmatrix} 1 & 0 \\ 0 & -1 \end{pmatrix}.$$

and whose boundary values $Y_\pm(z)$ satisfy equation:

$$Y_+(z) = Y_-(z) \begin{pmatrix} 1 & -2\pi i \omega(z) \\ 0 & 1 \end{pmatrix}, \qquad z \in L.$$

This Riemann-Hilbert problem can be also used to explain the appearance (see e.g.[ASM]) of the KP-type hierarchies in the matrix models. Indeed, let us assume that the weight function $\omega(\lambda)$ is of the form

$$\omega(z) = \exp \sum_{k=1}^{m} t_k z^k$$

and put

$$\Psi(z) = Y(z) \exp \left(\frac{1}{2} \sum_{k=1}^{m} t^k z^k \sigma_3 \right).$$

Then, the standard Liouville - type arguments yield (cf. (1.12-16) in [FIK4]) the system of linear differential and difference equations for function $\Psi(z) \equiv \Psi(z; n, t_1, t_2, t_3, ...)$,

$$\Psi(z; n+1) = U_n(z)\Psi(z; n)$$
$$\partial_z \Psi(z) = A_n(z)\Psi(z)$$
$$\partial_{t_k}\Psi(z) = V_n^k(z)\Psi(z), \qquad k = 1, 2, 3, ...,$$

where $U_n(z)$, $A_n(z)$, and $V_n^k(z)$ are polynomial on z (for their exact expressions in terms of the corresponding R_n see [FIK4]). The first two equations constitute the Lax pair for the relevant Freud equation:

$$U_n'(z) = A_{n+1}(z)U_n(z) - U_n(z)A_n(z).$$

The compatability conditions of the third equations with the different k generate the KP-type hierarchy of the integrable PDEs; the compatability condition of the second and the third equations produces the Virasoro-type constraints; the compatability condition of the first and the third equations is related to the Toda-type hierarchy and vertex operators.

Remark 3. Theorem 1 can be applied to proving the universality of the local distribution of eigenvalues in the matrix model with quartic potential. In fact, semiclassical asymptotics (1.15) leads to the Dyson (see [Dys]) sine-kernel for the local distribution of eigenvalues at a regular point z. In a completely different approach, the sine-kernel at regular points is proved in [PS]. At the endpoints of the spectrum we use the semiclassical asymptotics (1.18), and it leads to the Airy kernel. We refer the reader to the papers of Bowick and Brézin [BB], Forrester [For], Moore [Moo], and Tracy and Widom [TW], where the Airy kernel is discussed for the Gaussian matrix model and some other related models, and, in addition, some nonrigorous arguments are given for general matrix models.

References

[1] M. Adler, T. Shiota, and P. van Moerbeke, Random matrices, vertex operators and the Virasoro algebra, *Phys. Letters* bf A208, 101–112 (1995).

[2] D. Bessis, C. Itzykson, and J.-B. Zuber, Quantum field theory techniques in graphical enumeration, *Adv. Applied Math.* 1, 109–157 (1980).

[3] M. J. Bowick and E. Brézin, Universal scaling of the tail of the density of eigenvalues in random matrix models, *Phys. Letts.* **B268**, 21-28 (1991).

[4] E. Brézin, C. Itzykson, G. Parisi, and J. B. Zuber, *Commun. Math. Phys.* **59**, 35–51 (1978).

174 4. Asymptotic Analysis

[5] E. Brézin and V. A. Kazakov, Exactly solvable field theories of closed strings, *Phys. Lett. B* **236**, 144–150 (1990).

[6] E. Brézin and A. Zee, Universality of the correlations between eigenvalues of large random matrices, *Nuclear Physics B* **402**, 613–627 (1993).

[7] G. M. Cicuta, L. Molinari, and Montaldi, Large N phase transition in low dimensions, *Mod. Phys. Lett.* **A1**, 125, 1986.

[8] Č. Crnković and G. Moore, Multicritical multi-cut matrix models, *Phys. Lett. B* **257**, (1991).

[9] P. A. Deift, T. Kriecherbauer, and K. T-R. McLaughlin, New results for the asymptotics of orthogonal polynomials and related problems via inverse spectral method. Preprint, 1996.

[10] K. Demeterfi, Two-dimensional quantum gravity, matrix models and string theory, *Internat. J. Modern Phys. A* **8**, 1185–1244 (1993).

[11] M. R. Douglas, N. Seiberg, and S. H. Shenker, Flow and instability in quantum gravity, *Phys. Lett. B* **244**, 381–386 (1990).

[12] M. R. Douglas and S. H. Shenker, Strings in less than one dimension, *Nucl. Phys. B* **335**, 635–654 (1990).

[13] F. J. Dyson, Correlation between the eigenvalues of a random matrix, *Commun. Math. Phys.* **19**, 235–250 (1970).

[14] P. A. Deift and X. Zhou, A steepest descent method for oscillatory Riemann-Hilbert problems. Asymptotics for the MKdV equation, *Ann. of Math.* **137**, 295–368 (1995).

[15] P. A. Deift and X. Zhou, Asymptotics for the Painlevé ii equation, *Comm. Pure Appl. Math* **48**, 277–337 (1995).

[16] A. S. Fokas and A. R. Its, The isomonodromy method and the Painlevé equations, *in the book: Important Developments in Soliton Theory*, A. S. Fokas, V. E. Zakharov (eds.), Berlin, Heidelberg, New York, Springer (1993).

[17] H. Flaschka and A. Newell, Monodromy and spectral preserving deformations, *Commun. Math. Phys.* **76**, 67–116 (1980).

[18] A. S. Fokas, A. R. Its, and A. V. Kitaev, Isomonodromic approach in the theory of two-dimensional quantum gravity, *Usp. Matem. Nauk* **45**, 6, 135–136 (1990) (in Russian).

[19] A. S. Fokas, A. R. Its, and A. V. Kitaev, Discrete Painlevé equations and their appearance in quantum gravity, *Commun. Math. Phys.* **142**, 313–344 (1991).

[20] A. S. Fokas, A. R. Its, and A. V. Kitaev, The matrix model of the two-dimensional quantum gravity and isomonodromic solutions of the discrete Painlevé equations, *in the book : Zap. Nauch. Semin. LOMI* **187**, 12 (1991) (in Russian).

[21] A. S. Fokas, A. R. Its, and A. V. Kitaev, The isomonodromy approach to matrix models in 2D quantum gravity, *Commun. Math. Phys.* **147**, 395–430 (1992).

[22] A. S. Fokas, A. R. Its, and X. Zhou, Continuous and discrete Painlevé equations, *in the book: Painlevé Transcsndents. Their Asymptotics and Physical Applications. NATO ASI Series: Series B: Physics* **278** , Plenum Press, N.Y. (1992).

[23] P. J. Forrester, The spectrum edge of random matrix ensembles, *Nucl. Phys.* **B402**, 709-728 (1993).

[24] G. Freud, On the coefficients in the recursion formulae of orthogonal polynomials, *Proc. Royal Irish Acad.* **76A**, 1–6 (1976).

[25] D. J. Gross and A. A. Migdal, Nonperturbative two-dimensional quantum gravity, *Phys. Rev. Lett.* **64**, 127–130 (1990).

[26] B. Grammaticos, A. Ramani, V. Papageorgiou, Do integrable mappings have the Painlevé property? *Phys. Rev. Lett.* **67**, 1825–1828 (1991).

[27] A. R. Its and V. Yu. Novokshenov, The isomonodromic deformation method in the theory of Painlevé equations. *Lecture Notes in Math.* **1191**. Springer-Verlag, Berlin – New York, 1986.

[28] C. Itzykson and J.-B. Zuber, The planar approximation. II, *J. Math. Phys.*, **21**, 411–421 (1980).

[29] M. Jimbo, T. Miwa, and K. Ueno, Monodromy preserving deformation of linear ordinary differential equations with rational coefficients I, *Physica D2*, 306–352 (1981).

[30] A. L. Levin and D. S. Lubinsky, Orthogonal polynomials and Christoffel functions for $\exp(-|x|^\alpha)$, $\alpha \leq 1$, *J. Approx. Theory* **80**, 219–252 (1995).

[31] J. S. Lew and D. A. Quarles, Jr., Nonnegative solutions of a nonlinear recurrence, *J. Approx. Theory* **38**, 357–379 (1983).

176   *4. Asymptotic Analysis*

[32] D. S. Lubinsky, H. N. Mhaskar, and E. B. Saff, A proof of Freud's conjecture for exponential weights, *Constr. Approx.* **4**, 65–83 (1988).

[33] D. S. Lubinsky and E. B. Saff, Strong asymptotics for extremal polynomials associated with weights on $\mathbb{R}$, *Lect. Notes Math.* **1305** (1988).

[34] A. P. Magnus, A proof of Freud's conjecture about orthogonal polynomials related to $|x|^\rho \exp(-x^{2m})$ for integer m. In: *Polynômes Orthogonaux et Applications. Proceedings Bar-le-Duc 1984*, Eds. C. Brezinski e.a., *Lecture Notes Math.* **1171**, 362–372 (1985).

[35] A. P. Magnus, On Freud's equations for exponential weights, *J. Approx. Theory* **46**, 65–99 (1986).

[36] A. P. Magnus, Painlevé-type differential equations for the recurrence coefficients of semi-classical orthogonal polynomials, *Journal of Computational and Applied Mathematics* **57**, 215–237 (1995).

[37] A. P. Magnus, Freud's equations for orthogonal polynomials as discrete Painlevé equations, *in these Proceedings* (1996).

[38] A. Máté, P. Nevai, and T. Zaslavsky, Asymptotic expansion of ratios of coefficients of orthogonal polynomials with exponential weights, *Trans. Amer. Math. Soc.* **287**, 495-505 (1985).

[39] M. L. Mehta, Painlevé transcendents in the theory of random matrices. *An introduction to methods of complex analysis and geometry for classical mechanics and non-linear waves (Chamonix, 1993)*, 197–208, Frontières, Gif-sur-Yvette, 1994.

[40] G. Moore, Matrix models of 2D gravity and isomonodromic deformations, *Progr. Theor. Phys. Suppl.* **102**, 255-285 (1990).

[41] F. W. Nijhoff, V. G. Papageorgiou, H. W. Capel, and G. R. W. Quispel, The lattice Gel'fand-Dikii hierarchy, *Inv. Probl.* **8**, 597–621 (1992).

[42] P. Nevai, Asymptotics for orthogonal polynomials associated with $\exp(-x^4)$, *SIAM J. Math. Anal.* **15**, 1177–1187 (1984).

[43] P. Nevai, Géza Freud, orthogonal polynomials and Christoffel functions. A case study, *J. Approx. Theory* **48**, 3–167 (1986).

[44] L. Pastur, On the universality of the level spacing distribution for some ensembles of random matrices, *Lett. Math. Phys.* **25**, 259–265 (1992).

[45] L. Pastur and M. Shcherbina, Universality of the local eigenvalue statistics for a class of unitary invariant random matrix ensembles. Preprint, 1996.

[46] V. Periwal and D. Shevitz, Exactly solvable unitary matrix models: multicritical potentials and correlations, *Nucl. Phys. B* **333**, 731–746 (1990).

[47] M. Plancherel and W. Rotach, Sur les valeurs asymptotiques des polynomes d'Hermite $H_n(x) = (-1)^n e^{x^2/2} d^n (e^{-x^2/2})/dx^n$, *Comment. Math. Helvet.* **1**, 227–254 (1929).

[48] E. A. Rakhmanov, Strong asymptotics for orthogonal polynomials, *Methods of approximation theory in complex analysis and mathematical physics (leningrad, 1991)*, 71–97, *Lect. Notes Math.* **1550**, Springer, Berlin, 1993.

[49] R.-C. Sheen, Plancherel–Rotach-type asymptotics for orthogonal polynomials associated with $\exp(-x^6/6)$, *J. Approx. theory* **50**, 232–293 (1987).

[50] Y. Sibuya, Linear differential equations in the complex domain: problems of analytic continuation, *Transl. Math. Monogr.*, **82**, AMS, Providence, RI, 1990.

[51] G. Szegö, *Orthogonal Polynomials*, 3rd ed., AMS, Providence, 1967.

[52] C. A. Tracy and H. Widom, Level-spacing distribution and the Airy kernel, *Commun. Math. Phys.* **159**, 151–174 (1994).

[53] E. Witten, Two-dimensional gravity and intersection theory on moduli spaces, *Surveys in differential geometry (Cambridge, MA, 1990)*, 243–310, Lehigh Univ., Bethlehem, PA, 1991.

A new spectral transform for solving the continuous and spatially discrete heat equations on simple trees

P.C. Bressloff[†] and A.S. Fokas[‡]
[†] Department of Mathematical Sciences,
Loughborough University
Loughborough, Leics. LE11 3TU, U.K.
[‡] Department of Mathematics, Imperial College,
University of London
London, SW7 2BZ, U.K.

Abstract

A new spectral method for solving linear and integrable nonlinear PDE's in two variables has recently been developed. This unified method is based on the fact that linear and integrable nonlinear equations admit a Lax pair formulation. For linear equations this method generates an elegant integral representation of the solution to a given initial–boundary value problem. In this paper, we derive such an integral representation for the continuous and spatially discrete heat equations on a bounded domain. This representation is particularly useful for solving the heat equation on complex topologies such as trees.

1   Introduction

A major limitation of the traditional transform methods for solving initial–boundary value problems for linear evolution equations is that the solution is expressed in terms of time–dependent spectral data. Recently, Fokas [1] has used a Lax pair [2] formulation to derive a new spectral representation of the solution to a given initial–boundary value (IVB) problem that involves constant spectral data. This representation is of the form suggested by the Ehrenpreis principle [3], namely

$$q(x,t) = \int_L e^{-ikx-\omega(k)t}\rho(k)dk \qquad (1.1)$$

where $e^{-ikx-\omega(k)t}$ is a particular solution of the given PDE, and the concrete form of the spectral data $\rho(k)$ and of the contour L depends on the particular initial–boundary value problem. It was subsequently shown by Bressloff [4] that this spectral decomposition can also be achieved using more conventional Green's function techniques.

In this paper we will use this new spectral method to solve certain initial–boundary value problems for the continuous (section 2) and for the spatially discrete (section 3) heat equations. We will also indicate how to extend our results to more complex domains such as trees (section 4). For concreteness, we shall follow the Lax pair approach of Fokas [1].

2   Heat equation on a finite interval

Theorem 1 *Let $q(x,t)$ satisfy*

$$q_t - q_{xx} = 0, \quad t \in [0,\infty), \; x \in [0,l], \; l > 0, \tag{2.1}$$

with initial and boundary conditions

$$q(x,0) = g(x), \; q(0,t) = h_0(t), \; q(l,t) = h_l(t), \tag{2.2}$$

where $g(x)$ is twice differentiable for $x \in [0,l]$ and $h_0(t), h_l(t)$ are Schwartz functions for $t \in [0,\infty)$. There exists a unique solution of this IBV problem which decays to zero as $t \to \infty$. This solution is given by

$$q(x,t) = \int_L e^{-ikx-k^2t}\rho(k)\frac{dk}{2\pi} \tag{2.3}$$

where the directed contour L is depicted in figure 1 and $\rho(k)$ is uniquely determined in terms of $g(x), h_0(t), h_l(t)$ as follows:

$$\rho(k) = s(k), \quad k \in \mathbf{R}; \tag{2.4a}$$

$$\rho(k) = \nu(k), \quad \frac{5\pi}{4} \leq \arg k \leq \frac{7\pi}{4}; \tag{2.4b}$$

$$\rho(k) = e^{ikl}\lambda(k), \quad \frac{\pi}{4} \leq \arg k \leq \frac{3\pi}{4} \tag{2.4c}$$

where

$$s(k) = \hat{g}(k), \qquad 0 \leq \arg k \leq \pi \tag{2.5}$$

$$\nu(k) = \frac{\hat{g}(-k) - e^{-2ikl}\hat{g}(k) + 2ik(e^{-ikl}\hat{h}_l(k) - \hat{h}_0(k))}{1 - e^{-2ikl}},$$
$$\frac{5\pi}{4} \leq \arg k \leq \frac{7\pi}{4} \tag{2.6}$$

$$\lambda(k) = \frac{e^{ikl}(\hat{g}(k) - \hat{g}(-k)) + 2ik(e^{ikl}\hat{h}_0(k) - \hat{h}_l(k))}{1 - e^{2ikl}},$$
$$\frac{\pi}{4} \leq \arg k \leq \frac{3\pi}{4}, \tag{2.7}$$

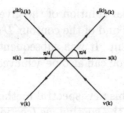

Figure 1: Contour L of theorem 1

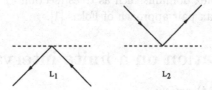

Figure 2: Contours L_1 and L_2

and

$$\hat{g}(k) = \int_0^l e^{ikx}g(x)dx, \quad \hat{h}_0(k) = \int_0^\infty e^{k^2t}h_0(t)dt, \qquad (2.8)$$
$$\hat{h}_l(k) = \int_0^\infty e^{k^2t}h_l(t)dt.$$

Proof The conditions on $g(x), h_0(t), h_l(t)$ imply that the integral in equation (2.3) is well defined. Furthermore, differentiation and integration can be interchanged, which implies that $q(x,t)$ solves equation (2.1). In order to verify that $q(x,t)$ satisfies $q(x,0) = g(x)$ we note that

$$q(x,0) = \int_{-\infty}^\infty e^{-ikx}\hat{g}(k)\frac{dk}{2\pi} + \int_{L_1} e^{-ikx}\nu(k)\frac{dk}{2\pi} + \int_{L_2} e^{ik(l-x)}\lambda(k)\frac{dk}{2\pi} \qquad (2.9)$$

where the contours L_1 and L_2 are shown in figure 2. The integrand of the second and third integrals in equation (2.8) are analytic in $5\pi/4 \leq \arg k \leq 7\pi/4$ and $\pi/4 \leq \arg k \leq 3\pi/4$ respectively. Thus, Cauchy's theorem implies that these integrals give zero contributions and equation (2.9) reduces to $q(x,0) = g(x)$.

In order to verify that $q(x,t)$ satisfies $q(0,t) = h_0(t)$ we note that

$$
\begin{aligned}
q(0,t) &= \int_{-\infty}^\infty e^{-k^2t}\hat{g}(k)\frac{dk}{2\pi} \\
&+ \int_{L_1} \frac{e^{-k^2t}}{1 - e^{-2ikl}}\left[\hat{g}(-k) - e^{-2ikl}\hat{g}(k) + 2ik(e^{-ikl}\hat{h}_l(k) - \hat{h}_0(k))\right]\frac{dk}{2\pi} \\
&+ \int_{L_2} \frac{e^{-k^2t}}{1 - e^{2ikl}}\left[e^{2ikl}(\hat{g}(k) - \hat{g}(-k)) + 2ik(e^{2ikl}\hat{h}_0(k) - e^{ikl}\hat{h}_l(k))\right]\frac{dk}{2\pi}
\end{aligned}
$$

Letting $k \to -k$ in the third integral, the above simplifies to

$$q(0,t) = \int_{-\infty}^{\infty} e^{-k^2 t} \hat{g}(k) \frac{dk}{2\pi} + \int_{L_1} e^{-k^2 t} \hat{g}(-k) \frac{dk}{2\pi} - \int_{L_1} 2ike^{-k^2 t} \hat{h}_0(k) \frac{dk}{2\pi} \quad (2.10)$$

This is precisely the expression that appears in the analysis of the heat equation with $x \in [0, \infty)$. Since $\hat{g}(k)$ is analytic for $0 \leq \arg k \leq \pi$, the combination of the first two terms in equation (2.10) gives zero contribution [1], and equation (2.10) reduces to $q(0,t) = h_0(t)$. A similar analysis yields $q(l,t) = h_l(t)$.

QED

Theorem 1 is a particular case of the following more general result.

Proposition 1 *There exists a solution of the heat equation on a finite interval that decays to zero as $t \to \infty$. This solution is given by equations (2.3) and (2.4) where the functions $s(k), \nu(k), \lambda(k)$ are defined by*

$$s(k) = \int_0^l e^{ikx} q(x,0) dx, \quad \Im(k) \geq 0 \quad (2.11)$$

$$\nu(k) = \int_0^\infty e^{k^2 t} [q_x(0,t) - ikq(0,t)] dt, \quad \frac{5\pi}{4} \leq \arg k \leq \frac{7\pi}{4} \quad (2.12)$$

$$\lambda(k) = \int_0^\infty e^{k^2 t} [q_x(l,t) - ikq(l,t)] dt, \quad \frac{\pi}{4} \leq \arg k \leq \frac{3\pi}{4}. \quad (2.13)$$

The boundary conditions $q(x,0), q(0,t)$ and $q(l,t)$ satisfy the constraints

$$\int_0^\infty e^{k^2 t} [q_x(0,t) - ikq(0,t)] dt$$
$$= e^{ikl} \int_0^\infty e^{k^2 t} [q_x(l,t) - ikq(l,t)] dt + \int_0^l e^{ikx} q(x,0) dx \quad (2.14)$$

for $\frac{\pi}{4} \leq \arg k \leq \frac{3\pi}{4}$ and

$$e^{-ikl} \int_0^\infty e^{k^2 t} [q_x(0,t) - ikq(0,t)] dt$$
$$= \int_0^\infty e^{k^2 t} [q_x(l,t) - ikq(l,t)] dt + e^{-ikl} \int_0^l e^{ikx} q(x,0) dx \quad (2.15)$$

for $\frac{5\pi}{4} \leq \arg k \leq \frac{7\pi}{4}$.

Proof   It is possible to verify directly that if the boundary values of q satisfy equations (2.14) and (2.15), then the function $q(x,t)$ defined by equation (2.3) satisfies the heat equation together with the boundary values $q(x,0), q(0,t)$, $q_x(0,t), q(l,t), q_x(l,t)$. Rather than give this proof, we use the method introduced in [1] to derive equations (2.3), (2.4) and (2.11)–(2.15). We emphasise that this derivation assumes *a priori* that $q(x,t)$ exists. However, this assumption can be eliminated, since *a posteriori*, one can verify directly that

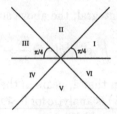

Figure 3: The definitions of the sectors I,...,VI

$q(x,t)$ defined by equation (2.3) solves the given IBV problem. An alternative derivation based on Green's functions is presented in [4].

Equation (2.1) admits the Lax pair [1]

$$\mu_x + ik\mu = q \qquad (2.16a)$$
$$\mu_t + k^2\mu = q_x - ikq \qquad (2.16b)$$

where k is a complex-valued parameter. Indeed, it is straightforward to verify that equations (2.16a) and (2.16b) are compatible if and only if q satisfies the heat equation. Let the sectors $I,...,VI$ of the complex k-plane be defined by $\arg k$ in the closed intervals

$$\left[0,\frac{\pi}{4}\right], \ \left[\frac{\pi}{4},\frac{3\pi}{4}\right], \ \left[\frac{3\pi}{4},\pi\right], \ \left[\pi,\frac{5\pi}{4}\right], \ \left[\frac{5\pi}{4},\frac{7\pi}{4}\right], \ \left[\frac{7\pi}{4},2\pi\right]$$

respectively (see figure 3). Let $\tilde{q}(x,t,k)$ be defined by

$$\tilde{q}(x,t,k) = q_x(x,t) - ikq(x,t) \qquad (2.17)$$

The general solution of equation (2.16a) is

$$\mu(x,t,k) = e^{-ikx}\mu(0,t,k) + \int_0^x e^{-ik(x-x')}q(x',t)dx'$$

Equations (2.16a) and (2.16b) are compatible, thus the solution $\mu(x,t,k)$ defined above is also a solution of (2.16b) if and only if $\mu(0,t,k)$ solves

$$\mu_t(0,t,k) + k^2\mu(0,t,k) = \tilde{q}(0,t,k)$$

A particular solution of this equation is

$$\mu(0,t,k) = \int_0^t e^{-k^2(t-t')}\tilde{q}(0,t',k)dt'$$

Thus, the function $\mu^{(46)}$ defined by

$$\mu^{(46)} = e^{-ikx}\int_0^t e^{-k^2(t-t')}\tilde{q}(0,t',k)dt' + \int_0^x e^{-ik(x-x')}q(x',t)dx', \qquad (2.18a)$$

for $k \in IV \cup VI$, solves both equations (2.16a,b). The superscript indicates the region where the above function is analytic and bounded. The boundedness of $\mu^{(46)}$ is a consequence of the following elementary facts. The function $e^{ik\xi}$ for $\xi \in \mathbf{R}^+$ is bounded in $I \cup II \cup III$, while the function $e^{k^2\tau}$ for $\tau \in R^+$ is bounded in $II \cup V$. Thus $\mu^{(46)}$ is bounded in the intersection of $I \cup III \cup IV \cup VI$ with the lower half plane, which is $IV \cup V \cup VI$.

Similarly, the functions $\mu^{(13)}$, $\mu^{(2)}$ and $\mu^{(5)}$ defined below also solve both equations (2.16a,b):

$$\mu^{(13)} = e^{-ik(x-l)} \int_0^t e^{-k^2(t-t')}\widetilde{q}(l,t',k)dt'$$

$$- \int_x^l e^{-ik(x-x')}q(x',t)dx', \quad k \in I \cup III \qquad (2.18b)$$

$$\mu^{(2)} = -e^{-ik(x-l)} \int_t^\infty e^{-k^2(t-t')}\widetilde{q}(l,t',k)dt''$$

$$- \int_x^l e^{-ik(x-x')}q(x',t)dx', \quad k \in II \qquad (2.18e)$$

$$\mu^{(5)} = -e^{-ikx} \int_t^\infty e^{-k^2(t-t')}\widetilde{q}(0,t',k)dt''$$

$$+ \int_0^x e^{-ik(x-x')}q(x',t)dx', \quad k \in V \qquad (2.18i)$$

In the domain of their overlap, the above functions are related by

$$\mu^{(13)} - \mu^{(46)} = -e^{-ikx-k^2t}s(k), \quad k \in \mathbf{R} \qquad (2.19a)$$

$$\mu^{(13)} - \mu^{(2)} = e^{-ikx-k^2t}e^{ikl}\lambda(k), \quad \frac{\pi}{4} \le \arg k \le \frac{3\pi}{4} \qquad (2.19c)$$

$$\mu^{(5)} - \mu^{(46)} = -e^{-ikx-k^2t}\nu(k), \quad \frac{5\pi}{4} \le \arg k \le \frac{7\pi}{4} \qquad (2.19f)$$

where the functions $s(k)$, $\nu(k)$, $\lambda(k)$ are defined by equations (2.11)–(2.13). Indeed, let $\Delta\mu$ denote the difference of any two solutions of equations (2.16). Since these solutions satisfy equation (2.16a), it follows that

$$\Delta\mu = e^{-ikx}f(t,k).$$

Similarly, since these solutions satisfy equation (2.16b), it follows that

$$\Delta\mu = e^{-k^2t}g(x,k).$$

Thus $\Delta\mu = e^{-ikx-k^2t}\rho(k)$, and the function $\rho(k)$ can be determined by evaluating this equation at any suitable x, t. For example, evaluating the equation $\mu^{(13)} - \mu^{(46)} = e^{-ikx-k^2t}\rho(k)$ at $x = t = 0$, it follows that $\rho(k) = -s(k)$. Similarly for equations (2.19b,c).

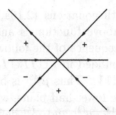

Figure 4: Directed contour L of the first Riemann–Hilbert problem

Integration by parts of equations (2.18) implies that if μ denotes $\mu^{(46)}$ or $\mu^{(13)}$ or $\mu^{(2)}$ or $\mu^{(5)}$, then for μ off the contour depicted in figure 1,

$$\mu(x, t, k) = \frac{q(x,t)}{ik} + O\left(\frac{1}{k^2}\right), \quad k \to \infty \tag{2.20}$$

Equations (2.19) and (2.20) define a Riemann–Hilbert (RH) problem with respect to the directed contour depicted in figure 4. The unique solution of the RH problem is given by [5]

$$\mu = -\frac{1}{2\pi i} \int_L e^{-ik'x - k'^2 t} \rho(k') \frac{dk'}{k' - k} \tag{2.21}$$

(The contour L differs from the contour of figure 1 in the direction of the rays defined by $\arg k = \pi/4$, $\arg k = 3\pi/4$, since $\rho(k)$ is defined by $e^{ikl}\lambda(k)$ and not by $-e^{ikl}\lambda(k)$). Equation (2.21) together with the equation obtained by considering the large k behaviour of equation (2.21) yields equations (2.3) and (2.4).

It will now be shown that if $q(x,t) \to 0$ as $t \to \infty$ then the boundary values of $q(x,t)$ satisfy equations (2.14) and (2.15). To derive equation (2.14) we note that $\mu^{(2)}$, in addition to satisfying equation (2.18c), also satisfies the equation

$$\mu^{(2)} = -\int_t^\infty e^{-k^2(t-t')} \tilde{q}(x, t', k) dt', \quad k \in II \tag{2.22}$$

This equation follows from equation (2.16b) and the assumption that

$$q(x, t) \to 0 \quad \text{as} \quad t \to \infty.$$

Equation (2.14) follows by evaluating equations (2.18c) and (2.22) at $x = 0$ and $t = 0$. Similarly, $\mu^{(5)}$, in addition to satisfying equation (2.18d), also satisfies the equation

$$\mu^{(5)} = -\int_t^\infty e^{-k^2(t-t')} \tilde{q}(x, t', k) dt', \quad k \in V \tag{2.23}$$

Equation (2.15) follows by evaluating equations (2.18d) and (2.23) at $x = l$ and $t = 0$.

QED

Remark: Equations (2.14) and (2.15) involve the five functions $q(x,0), q(0,t)$, $q_x(0,t), q(l,t), q_x(l,t)$. Consider the following IBV problem for the heat equation

$$q(x,0) = g(x), \tag{2.24a}$$

$$\alpha_1 q_x(0,t) + \alpha_2 q(0,t) = h_0(t), \tag{2.24b}$$

$$\beta_1 q_x(l,t) + \beta_2 q(l,t) = h_l(t) \tag{2.24c}$$

where $\alpha_1, \alpha_2, \beta_1, \beta_2$ are given constants and $g(x), h_0(t), h_l(t)$ are given functions. Equations (2.14), (2.15) and (2.24) are five algebraic equations for the following five unknown functions:

$$\int_0^l e^{ikl} q(x,0)dx, \quad \int_0^\infty e^{k^2 t} q(0,t)dt, \quad \int_0^\infty e^{k^2 t} q_x(0,t)dt,$$

$$\int_0^\infty e^{k^2 t} q(l,t)dt, \quad \int_0^\infty e^{k^2 t} q_x(l,t)dt$$

The above functions uniquely specify $s(k), \nu(k), \lambda(k)$ (see equations (2.11)–(2.13)), and hence $q(x,t)$ (see equations (2.3) and (2.4)).

As a particular example of the above general case, consider the case that $\alpha_1 = \beta_1 = 0$ and $\alpha_2 = \beta_2 = 1$. Then equation (2.14) and the equation resulting from equation (2.15) by replacing k with $-k$ yield

$$\left[1 - e^{2ikl}\right] \int_0^\infty e^{k^2 t} q_x(l,t)dt$$

$$= e^{ikl}[\hat{g}(k) - \hat{g}(-k)] + 2ik e^{ikl}\hat{h}_0(k) - ik(1 + e^{2ikl})\hat{h}_l(k)$$

for $\frac{\pi}{4} \leq \arg k \leq \frac{3\pi}{4}$ and

$$\left[1 - e^{-2ikl}\right] \int_0^\infty e^{k^2 t} q_x(0,t)dt$$

$$= \hat{g}(k) - e^{2ikl}\hat{g}(-k) - 2ik e^{ikl}\hat{h}_l(k) + ik(1 + e^{-2ikl})\hat{h}_0(k),$$

for $\frac{\pi}{4} \leq \arg k \leq \frac{3\pi}{4}$. Using the above equations in (2.12) and (2.13) to replace $\int_0^\infty e^{k^2 t} q_x(0,t)dt$ and $\int_0^\infty e^{k^2 t} q_x(l,t)dt$, equations (2.12) and (2.13) become equations (2.6) and (2.7)

3 Spatially discrete heat equation

In this section we shall illustrate how the new spectral method can be extended to discrete evolution equations. For the sake of concreteness, we shall consider a discretized version of the heat equation. Partition the x-axis into segments of equal length Δx and set $x = n\Delta x$, integer n. Let $q_n(t) := q(n\Delta x, t)$. Using a standard finite difference scheme, the heat equation on $-\infty < x < \infty$ becomes

$$\dot{q}_n(t) = \frac{1}{\Delta x^2}\left[q_{n+1}(t) + q_{n-1}(t) - 2q_n(t)\right]$$

where $\dot{q} := dq/dt$. We shall now consider the analog of theorem 1 and proposition 1 for the discrete heat equation on a finite interval. In the following we set $\Delta x = 1$.

Theorem 2 *Let $q_n(t)$, $n = 0, ..., N$, satisfy the spatially discrete heat equation*

$$\dot{q}_n(t) = [q_{n+1}(t) + q_{n-1}(t) - 2q_n(t)], \quad 1 \leq n \leq N - 1 \tag{3.1}$$

$$q_n(0) = g_n, \quad 0 \leq n \leq N, \quad q_0(t) = h_0(t), \quad q_N(t) = h_N(t) \tag{3.2}$$

where $h_0(t), h_N(t)$ are Schwartz functions for $t \in [0, \infty)$. There exists a unique solution of this IVB problem that decays to zero as $t \to \infty$. This solution is given by

$$q_n(t) = \int_C z^n e^{-\epsilon(z)t}\rho(z)\frac{dz}{2\pi i z} \tag{3.3}$$

where the directed contour $C = C_1 \cup C_2 \cup C_3$ is depicted in figure 5, $\epsilon(z) = [2 - z - z^{-1}]$ and $\rho(z)$ is uniquely determined in terms of g_n, $h_0(t)$, $h_N(t)$ as follows:

$$\rho(z) = s(z), \quad z \in C_1; \tag{3.4a}$$

$$\rho(z) = \nu(z), \quad z \in C_2; \tag{3.4b}$$

$$\rho(z) = z^{-N}\lambda(z), \quad z \in C_3 \tag{3.4c}$$

with

$$C_1 = \{z; |z| = 1\}, \tag{3.5a}$$

$$C_2 = \{z; \Re(\epsilon(z)) = 0, |z| \leq 1\}, \tag{3.5b}$$

$$C_3 = \{z; \Re(\epsilon(z)) = 0, |z| \geq 1\} \tag{3.5c}$$

and

$$s(z) = \hat{g}(z) \tag{3.6}$$

$$\nu(z) = \frac{\hat{g}(1/z) - z^{2N}\hat{g}(z) + (z - z^{-1})\left(\hat{h}_0(z) - z^N\hat{h}_N(z)\right)}{1 - z^{2N}} \tag{3.7}$$

$$\lambda(z) = \frac{z^{-N}\left(\hat{g}(z) - \hat{g}(1/z)\right) + (z - z^{-1})\left(\hat{h}_N(z) - z^{-N}\hat{h}_0(z)\right)}{1 - z^{-2N}} \tag{3.8}$$

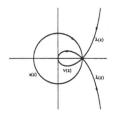

Figure 5: The directed contour C of theorem 2

and

$$\hat{g}(z) = \sum_{m=1}^{N-1} z^{-m} g_m, \quad \hat{h}_0(z) = \int_0^\infty e^{\epsilon(z)t} h_0(t) dt, \qquad (3.9)$$

$$\hat{h}_N(z) = \int_0^\infty e^{\epsilon(z)t} h_N(t) dt$$

Proof   It is straightforward to establish that the integral in equation (3.3) is well defined and that $q_n(t)$ solves equation (3.1). In order to verify that $q_n(t)$ satisfies $q_n(0) = g_n$ we note that

$$q_n(0) = \int_{C_1} z^n \hat{g}(z) \frac{dz}{2\pi i z} + \int_{C_2} z^n \nu(z) \frac{dz}{2\pi i z} + \int_{C_3} z^{n-N} \lambda(z) \frac{dz}{2\pi i z} \qquad (3.10)$$

with the contours C_1, C_2, C_3 defined by equation (3.5). The integrand of the second and third integrals in equation (3.10) are analytic in the domains $\{z; \Re(\epsilon(z)) \le 0, |z| \le 1\}$ and $\{z; \Re(\epsilon(z)) \le 0, |z| \ge 1\}$ respectively. Thus, Cauchy's theorem implies that these integrals give zero contributions and equation (3.10) reduces to $q_n(0) = g_n(0)$.

In order to verify that $q_n(t)$ satisfies $q_0(t) = h_0(t)$ we note that

$$q_0(t) = \int_{C_1} e^{-\epsilon(z)t} \hat{g}(z) \frac{dz}{2\pi i z}$$

$$+ \int_{C_2} \frac{e^{-\epsilon(z)t}}{1 - z^{2N}} \left[\hat{g}(1/z) - z^{2N} \hat{g}(z) + (z - z^{-1}) \left(\hat{h}_0(z) - z^N \hat{h}_N(z) \right) \right] \frac{dz}{2\pi i z}$$

$$+ \int_{C_3} \frac{e^{-\epsilon(z)t}}{1 - z^{-2N}} \left[z^{-2N} \left(\hat{g}(z) - \hat{g}(1/z) \right) \right.$$

$$\left. + (z - z^{-1}) \left(z^{-N} \hat{h}_N(z) - z^{-2N} \hat{h}_0(z) \right) \right] \frac{dz}{2\pi i z}$$

Letting $z \to 1/z$ in the third integral, the above simplifies to

$$q_0(t) = \int_{C_1} e^{-\epsilon(z)t} \hat{g}(z) \frac{dz}{2\pi i z} + \int_{C_2} e^{-\epsilon(z)t} \hat{g}(1/z) \frac{dz}{2\pi i z}$$

$$+ \int_{C_2} e^{-\epsilon(z)t} (z - z^{-1}) \hat{h}_0(z) \frac{dz}{2\pi i z} \qquad (3.11)$$

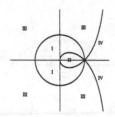

Figure 6: The definitions of the sectors I,...,IV

Since $\hat{g}(z)$ is analytic for $|z| \geq 1$, the combination of the first two terms in (3.11) gives zero contribution, and equation (3.11) reduces to $q_0(t) = h_0(t)$. A similar analysis yields $q_N(t) = h_N(t)$.

QED

Proposition 2 *There exists a solution of the discrete heat equation on a finite interval that decays to zero as $t \to \infty$. This solution is given by equations (3.3) and (3.4) where the functions $s(z), \nu(z), \lambda(z)$ are defined by*

$$s(z) = \sum_{m=1}^{N-1} z^{-m} g_m, \quad |z| \geq 1 \tag{3.12}$$

$$\nu(z) = \int_0^\infty e^{\epsilon(z)t'} \left[q_1(t') - z^{-1} q_0(t') \right] dt', \quad \Re(\epsilon(z)) \leq 0 \tag{3.13}$$

$$\lambda(z) = \int_0^\infty e^{\epsilon(z)t'} \left[q_N(t') - z^{-1} q_{N-1}(t') \right] dt', \quad \Re(\epsilon(z)) \leq 0. \tag{3.14}$$

The boundary conditions $q_n(0)$, $q_0(t)$, $q_N(t)$ satisfy the constraints

$$\int_0^\infty e^{\epsilon(z)t'} \left[q_1(t') - z^{-1} q_0(t') \right] dt'$$
$$= z^{1-N} \int_0^\infty e^{\epsilon(z)t'} \left[q_N(t') - z^{-1} q_{N-1}(t') \right] dt' + \sum_{m=1}^{N-1} z^{-m} q_m(t) \tag{3.15}$$

for $\Re(\epsilon(z)) \leq 0$, $|z| \geq 0$ and

$$z^{N-1} \int_0^\infty e^{\epsilon(z)t'} \left[q_1(t') - z^{-1} q_0(t') \right] dt'$$
$$= \int_0^\infty e^{\epsilon(z)t'} \left[q_N(t') - z^{-1} q_{N-1}(t') \right] dt' + \sum_{m=1}^{N-1} z^{N-m-1} q_m(t) \tag{3.16}$$

for $\Re(\epsilon(z)) \leq 0$, $|z| \leq 0$.

Proof Equation (3.1) admits the Lax pair [4]

$$\mu_{n+1} - z\mu_n = q_n \tag{3.17a}$$
$$\dot{\mu}_n + \epsilon(z)\mu_n = q_n - z^{-1} q_{n-1}, \quad 1 \leq n \leq N-1 \tag{3.17b}$$

where z is a complex-valued parameter. One can easily verify that equations (3.17a) and (3.17b) are compatible if and only if q_n satisfies the discrete heat equation. Let the sectors I,...,IV of the complex z-plane be defined as follows (see figure 6):

$$\{z : |z| \leq 1, \Re(\epsilon(z)) \geq 0\}, \quad \{z : |z| \leq 1, \Re(\epsilon(z)) \leq 0\},$$
$$\{z : |z| \geq 1, \Re(\epsilon(z)) \geq 0\}, \quad \{z : |z| \geq 1, \Re(\epsilon(z)) \leq 0\}$$

Let $\tilde{q}_n(t, z)$ be defined by

$$\tilde{q}_n(t, z) = q_n(t) - z^{-1}q_{n-1}(t) \tag{3.18}$$

The general solution of equation (3.17a) is

$$\mu_n(t, z) = \sum_{m=1}^{n-1} z^{n-m-1}q_m(t) + z^{n-1}\mu_1(t), \quad 1 \leq n \leq N - 1$$

Equations (3.17a) and (3.17b) are compatible, thus the solution $\mu_n(t, z)$ defined above is also a solution of (3.17b) if and only if

$$\dot{\mu}_1(t) + \epsilon(z)\mu_1(t) = \tilde{q}_1(t, z)$$

A particular solution of this equation is

$$\mu_1(t) = \int_0^t e^{-\epsilon(z)(t-t')}\tilde{q}_1(t', z)dt',$$

Thus the function $\mu^{(1)}$ defined by

$$\mu^{(1)} = z^{n-1}\int_0^t e^{-\epsilon(z)(t-t')}\tilde{q}_1(t', z)dt' + \sum_{m=1}^{n-1} z^{n-m-1}q_m(t), \quad z \in I \tag{3.19a}$$

solves both equations (3.17a) and (3.17b). The superscript indicates the region where the above function is analytic in the extended complex z-plane.

Similarly, the functions $\mu^{(2)}, \mu^{(3)}, \mu^{(4)}$ defined below also solve both equations (3.17)

$$\mu^{(2)} = -z^{n-1}\int_t^\infty e^{-\epsilon(z)(t-t')}\tilde{q}_1(t', z)dt'$$
$$+ \sum_{m=1}^{n-1} z^{n-m-1}q_m(t), \quad z \in II \tag{3.19b}$$

$$\mu^{(3)} = z^{n-N}\int_0^t e^{-\epsilon(z)(t-t')}\tilde{q}_N(t', z)dt'$$

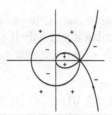

Figure 7: Directed contour C of the second Riemann–Hilbert problem

$$- \sum_{m=n}^{N-1} z^{n-m-1} q_m(t), \quad z \in III \qquad (3.19c)$$

$$\mu^{(4)} = -z^{n-N} \int_t^\infty e^{-\epsilon(z)(t-t')} \widetilde{q}_N(t', z) dt'$$

$$- \sum_{m=n}^{N-1} z^{n-m-1} q_m(t), \quad z \in IV \qquad (3.19d)$$

In the domain of their overlap, the above functions are related by

$$\mu^{(3)} - \mu^{(1)} = -z^{n-1} e^{-\epsilon(z)t} s(z), \quad z \in C_1 \qquad (3.20a)$$

$$\mu^{(3)} - \mu^{(4)} = z^{n-1} e^{-\epsilon(z)t} \lambda(z), \quad z \in C_3 \qquad (3.20b)$$

$$\mu^{(2)} - \mu^{(1)} = -z^{n-1} e^{-\epsilon(z)t} \nu(z), \quad z \in C_2 \qquad (3.20c)$$

where the functions $s(z)$, $\nu(z)$, $\lambda(z)$ are defined by equations (3.12)–(3.14). Let $\Delta\mu$ denote the difference of any two solutions of equations (3.17). Since these solutions satisfy equation (3.17a), it follows that $\Delta\mu = z^{n-1} f(t, z)$. Similarly, since these solutions satisfy equation (3.17b), it follows that $\Delta\mu = e^{-k^2 t} g_n(z)$. Thus $\Delta\mu = z^{n-1} e^{-k^2 t} \rho(z)$, and the function $\rho(z)$ can be determined by evaluating this equation at any suitable x, t. For example, evaluating the equation $\mu^{(3)} - \mu^{(1)} = z^{n-1} e^{-k^2 t} \rho(z)$ at $z = 1$, $t = 0$, it follows that $\rho(z) = -s(z)$. Similarly for equations (3.20b) and (3.20c).

We deduce from the form of $\mu^{(3)}$ or $\mu^{(4)}$ that for μ off the contour depicted in figure 5,

$$\mu_n(t, z) = \frac{q_n(t)}{z} + O\left(\frac{1}{z^2}\right), \quad t \to \infty \qquad (3.21)$$

Equations (3.20) and (3.21) define a RH problem with respect to the directed contour in figure 7. The unique solution of the RH problem is given by [5]

$$\mu = -\frac{1}{2\pi i} \int_C e^{-\epsilon(z)} z^{n-1} \rho(z) \frac{dz'}{z' - z} \qquad (3.22)$$

(The contour C differs from the contour of figure 5 in the direction of C_3 since $\rho(z)$ is defined by $z^{-N} \lambda(z)$ and not by $-z^{-N} \lambda(z)$). Equation (3.22) together

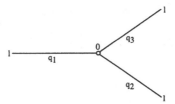

Figure 8: Simple tree consisting of three finite segments joined at a common vertex

with the equation obtained by considering the large z behaviour of (3.22) yields equations (3.3) and (3.4).

Finally, we shall show that if $q_n(t) \to 0$ as $t \to \infty$ then the boundary values of $q_n(t)$ satisfy equations (3.15) and (3.16). To derive equation (3.15) we note that $\mu^{(4)}$, in addition to satisfying equation (3.19d), also satisfies the equation

$$\mu^{(4)} = -\int_t^\infty e^{-\epsilon(z)(t-t')} \tilde{q}_n(t', z) dt', \quad z \in IV \tag{3.23}$$

This equation follows from equation (3.17b) and the assumption that

$$q(x, t) \to 0 \quad \text{as} \quad t \to \infty.$$

Equation (3.15) follows by evaluating equations (3.19d) and (3.23) at $n = 1$ and $t = 0$. Similarly $\mu^{(2)}$, in addition to satisfying equation (3.19b), also satisfies the equation

$$\mu^{(2)} = -\int_t^\infty e^{-\epsilon(z)(t-t')} \tilde{q}_n(t', z) dt', \quad z \in II \tag{3.24}$$

Equation (3.16) follows by evaluating equations (3.19b) and (3.24) at $n = N$ and $t = 0$.

QED

4 Heat equation on a tree

It is straightforward to extend the spectral transform method to the case of more complex topologies, a classical example of which is the one–dimensional heat equation on a branching structure or tree. Many physical systems can be reduced to this problem after suitable approximations. For example, the heat equation (with damping) has been used to model the passive membrane properties of a neuron's dendritic tree [6]. To illustrate the basic idea we shall consider the heat equation on the simple tree depicted in figure 8, which consists of P finite segments of length l labelled $p = 1, ..., P$ connected at a common vertex. Let $q_p(x,t)$, $0 \leq x \leq l$ denote the field on the pth segment and take the coordinate value of the branching node to be zero for all p. Let each q_p satisfy

$$q_{p,t} = q_{p,xx}, \quad t \in [0,\infty), \quad x \in [0,l] \tag{4.1}$$

Equation (4.1) is supplemented by two conditions ensuring (a) continuity at the node,

$$q_p(0,t) = h(t), \quad \text{for all} \quad p = 1, ..., P \tag{4.2}$$

and (b) conservation of current through the node,

$$\sum_{p=1}^{P} q_{p,x}(0,t) = 0, \tag{4.3}$$

The unknown function $h(t)$ is determined self-consistently.

We can now solve equation (4.1) for each segment p using proposition 1 of section 2. Thus

$$q_p(x,t) = \int_L e^{-ikx-k^2t} \rho_p(k) \frac{dk}{2\pi}, \quad p = 1, ..., P \tag{4.4}$$

where the directed contour L is depicted in figure 1, $\rho_p(k)$ is given by

$$\rho_p(k) = s_p(k), \quad k \in R; \tag{4.5a}$$

$$\rho_p(k) = \nu_p(k), \quad \frac{5\pi}{4} \leq \arg k \leq \frac{7\pi}{4}; \tag{4.5b}$$

$$\rho_p(k) = e^{ikl}\lambda_p(k), \quad \frac{\pi}{4} \leq \arg k \leq \frac{3\pi}{4} \tag{4.5c}$$

and the functions $s_p(k), \nu_p(k), \lambda_p(k)$ are defined by

$$s_p(k) = \int_0^l e^{ikx} q_p(x,0)dx, \quad \Im(k) \geq 0 \tag{4.6}$$

$$\nu_p(k) = \int_0^\infty e^{k^2t}[q_{p,x}(0,t) - ikq_p(0,t)]\,dt, \quad \frac{5\pi}{4} \leq \arg k \leq \frac{7\pi}{4} \tag{4.7}$$

$$\lambda_p(k) = \int_0^\infty e^{k^2t}[q_{p,x}(l,t) - ikq_p(l,t)]\,dt, \quad \frac{\pi}{4} \leq \arg k \leq \frac{3\pi}{4} \tag{4.8}$$

Furthermore, the boundary conditions $q_p(x,0), q_p(0,t)$ and $q_p(l,t)$ satisfy the constraints

$$\int_0^\infty e^{k^2t}[q_{p,x}(0,t) - ikq_p(0,t)]\,dt$$

$$= e^{ikl}\int_0^\infty e^{k^2t}[q_{p,x}(l,t) - ikq_p(l,t)]\,dt + \int_0^l e^{ikx}q_p(x,0)dx \tag{4.9}$$

for $\frac{\pi}{4} \leq \arg k \leq \frac{3\pi}{4}$ and

$$e^{-ikl} \int_0^\infty e^{k^2t} \left[q_{p,x}(0,t) - ikq_p(0,t) \right] dt$$

$$= \int_0^\infty e^{k^2t} \left[q_{p,x}(l,t) - ikq_p(l,t) \right] dt + e^{-ikl} \int_0^l e^{ikx} q_p(x,0) dx \quad (4.10)$$

for $\frac{5\pi}{4} \leq \arg k \leq \frac{7\pi}{4}$.

A typical IVB problem for the given configuration will specify the initial data on each segment together with a boundary condition at each end-point $x = l$. This, together with equations (4.9) and (4.10), yields $4P$ algebraic equations for the $5P$ unknowns

$$\int_0^l e^{ikl} q_p(x,0) dx, \quad \int_0^\infty e^{k^2t} q_p(0,t) dt, \quad \int_0^\infty e^{k^2t} q_{p,x}(0,t) dt,$$

$$\int_0^\infty e^{k^2t} q_p(l,t) dt, \quad \int_0^\infty e^{k^2t} q_{p,x}(l,t) dt.$$

The additional P algebraic equations necessary for obtaining a unique solution are obtained from continuity and current conservation at the vertex, equations (4.2) and (4.3). As a particular example, consider the case

$$q_p(x,0) = g_p(x), \quad q_p(l,t) = h_p(t) \quad (4.11)$$

and define

$$\hat{g}_p(k) = \int_0^l e^{ikx'} g_p(x') dx', \quad \hat{h}(k) = \int_0^\infty e^{k^2t'} h(t') dt', \quad (4.12)$$

$$\hat{h}_p(k) = \int_0^\infty e^{k^2t'} h_p(t') dt'$$

Then equation (4.9) and the equation resulting from equation (4.10) by replacing k with $-k$ yield (for $\frac{\pi}{4} \leq \arg k \leq \frac{3\pi}{4}$)

$$\left[1 - e^{2ikl} \right] \int_0^\infty e^{k^2t} q_{p,x}(l,t) dt$$

$$= e^{ikl} [\hat{g}_p(k) - \hat{g}_p(-k)] + 2ike^{ikl}\hat{h}(k) - ik(1 + e^{2ikl})\hat{h}_p(k) \quad (4.13)$$

$$\left[1 - e^{-2ikl} \right] \int_0^\infty e^{k^2t} q_{p,x}(0,t) dt$$

$$= \hat{g}_p(k) - e^{2ikl}\hat{g}_p(-k) - 2ike^{ikl}\hat{h}_p(k) + ik(1 + e^{-2ikl})\hat{h}(k) \quad (4.14)$$

In deriving equations (4.13) and (4.14) we have used the continuity condition (4.2). Equations (4.13) and (4.14) express

$$\int_0^\infty e^{k^2t} q_{p,x}(0,t) dt \quad \text{and} \quad \int_0^\infty e^{k^2t} q_{p,x}(l,t) dt$$

in terms of the known functions $\hat{h}_p(k)$, $\hat{g}_p(k)$ and the unknown function $\hat{h}(k)$. The latter may be obtained in terms of the former by summing equation (4.14) with respect to p and using the current conservation condition (4.3). The result is

$$\hat{h}(k) = \frac{i}{kP(1 + e^{-2ikl})} \sum_{p=1}^{P} \left[\hat{g}_p(k) - e^{2ikl}\hat{g}_p(-k) - 2ike^{ikl}\hat{h}_p(k) \right] \qquad (4.15)$$

We conclude that the spectral transform method presented in this paper reduces the problem of prescribed boundary data for the heat equation on a complex network to the simple problem of solving a system of algebraic equations. These equations determine the spectral data $\rho(k)$ appearing in the separable form of the solution $q_p(x, t)$ on each branch, equation (4.4).

Acknowledgement

A. S. Fokas was partially supported by the EPSRC (GR/J71885) and by the National Science Foundation (DMS – 9500311).

References

[1] A.S. Fokas, "Lax pairs and a new spectral method for solving linear PDE's in two variables" Proc. Roy. Soc. A (in press).

[2] P.D. Lax, Comm. Pure and Appl. Math., **21** 467 (1968).

[3] L. Ehrenpreis, *Fourier analysis in several complex variables*, Wiley–Interscience, New York, 1970.

[4] P.C. Bressloff, "A new Green's function method for solving linear PDE's in two variables" J. Math. Anal. (in press).

[5] M.J. Ablowitz and A.S. Fokas, *Complex variables and applications*, Cambridge University Press, 1997.

[6] W. Rall, "Core conductor theory and cable properties of neurons". In *Handbook of neurophysiology: the nervous system* (ed. E.R. Kandel) pp. 39-97, American Physiological Society, Bethesda, Maryland, 1977.

Chapter 5

Discrete Painlevé Equations

The Discrete Painlevé I Hierarchy

Clio Cresswell[†] and Nalini Joshi[‡]
† School of Mathematics
University of New South Wales
Sydney NSW 2052, Australia

‡ Department of Pure Mathematics
University of Adelaide
Adelaide SA 5005, Australia

Abstract

The discrete Painlevé I equation (dP$_I$) is an integrable difference equation which has the classical first Painlevé equation (P$_I$) as a continuum limit. dP$_I$ is believed to be integrable because it is the discrete isomonodromy condition for an associated (single-valued) linear problem. In this paper, we derive higher-order difference equations as isomonodromy conditions that are associated to the same linear deformation problem. These form a hierarchy that may be compared to hierarchies of integrable ordinary differential equations (ODEs). We strengthen this comparison by continuum limit calculations that lead to equations in the P$_I$ hierarchy. We propose that our difference equations are discrete versions of higher-order Painlevé equations.

1 Introduction

Our aim is to derive higher-order versions of the equation

$$y_{n+1} + y_n + y_{n-1} = \frac{\alpha n + \beta}{y_n} + \gamma. \tag{1}$$

Equation (1) is known as the first discrete Painlevé equation or dP$_I$, because the scaled continuum limit $y_n = 1 + h^2 u(x)$, $x = nh$, with $\alpha = r_1 h^5$, $\beta = -3 + r_2 h^2$, $\gamma = 3 - \beta - r_3 h^4$, where the r_i $(i = 1, 2, 3)$ are constants, yields a scaled and translated version of the classical first Painlevé equation (P$_I$)

$$u'' = 6u^2 + x,$$

as $h \to 0$.

P_I is the simplest of six well known nonlinear second-order ODEs in the complex plane called the Painlevé equations. Their characteristic property that all movable singularities of all solutions are poles is called the Painlevé property. Painlevé [1], Gambier [2], and Fuchs [3] identified them (under some mild conditions) as the only such equations with the Painlevé property whose general solutions are new transcendental functions.

P_I is believed to be integrable because it is the isomonodromy condition for an associated (single-valued) linear system of differential equations [4]. dP_I is also an isomonodromy condition. Moreover, it possesses a discrete version of the Painlevé property called the singularity confinement property proposed by Grammaticos, Ramani et al[5, 6].

Joshi et al [7] derived the linear problem associated with dP_I by starting with the Ablowitz-Ladik [8, 9] scattering problem. The latter authors in turn based their scattering problem on the general AKNS problem given by Ablowitz, Kaup, Newell, and Segur [10].

It is well known that a single linear problem can give rise to a hierarchy of integrable ODEs [11]. In this paper, we announce an extension of such hierarchies to the discrete realm by deriving hierarchies of integrable difference equations.

The plan of the paper is as follows. We recall the isomonodromy problem for dP_I in section 2 for completeness and then show how higher-order difference equations arise from it, giving examples up to the sixth-order.

In section 3, we give continuum limits of the second and fourth-order dP_I equations. In section 4, we examine a linear problem associated with P_I and show that the continuum limit found for the fourth-order dP_I is the next equation in the P_I hierarchy. We propose that our hierarchy of difference equations is an integrable discrete version of such a hierarchy.

2 Construction of the dP_I hierarchy

The linear problem associated with dP_I [7] is

$$\alpha_n w_{n+1} = \lambda w_n - w_{n-1}, \tag{2}$$

$$\frac{\partial w_n}{\partial \lambda} = a_n w_{n+1} + b_n w_n, \tag{3}$$

where w, α, a, b depend on a discrete variable n and w, a, b also depend on the continuous variable λ (a, b rational in λ).

The aim of this section is to show that dP_I is one of a whole range of higher order compatibility conditions arising from this system.

The compatibility conditions of equations (2) and (3) are

$$b_{n+1} - b_{n-1} + \lambda\left(\frac{a_{n+1}}{\alpha_{n+1}} - \frac{a_n}{\alpha_n}\right) = 0,$$

$$\frac{\lambda^2}{\alpha_n}\left(\frac{a_{n+1}}{\alpha_{n+1}} - \frac{a_n}{\alpha_n}\right) + \frac{\lambda}{\alpha_n}(b_{n+1} - b_n) - \frac{1}{\alpha_n}\left(\alpha_n\frac{a_{n+1}}{\alpha_{n+1}} - a_{n-1} + 1\right) = 0.$$

Define

$$p_n = \frac{a_n}{\alpha_n} \qquad \text{and} \qquad q_n = b_n + b_{n-1}.$$

We concentrate on results for p_n. Corresponding results for q_n can be obtained from the first compatibility condition above.

The two compatibility conditions collapse down into a single condition in terms of p_n

$$\alpha_{n+1}p_{n+2} + (\alpha_n - \lambda^2)p_{n+1} + (\lambda^2 - \alpha_n)p_n - \alpha_{n-1}p_{n-1} + 2 = 0. \qquad (4)$$

Take p_n to be a finite Laurent expansion in λ

$$p_n = \sum_{k=0}^{l} P_{k,n}\lambda^k,$$

and substitute this expansion into (4). (Wherever convenient below we refer to $P_{k,n}$ as P_k.) We then find the following set of simultaneous equations for $P_{0,n}, P_{1,n}, ... P_{l,n}$

$$P_{l,n} - P_{l,n+1} = 0,$$
$$P_{l-1,n} - P_{l-1,n+1} = 0,$$
$$\alpha_{n+1}P_{k,n+2} + \alpha_n P_{k,n+1} - \alpha_n P_{k,n} - \alpha_{n-1}P_{k,n-1} + P_{k-2,n} - P_{k-2,n+1} = 0,$$

for $2 \leq k \leq l$, and

$$\alpha_{n+1}P_{1,n+2} + \alpha_n P_{1,n+1} - \alpha_n P_{1,n} - \alpha_{n-1}P_{1,n-1} = 0$$
$$\alpha_{n+1}P_{0,n+2} + \alpha_n P_{0,n+1} - \alpha_n P_{0,n} - \alpha_{n-1}P_{0,n-1} + 2 = 0$$

for $k = 0, 1$.

Note that the equation for P_0 is inhomogeneous (i.e. it has a nonzero forcing term) whereas the equation for P_1 is homogeneous. Also note that the equations defining P_k for even and odd k may be separated into otherwise identical equations. These lead to two inconsistent compatibility conditions for α unless one of these two subsequences of P_k vanishes identically. In fact, $P_l = 0$ for odd l is the only possibility due to the inhomogeneity of the equation for P_0.

We are therefore left with the following system

$$P_{2m,n} - P_{2m,n+1} = 0,$$

$$\alpha_{n+1}P_{2k,n+2} + \alpha_n P_{2k,n+1} - \alpha_n P_{2k,n} - \alpha_{n-1}P_{2k,n-1} + P_{2k-2,n} - P_{2k-2,n+1} = 0$$

for $1 \le k \le m$, and

$$\alpha_{n+1}P_{0,n+2} + \alpha_n P_{0,n+1} - \alpha_n P_{0,n} - \alpha_{n-1}P_{0,n-1} + 2 = 0$$

for $k = 0$, where we have taken $l = 2m$ $(m \in \mathbf{N})$.

Below we list the solutions for different m up to $m = 3$. The c_i's, $i = 1, 2, ..., 5$ are constants.

- $m = 0$ yields the trivial linear nonautonomous equation

$$\alpha_n c_2 = c_1 + c_0(-1)^n - n$$

with

$$P_{0,n} = c_2 \neq 0.$$

- $m = 1$ yields the second-order equation

$$\alpha_n c_3(\alpha_{n-1} + \alpha_n + \alpha_{n+1}) + \alpha_n c_2 = c_1 + c_0(-1)^n - n \qquad (5)$$

with

$$P_{0,n} = c_2 + c_3(\alpha_n + \alpha_{n-1}),$$
$$P_{2,n} = c_3 \neq 0.$$

In this case, we have recovered the more general version of dP$_I$ [5].

- $m = 2$ yields the fourth-order equation

$$\alpha_n c_4(\alpha_{n+1}\alpha_{n+2} + \alpha_{n+1}^2 + 2\alpha_n\alpha_{n+1} + \alpha_{n-1}\alpha_{n-2} + \alpha_{n-1}^2 + 2\alpha_n\alpha_{n-1}$$
$$+\alpha_n^2 + \alpha_{n-1}\alpha_{n+1}) + \alpha_n c_3(\alpha_{n-1} + \alpha_n + \alpha_{n+1}) + \alpha_n c_2$$
$$= c_1 + c_0(-1)^n - n \qquad (6)$$

with

$$P_{0,n} = c_2 + c_3(\alpha_n + \alpha_{n-1})$$
$$+c_4(2\alpha_n\alpha_{n-1} + \alpha_{n-1}^2 + \alpha_{n-1}\alpha_{n-2} + \alpha_n\alpha_{n+1} + \alpha_n^2),$$
$$P_{2,n} = c_3 + c_4(\alpha_n + \alpha_{n-1}),$$
$$P_{4,n} = c_4 \neq 0.$$

- $m = 3$ yields the sixth-order equation

$$\alpha_n c_5 (\alpha_{n+1}\alpha_{n+2}\alpha_{n+3} + 2\alpha_n\alpha_{n+1}\alpha_{n+2} + \alpha_{n-1}\alpha_{n+1}\alpha_{n+2} + \alpha_{n+1}^3$$
$$+\alpha_{n+1}\alpha_{n+2}^2 + 2\alpha_{n+2}\alpha_{n+1}^2 + 3\alpha_{n+1}\alpha_n^2 + 3\alpha_n\alpha_{n+1}^2 + \alpha_{n-1}\alpha_{n-2}\alpha_{n-3}$$
$$+2\alpha_n\alpha_{n-1}\alpha_{n-2} + \alpha_{n-2}\alpha_{n-1}\alpha_{n+1} + \alpha_{n-1}^3 + \alpha_{n-2}^2\alpha_{n-1} + 2\alpha_{n-1}^2\alpha_{n-2}$$
$$+3\alpha_{n-1}\alpha_n^2 + 3\alpha_{n-1}^2\alpha_n + 4\alpha_{n-1}\alpha_n\alpha_{n+1} + \alpha_n^3 + \alpha_{n-1}\alpha_{n+1}^2$$
$$+\alpha_{n-1}^2\alpha_{n+1}) + \alpha_n c_4(\alpha_{n+1}\alpha_{n+2} + \alpha_{n+1}^2 + 2\alpha_n\alpha_{n+1} + \alpha_{n-1}\alpha_{n-2}$$
$$+\alpha_{n-1}^2 + 2\alpha_n\alpha_{n-1} + \alpha_n^2 + \alpha_{n-1}\alpha_{n+1}) + \alpha_n c_3(\alpha_{n-1} + \alpha_n + \alpha_{n+1})$$
$$+\alpha_n c_2 = c_1 + c_0(-1)^n - n$$

with

$$P_{0,n} = c_2 + c_3(\alpha_n + \alpha_{n-1}) + c_4(2\alpha_n\alpha_{n-1} + \alpha_{n-1}^2 + \alpha_{n-1}\alpha_{n-2}$$
$$+\alpha_n\alpha_{n+1} + \alpha_n^2) + c_5(\alpha_n\alpha_{n+1}\alpha_{n+2} + 2\alpha_{n-1}\alpha_n\alpha_{n+1} + 3\alpha_n^2\alpha_{n-1}$$
$$+3\alpha_{n-1}^2\alpha_n + \alpha_{n-2}^2\alpha_{n-1} + 2\alpha_{n+1}\alpha_n^2 + \alpha_n^3 + \alpha_{n-1}^3 + \alpha_n\alpha_{n+1}^2$$
$$+\alpha_{n-1}\alpha_{n-2}\alpha_{n-3} + 2\alpha_n\alpha_{n-1}\alpha_{n-2} + 2\alpha_{n-1}^2\alpha_{n-2}),$$
$$P_{2,n} = c_3 + c_4(\alpha_n + \alpha_{n-1})$$
$$+c_5(2\alpha_n\alpha_{n-1} + \alpha_{n-1}^2 + \alpha_{n-1}\alpha_{n-2} + \alpha_n\alpha_{n+1} + \alpha_n^2),$$
$$P_{4,n} = c_4 + c_5(\alpha_n + \alpha_{n-1}),$$
$$P_{6,n} = c_5 \neq 0.$$

This process of solving the system for increasing m continues indefinitely. At each stage, the order of the compatibility condition increases by 2.

3   Continuum limits

In this section, we derive continuum limits of difference equations found in the previous section.

3.1   The case $m = 1$

Consider equation (5). If $c_0 = 0$, we recover the equation often referred to as dP_I, with P_I as one of its continuum limits. If $c_0 \neq 0$ however, the term $(-1)^n$ suggests an odd-even dependence in α_n. This dependence must be taken into account to obtain a meaningful continuum limit. This leads to the transformation

$$\alpha_{2k-1} = u_k, \qquad \alpha_{2k} = v_k.$$

(This is similar to the limit pointed out by Grammaticos *et al* [12] for a discrete version of the second Painlevé equation generalised with an additional $(-1)^n$ term.)

In our search for a continuum limit, we use the substitutions

$$u_k = 1 + hy(kh), \quad v_k = z(kh), \quad t = kh, \quad c_3 = -\frac{2}{r_1}h^{-3},$$

and for ease of notation rename

$$\mu = \frac{c_1 + c_0}{c_3}, \quad \nu = \frac{c_1 - c_0 + 1}{c_3}, \quad \sigma = -\frac{c_2}{c_3} \quad \text{and} \quad \rho = -\frac{2}{c_3}.$$

Then we find

$$z = \frac{\sigma + \nu - 1}{2} - \frac{\nu + 1}{2}yh + \frac{1}{2}\left(r_1 t + \nu y^2 - \frac{\nu}{2}y_t - \frac{1}{2}y_t\right)h^2$$
$$+ \frac{1}{2}\left(\frac{r_1}{2} + \nu y y_t - r_1 t y - \nu y^3\right)h^3 + O(h^4).$$

Using the scalings

$$\mu = -\frac{\sigma^2}{4} + \frac{\nu^2}{4} + \frac{\nu}{2} + \sigma - \frac{3}{4} - r_2 h^3,$$
$$\sigma = \frac{3}{2} + \frac{\nu^2}{2} + r_3 h^2,$$
$$\nu = 1 + \frac{2}{3}r_4 h,$$

in the limit $h \to 0$ we are left with

$$y_{tt} - 2y^3 + 2r_4 y^2 - 2r_1 t y + 2r_3 y + \frac{2}{3}r_1 r_4 t + r_1 + 2r_2 = 0.$$

This is a scaled and translated version of the second Painlevé equation (P_{II}).

3.2   The case $m = 2$

For simplicity, we restrict ourselves to the case $c_0 = 0$. First, we illustrate the fact that second-order continuum limits are possible even though difference equation (6) is fourth order. Under the substitution

$$\alpha_n = 1 + h^2 u(nh), \quad t = nh,$$

and the scalings

$$10 + \frac{3c_3}{c_4} - \frac{c_1}{c_4} + \frac{c_2}{c_4} = r_1 h^4, \quad 20 + \frac{3c_3}{c_4} + \frac{c_1}{c_4} = r_2 h^2, \quad \frac{1}{c_4} = r_3 h^5,$$

equation (6) becomes

$$\left(10 + \frac{c_3}{c_4}\right)u_{tt} + \left(10 - \frac{c_1}{c_4}\right)u^2 + r_2 u + r_1 + r_3 t = 0$$

in the limit $h \to 0$. This is a scaled and translated version of P_I. However, this case restricts the four degrees of freedom contained in the parameters of equation (6) to three.

Scalings that maintain the full four degrees of freedom, i.e.

$$10 + \frac{3c_3}{c_4} - \frac{c_1}{c_4} + \frac{c_2}{c_4} = r_1 h^6, \qquad 20 + \frac{3c_3}{c_4} + \frac{c_1}{c_4} = r_2 h^4,$$

$$10 + \frac{c_3}{c_4} = r_3 h^2, \qquad \frac{1}{c_4} = r_4 h^7,$$

lead, in the limit $h \to 0$, to the fourth-order equation

$$u_{tttt} + 5(u_t)^2 + 10 u u_{tt} + r_3 u_{tt} + 10 u^3 + 3 r_3 u^2 + r_2 u + r_1 + r_4 t = 0. \qquad (7)$$

We discuss the significance of this equation in the next section.

4 Discussion

The aim of this section is to show that equation (7), the continuum limit for the case $m = 2$, $c_0 = 0$, is the next equation in the P_I hierarchy.

The continuum limit equations found for the case $m = 1$ are P_I and P_{II}. These equations are considered integrable as they are isomonodromy conditions for a single-valued linear problem. Furthermore, they may be viewed as equations lying at the base of two integrable hierarchies of equations. Consider the following isomonodromy problem associated with P_I

$$\mathbf{v_x} = \begin{pmatrix} 0 & 1 \\ -u(x) - k^2 & 0 \end{pmatrix} \mathbf{v}, \qquad \mathbf{v_k} = \begin{pmatrix} A & B \\ C & -A \end{pmatrix} \mathbf{v}, \qquad (8)$$

where u is the potential, k is the eigenvalue and

$$\begin{aligned}
A &= -2u_x k, \\
B &= (2r_2 + 4u)k - 8k^3, \\
C &= -(2u_{xx} + 2r_2 u + 4u^2)k + (-2r_2 + 4u)k^3 + 8k^5.
\end{aligned}$$

P_I is a compatibility condition for system (8) with these choices of A, B, C. A hierarchy of compatibility conditions arises when A, B and C are expanded to higher degree in k. The next self-consistent expansions are

$$\begin{aligned}
A &= -\left(\frac{s_0}{16} u_{xxx} + \frac{3s_0}{8} u u_x + \frac{s_1}{4} u_x \right) k + \frac{s_0}{4} u_x k^3, \\
B &= \left(\frac{s_0}{8} u_{xx} + \frac{3s_0}{8} u^2 + \frac{s_1}{2} u - s_2 \right) k - \left(\frac{s_0}{2} u + s_1 \right) k^3 + s_0 k^5, \qquad (9) \\
C &= \left(\frac{s_0}{8} u u_{xx} - \frac{s_0}{16} u_x^2 + \frac{s_0}{4} u^3 + \frac{s_1}{4} u^2 - 2x + s_3 \right) k \\
&\quad + \left(\frac{s_0}{8} u_{xx} + \frac{s_0}{8} u^2 + \frac{s_1}{2} u + s_2 \right) k^3 + \left(-\frac{s_0}{2} u + s_1 \right) k^5 - s_0 k^7,
\end{aligned}$$

where s_0, s_1, s_2, s_3 are constants. A straight forward calculation shows that equation (7) is a compatibility condition of (8) with A, B, C expanded as in (9) (renaming $s_0 = -\dfrac{32}{r_4}$, $s_1 = -\dfrac{8r_3}{r_4}$, $s_2 = \dfrac{2r_2}{r_4}$, $s_3 = -\dfrac{2r_1}{r_4}$).

5   Acknowledgements

The research reported in this paper was made possible through an Australian Postgraduate Award and supported by the Australian Research Council.

References

[1] P. Painlevé. Sur les équations différentielles du second ordre et d'ordre supérieur dont l'intégrale générale est uniforme. *Acta Math.*, 25:1–85, 1902.

[2] B. Gambier. Sur les équations différentielles du second ordre et du premier degré dont l'intégral général est à points critiques fixes. *Acta Math.*, 33:1–55, 1910.

[3] R. Fuchs. Über lineare homogene differentialgleichungen zweiter or ordnung mit drei im endlich gelegene wesentlich singulären stellen. *Math. Annalen*, 63:301–321, 1907.

[4] K. Okamoto. Isomonodromic deformation and Painlevé equations, and the Garnier system. *J. Fac. Sci. Univ. Tokyo Sec. IA Math*, 33:575–618, 1986.

[5] B. Grammaticos, A. Ramani, and V. Papageorgiou. Do integrable mappings have the Painlevé property? *Phys. Rev. Lett.*, 67:1825–1828, 1991.

[6] A. Ramani, B. Grammaticos, and J. Hietarinta. Discrete versions of the Painlevé equations. *Phys. Rev. Lett.*, 67:1829–1832, 1991.

[7] N. Joshi, D. Burtonclay, and R. Halburd. Nonlinear nonautonomous discrete dynamical systems from a general discrete isomonodromy problem. *Lett. Math. Phys.*, 26:123–131, 1992.

[8] M.J. Ablowitz and J.F. Ladik. Nonlinear differential-difference equations. *J. Math. Phys.*, 16:598–603, 1975.

[9] M.J. Ablowitz and J.F. Ladik. Nonlinear differential-difference equations and fourier analysis. *J. Math. Phys.*, 17:1011–1018, 1976.

[10] M.J. Ablowitz, D.J. Kaup, A.C. Newell, and H. Segur. The inverse scattering transform — Fourier analysis for nonlinear problems. *Stud. Appl. Math*, 53:249–315, 1974.

[11] H. Flaschka and A.C. Newell. Monodromy- and spectrum-preserving deformations I. *Commun. Math. Phys.*, 76:65–116, 1980.

[12] B. Grammaticos, F.W. Nijhoff, V. Papageorgiou, A. Ramani, and J. Satsuma. Linearization and solutions of the discrete Painlevé III equation. *Phys. Lett. A*, 185:446–452, 1994.

Rational solutions to d-P$_{IV}$

Jarmo Hietarinta[1,2*] and Kenji Kajiwara[2†]

^{1} Department of Physics, University of Turku
FIN-20014 Turku, Finland
^{2} Department of Electrical Engineering
Doshisha University, Tanabe, Kyoto 610-03, Japan

Abstract

We study the rational solutions of the discrete version of Painlevé's fourth equation (d-P$_{IV}$). The solutions are generated by applying Schlesinger transformations on the seed solutions $-2z$ and $-1/z$. After studying the structure of these solutions we are able to write them in a determinantal form that includes an interesting parameter shift that vanishes in the continuous limit.

1 Introduction

One important question in the study of discrete versions of continuous differential equations concerns the existence of corresponding special solutions. For continuous Painlevé equations rational and special function solutions are known [1, 2], and in many cases even a rigorous classification has been done [3]. If one proposes a discrete version of a Painlevé equation it is not enough that in some continuous limit the continuous Painlevé equation is obtained, but in addition the proposed equation should share some further properties of the original equation. One of these properties should be the equivalent of the Painlevé property, called "singularity confinement" [4]. This has already been used to propose discrete forms of the Painlevé equations [5]. Other structures of the continuous Painlevé equations that have been shown to exist for the discrete ones include their relationships by coalescence limits [6] and the existence of Hirota forms [7] for these equations. What is still largely an open question is the fate of the special solutions (rational, algebraic, special function) known for the continuous case.

---

*Email: hietarin@newton.tfy.utu.fi
†Email: kaji@elrond.doshisha.ac.jp

Here we discuss the determinantal structure of the *rational solutions* to the discrete fourth Painlevé equation, (d-P_{IV}), given by [5]

$$
(x_{n+1} + x_n)(x_n + x_{n-1}) =
$$
$$
\frac{(x_n + \alpha + \beta)(x_n + \alpha - \beta)(x_n - \alpha + \beta)(x_n - \alpha - \beta)}{(x_n + z_n + \gamma)(x_n + z_n - \gamma)} \quad (1)
$$

where $z_n = \delta n + \zeta$. [Note that this equation is invariant under the change sign of any of parameters α, β, γ.]

One reason for calling (1) d-P_{IV} is that if we put $\alpha^2 = -a + \sqrt{-b/2} + \delta^{-2}$, $\beta^2 = -a - \sqrt{-b/2} + \delta^{-2}$, $\gamma = \delta^{-1}$ and then take the limit $\delta \to 0$ we get P_{IV}:

$$
\frac{d^2 w}{dz^2} = \frac{1}{2w}\left(\frac{dw}{dz}\right)^2 + \frac{3}{2}w^3 + 4zw^2 + 2(z^2 - a)w + \frac{b}{w} . \quad (2)
$$

The comparison of (1) and (2) reveals the first interesting difference between the discrete and continuous versions: the discrete one has more parameters.

But why should one insist on writing the solution in determinantal form? It is well known that most integrable systems have multisoliton (and rational) solutions in determinantal form, and the same holds for many special solutions of continuous Painlevé equations. This determinantal structure actually carries fundamental information and reveals the "basic" object hidden in the solutions. For example it is well known that P_{II},

$$
\frac{d^2 u}{dz^2} = 2u^3 - 4zu + 4\alpha, \quad (3)
$$

where α is a parameter, admits rational solutions for $\alpha = N + 1$. It has also been shown that these solutions can be are expressed as [8]

$$
u = \frac{d}{dz} \log \frac{\tau_{N+1}}{\tau_N}, \quad (4)
$$

where τ_N's are polynomials in z (Vorob'ev–Yablonski polynomials). However, it has been shown only recently that τ_N can be expressed in determinantal form [9],

$$
\tau_N = \begin{vmatrix} q_N(z,t) & q_{N+1}(z,t) & \cdots & q_{2N-1}(z,t) \\ q_{N-2}(z,t) & q_{N-1}(z,t) & \cdots & q_{2N-3}(z,t) \\ \vdots & \vdots & \ddots & \vdots \\ q_{-N+2}(z,t) & q_{-N+3}(z,t) & \cdots & q_1(z,t) \end{vmatrix}, \quad (5)
$$

where q_k's are the so called Devisme polynomials defined by

$$
\sum_{k=0}^{\infty} q_k(z,t)\lambda^k = \exp\left(z\lambda + t\lambda^2 + \frac{1}{3}\lambda^3\right), \quad \text{and} \quad q_k(z,t) = 0, \text{ for } k < 0 . \quad (6)
$$

Thus, these Devisme polynomials may be considered as basic polynomials in the rational solutions of P_{II}. Similar determinantal structure of the rational solutions is observed for the discrete case: in fact, it has been shown that in the case of d-P_{II}, the rational solutions can be expressed by determinants whose entries are given by Laguerre polynomials [10, 11].

In the case of P_{IV}, it is well known [1] that the continuous P_{IV} has three rational solution hierarchies, whose "seed" solutions are

$$ y(z) = -2z, \quad -\frac{1}{z}, \quad \text{or} \quad -\frac{2}{3}z, \tag{7} $$

and determinantal forms of the first two hierarchies have been discussed in general terms, e.g., in [2]. The detailed results for the first two hierarchies are as follows [12]. Let τ_N^ν be an $N \times N$ determinant of Hankel type given by

$$ \tau_N^\nu = \det |H_{\nu+i+j-2}|_{1 \le i,j \le N} \ , \quad \tau_0^\nu = 1 \ , \tag{8} $$

where H_n, $n = 0, 1, 2 \cdots$ are Hermite polynomials in x characterized by the recursion relations

$$ H_{n+1} = xH_n - nH_{n-1}, \quad \frac{d}{dx}H_n = nH_{n-1}, \quad H_0 = 1, \ H_1 = x. \tag{9} $$

Then

$$ w = -\sqrt{2}\frac{d}{dx}\log\left(\frac{\tau_{N+1}^\nu}{\tau_N^\nu}\right) \ , \quad z = \frac{1}{\sqrt{2}}x \ , \tag{10} $$

are rational solutions of $P_{IV}(2)$ for parameter values

$$ (a,b) = (-(\nu + 2N + 1), -2\nu^2), \quad \nu, N \in \mathbb{Z}, \quad \nu \ge 1, \quad N \ge 0. \tag{11} $$

Moreover,

$$ w = \sqrt{2}\left(\frac{d}{dx}\log\left(\frac{\tau_{N+1}^\nu}{\tau_N^{\nu+1}}\right) - x\right) \ , \quad z = \frac{1}{\sqrt{2}}x \ , \tag{12} $$

yield rational solutions of $P_{IV}(2)$ with

$$ (a,b) = (N - \nu, -2(\nu + N + 1)^2), \quad \nu, N \in \mathbb{Z}, \quad \nu \ge 0, \quad N \ge 0. \tag{13} $$

The solutions given in (10) and (12) correspond to the "$-1/z$" and "$-2z$" hierarchies, respectively[1].

Here our object is to find the discrete versions of these results for the $-2z$ and $-1/z$ hierarchies. (We hope to return the more complicated $-\frac{2}{3}z$ case elsewhere.) We cannot simply discretize the continuous results, because

---

[1]Equations (10) and (12) actually give half of the solutions in $-1/z$ and $-2z$ hierarchies obtained by Murata [3]. The other half is given in terms of polynomials similar to the Hermite polynomials, but with some different signs in the recursion relations above.

the discrete equation has more parameters, and the way the new parameters modify the continuous results is indeed one of the interesting questions. In approaching this problem we will not use any detailed properties of the continuous case, we just assume that that such rational solutions should arise from determinants of polynomials, which always happens in the continuous case. Starting with the discrete version of the usual seed solutions, we first construct a set of solutions in both hierarchies using Bäcklund–Schlesinger transformations (discussed in the next section) and then search for the determinantal structure by studying the properties of these rational solutions, the main clue being the factorization of the denominator.

2   Bäcklund-Schlesinger transformation

In order to generate rational solutions we use the Bäcklund–Schlesinger transformations on the seed solutions. These transformations were given in [13]. We write them as follows:

Bäcklund transformation: Let us assume that $x(n)$ solves d-P_{IV} with parameter values (α, β, γ) (in which case we often write $x = x(n; \alpha, \beta, \gamma)$ or $x = \{f(n); \alpha, \beta, \gamma\}$). Using these let us define

$$BT_{a,b}(x) := \left[\frac{x\bar{x} + \bar{x}(\tilde{z} + a) + x(\tilde{z} - a) + b^2 - a^2)}{x + \bar{x}} \right]_{n \to n + \frac{1}{2}} \tag{14}$$

where $\tilde{z} = (n + \frac{1}{2})\delta + \zeta$, and $\bar{x} = x(n+1)$. Note the shifts in n. Then we get new solutions from the elementary Bäcklund transformations BT_i as follows:

$$\begin{aligned}
x(n; \alpha + \tfrac{1}{2}\delta, \gamma, \beta) &= BT_1 x(n) = BT_{\alpha,\beta}(x(n; \alpha, \beta, \gamma)), \\
x(n; \alpha - \tfrac{1}{2}\delta, \gamma, \beta) &= BT_2 x(n) = BT_{-\alpha,\beta}(x(n; \alpha, \beta, \gamma)), \\
x(n; \beta + \tfrac{1}{2}\delta, \gamma, \alpha) &= BT_3 x(n) = BT_{\beta,\alpha}(x(n; \alpha, \beta, \gamma)), \\
x(n; \beta - \tfrac{1}{2}\delta, \gamma, \alpha) &= BT_4 x(n) = BT_{-\beta,\alpha}(x(n; \alpha, \beta, \gamma)).
\end{aligned} \tag{15}$$

These transformations jump too much in the parameter space and therefore it is useful to define **Schlesinger transformations** that generate new solutions by changing only one of the parameters

$$\begin{aligned}
x(n; \alpha + \delta, \beta, \gamma) &= S_1(x) = BT_1 BT_1 \, x(n; \alpha, \beta, \gamma), \\
x(n; \alpha - \delta, \beta, \gamma) &= S_2(x) = BT_2 BT_2 \, x(n; \alpha, \beta, \gamma), \\
x(n; \alpha, \beta + \delta, \gamma) &= S_3(x) = BT_2 BT_3 BT_1 BT_3 \, x(n; \alpha, \beta, \gamma), \\
x(n; \alpha, \beta - \delta, \gamma) &= S_3(x) = BT_2 BT_3 BT_2 BT_4 \, x(n; \alpha, \beta, \gamma).
\end{aligned} \tag{16}$$

It should be noted that sometimes there are barriers over which Schlesinger transformations cannot cross. For example if $\alpha = \gamma + (m + \frac{1}{2})\delta$ and $\beta = \gamma - (n + \frac{1}{2})\delta$ (which is relevant for some rational hierarchies) then the barrier $n = 0$ cannot be crossed by Schlesinger transformations because BT_3 yields the 0 solution, similarly for $m = 0$ with BT_2.

3   Some solutions

Using the above Schlesinger transformations we can construct other rational solutions from the seed solutions. The seed for the $-2z$ hierarchy is given by

$$x_{00} = \{-2z, \gamma + \tfrac{1}{2}\delta, \gamma - \tfrac{1}{2}\delta, \gamma\}. \tag{17}$$

(here and in the following $z = n\delta + \zeta$.) Since equation (1) is invariant under the sign changes in α, β and γ we may always assume the above signs in front of γ. With Schlesinger transformations we can reach the parameter values

$$\alpha = \gamma + (M + \tfrac{1}{2})\delta, \quad \beta = \gamma - (N + \tfrac{1}{2})\delta, \tag{18}$$

where M and N are nonnegative integers, and we may also assume that $M \geq N$ because of the $\alpha \leftrightarrow \beta$ symmetry. This then defines the extent of the hierarchy. For particular solutions we give N, M as subscripts. Some further solutions obtained this way are (see also [14])

$$x_{10} = \left\{ \frac{-2z^2 + \delta(\gamma + \delta)}{z}, \gamma + \tfrac{3}{2}\delta, \gamma - \tfrac{1}{2}\delta, \gamma \right\},$$

$$x_{20} = \left\{ \frac{4z^3 + 2z\delta(-3\gamma - 5\delta)}{-2z^2 + \delta(\gamma + 2\delta)}, \gamma + \tfrac{5}{2}\delta, \gamma - \tfrac{1}{2}\delta, \gamma \right\},$$

$$x_{30} = \left\{ \frac{4z^4 + 4z^2\delta(-3\gamma - 7\delta) + 3\delta^2(\gamma^2 + 5\gamma\delta + 6\delta^2)}{-2z^3 + z\delta(3\gamma + 8\delta)}, \gamma + \tfrac{7}{2}\delta, \gamma - \tfrac{1}{2}\delta, \gamma \right\},$$

$$x_{31} = \left\{ \frac{8z^5 - 16z^3\delta^2 + 2z\delta^2(3\gamma^2 + \delta^2)}{-4z^4 + 4z^2\delta^2 + \delta^2(\gamma^2 - \delta^2)}, \gamma + \tfrac{3}{2}\delta, \gamma - \tfrac{3}{2}\delta, \gamma \right\}.$$

The $-1/z$-hierarchy is not connected to the $-2z$ hierarchy by a Schlesinger transformation, but only by a Bäcklund transformation followed by a redefinition of γ. In any case the seed is

$$x_{10} = \left\{ \frac{-\delta(\gamma + \delta)}{z}, \gamma + \tfrac{3}{2}\delta, \gamma + \tfrac{1}{2}\delta \right\}, \tag{19}$$

and in general the parameter values in this hierarchy are

$$\alpha = \gamma + (M + \tfrac{1}{2})\delta, \quad \beta = \gamma + (N + \tfrac{1}{2})\delta, \tag{20}$$

with $0 \leq N < M$. Some further solutions are given by

$$x_{20} = \left\{ \frac{2z\delta(2\gamma + 3\delta)}{-2z^2 + \delta(\gamma + 2\delta)}, \gamma + \tfrac{5}{2}\delta, \gamma + \tfrac{1}{2}\delta, \gamma \right\},$$

$$x_{21} = \left\{ \frac{\delta(\gamma + 2\delta)(-2z^2 + \delta(\gamma + \delta))}{2z^3 + z\delta\gamma}, \gamma + \tfrac{5}{2}\delta, \gamma + \tfrac{3}{2}\delta, \gamma \right\},$$

	M=0	M=1	M=2	M=3	M=4	N	β
				0	$\frac{[6]}{[3][4]}$	3	$\gamma+\frac{7}{2}\delta$
$-1/z$			0	$\frac{[4]}{[3][2]}$	$\frac{[9]}{[4][6]}$	2	$\gamma+\frac{5}{2}\delta$
		0	$\frac{[2]}{[1][2]}$	$\frac{[5]}{[2][4]}$	$\frac{[8]}{[3][6]}$	1	$\gamma+\frac{3}{2}\delta$
	0	$\frac{[0]}{[0][1]}$	$\frac{[1]}{[0][2]}$	$\frac{[2]}{[0][3]}$	$\frac{[3]}{[0][4]}$	0	$\gamma+\frac{1}{2}\delta$

$$M = \quad 0 \qquad 1 \qquad 2 \qquad 3 \qquad 4 \qquad\qquad N \qquad \beta$$
$$\alpha = \; \gamma+\tfrac{1}{2}\delta \;\; \gamma+\tfrac{3}{2}\delta \;\; \gamma+\tfrac{5}{2}\delta \;\; \gamma+\tfrac{7}{2}\delta \;\; \gamma+\tfrac{9}{2}\delta$$

	M=0	M=1	M=2	M=3	M=4	N	β
	0	$\frac{[0]}{[0][1]}$	$\frac{[1]}{[0][2]}$	$\frac{[2]}{[0][3]}$	$\frac{[3]}{[0][4]}$	0	$\gamma-\frac{1}{2}\delta$
		$\frac{[3]}{[2][2]}$	$\frac{[6]}{[3][4]}$	$\frac{[9]}{[4][6]}$	$\frac{[12]}{[5][8]}$	1	$\gamma-\frac{3}{2}\delta$
$-2z$			$\frac{[11]}{[6][6]}$	$\frac{[16]}{[9][8]}$		2	$\gamma-\frac{5}{2}\delta$
				$\frac{[23]}{[12][12]}$		3	$\gamma-\frac{7}{2}\delta$

Figure 1: Observed degrees in n (in square brackets) of various solutions. The denominator always factorizes. For the $-2z$ hierarchy we have subtracted $-2z$ from each solution.

$$x_{30} = \left\{ \frac{3\delta(-\gamma-2\delta)(-2z^2+\delta(\gamma+3\delta))}{-2z^3+z\delta(3\gamma+8\delta)}, \gamma+\tfrac{7}{2}\delta, \gamma+\tfrac{1}{2}\delta, \gamma \right\},$$

$$x_{32} = \left\{ \frac{(-\gamma-3\delta)\delta(4z^4-4z^2\delta^2+3\delta^2(\gamma^2+2\gamma\delta+\delta^2))}{4z^5+4z^3\delta(2\gamma+\delta)+z\delta^2(3\gamma^2+4\gamma\delta+\delta^2)}, \gamma+\tfrac{7}{2}\delta, \gamma+\tfrac{5}{2}\delta, \gamma \right\}.$$

It is obvious from these examples (and from the equation) that there is an overall scaling invariance, and because of this we will in the following simplify expressions by scaling out δ by writing $\gamma = c\delta$, $z = n\delta$.

4   The elementary polynomials

A common property for rational and other special function solutions of the continuous case is that the denominator factorizes into two determinants. Thus after constructing a set of solutions (all computations were done using REDUCE [15]) we studied the factorization of their denominators, the results are given in Figure 1. From it we can see that for the $-2z$ hierarchy the denominators of x_{MN} factorize with factors of degree $N(M+1)$ and $(N+1)M$ in n, and for the $-1/z$ hierarchy the degrees are $N(M-N)$ and $(N+1)(M-N)$.

Next observe that when $N = 0$ one of the factors in the denominator is constant and the other grows *linearly* with M. (For this value $N = 0$ the denominators are the same for the $-2z$ and $-1/z$ hierarchies.) The denominator is interpreted as the product of an 0×0 and 1×1 matrix, and this suggests that the denominators can provide us the basic polynomials that are to be used as matrix elements. The first few polynomials obtained this way are (when normalized to be monic):

$$
\begin{aligned}
p_0(n, c) &:= 1, \\
p_1(n, c) &:= n, \\
p_2(n, c) &:= n^2 - \tfrac{1}{2}c - 1, \\
p_3(n, c) &:= n^3 + n(-\tfrac{3}{2}c - 4), \\
p_4(n, c) &:= n^4 + n^2(-3c - 10) + \tfrac{3}{4}c^2 + \tfrac{21}{4}c + 9, \\
p_5(n, c) &:= n^5 + 5n^3(-c - 4) + n(\tfrac{15}{4}c^2 + \tfrac{125}{4}c + 64), \\
p_6(n, c) &:= n^6 + n^4(-\tfrac{15}{2}c - 35) + n^2(\tfrac{45}{4}c^2 + \tfrac{435}{4}c + 259) \\
&\quad - \tfrac{15}{8}c^3 - \tfrac{225}{8}c^2 - \tfrac{555}{4}c - 225, \\
p_7(n, c) &:= n^7 + n^5(-\tfrac{21}{2}c - 56) + n^3(\tfrac{105}{4}c^2 + \tfrac{1155}{4}c + 784) \\
&\quad + n(-\tfrac{105}{8}c^3 - \tfrac{1785}{8}c^2 - \tfrac{2499}{2}c - 2304)
\end{aligned}
$$

One indication that we are on the right track is obtained when we observe that the $p_N(n, c)$ satisfy recursion relations:

$$
\begin{aligned}
p_{N+1}(n, c) &= n\, p_N(n, c) - \tfrac{N}{2}(c + N + 1)\, p_{N-1}(n, c + 1), \\
p_N(n + \tfrac{1}{2}, c) - p_N(n - \tfrac{1}{2}, c) &= N\, p_{N-1}(n, c + \tfrac{1}{2}).
\end{aligned}
\tag{21}
$$

As was noted in Sec. 1 the matrix elements of the corresponding continuous case are given in terms of Hermite polynomials. Now we note that in the continuous limit (which means $n = x/(\sqrt{2}\delta)$, $c = 1/\delta^2$, $\delta \to 0$) we get for

$$
H_N(x) := \lim_{\delta \to 0} (\sqrt{2}\delta)^N p_N(x/(\sqrt{2}\delta), 1/\delta^2),
\tag{22}
$$

the recursions relations of Hermite polynomials (9).

5   Matrix form of the denominators

The above indicates what the polynomial matrix entries should be, and determining the structure of these matrices in the denominator is the next problem.

For $N = 0$ the denominator was interpreted as the product of an 0×0 and 1×1 matrix. For $N = 1$ we can see from the table that one of the factors grows by 1 as M increases by 1 and the other factor increases by 2.

These factors were interpreted as 1×1 and 2×2 matrices, respectively. The 1×1 part for $N = 1$ was then found to be proportional to the *shifted* basic polynomials given above: for a given M this factor is $p_{M+1}(c-1)$ in the $-2z$ hierarchy and $p_{M+1}(c+1)$ in the $-1/z$ hierarchy. The necessity of shifts in c is an important new ingredient, and something that exists only in the discrete version.

The next problem was to find a suitable matrix structure for the factors of the denominator. Our working assumption was that for any M the denominator should be a product of a $N \times N$ and a $(N+1) \times (N+1)$ matrix. The degrees of these factors led us to try the determinantal structure of Hankel type

$$T_{L,K} := \begin{vmatrix} q_K & q_{K+1} & \cdots & q_{K+L-1} \\ q_{K+1} & q_{K+2} & \cdots & q_{K+L} \\ \vdots & \vdots & \ddots & \vdots \\ q_{K+L-1} & q_{K+L} & \cdots & q_{K+2L-2} \end{vmatrix},$$

because now if the entries q_K are non-monic polynomials of degree K then $deg(T_{L,K}) = L(K + L - 1)$.

A more detailed study with some trial and error revealed that q's are not simply proportional to the p's obtained before, but that different shifts in c are also needed for the matrix elements. Our final result is as follows:

For the $-2z$ hierarchy (for which $\alpha = \gamma + (M + \frac{1}{2})\delta$, $\beta = \gamma - (N + \frac{1}{2})\delta$ where $0 \le N \le M$) the denominator can be expressed as

$$den(x_{M,N}) = \tau_{N,M-N+2}(-1)\, \tau_{N+1,M-N}(0), \tag{23}$$

where

$$\tau_{L,K}(s) :=$$
$$\begin{vmatrix} q_K(s) & q_{K+1}(s) & \cdots & q_{K+L-1}(s) \\ q_{K+1}(s-1) & q_{K+2}(s-1) & \cdots & q_{K+L}(s-1) \\ \vdots & \vdots & \ddots & \vdots \\ q_{K+L-1}(s-L+1) & q_{K+L}(s-L+1) & \cdots & q_{K+2L-2}(s-L+1) \end{vmatrix}, \tag{24}$$

and

$$q_M(s) := (c+s)^M\, p_M(c+s)/M!\,. \tag{25}$$

For the $-1/z$ hierarchy ($\alpha = \gamma + (M + \frac{1}{2})\delta$, $\beta = \gamma + (N + \frac{1}{2})\delta$, where $0 \le N < M$) we found same matrix form, but with different degrees and shifts:

$$den(x_{M,N}) = \tau_{N,M-2N+1}(N)\, \tau_{N+1,M-2N}(N). \tag{26}$$

6   The numerator

Finding a determinantal form for the numerator was more difficult, because
it did not factorize. However this was expected, because usually it turns out
that the numerator is the sum of two products of τ-functions, c.f., (4,10,12).

Thus we tried to express the numerator as a sum of two products of two τ-
functions (24), with possible shifts not only in c, but also in n, corresponding
to the derivatives in the continuous cases. Furthermore it seemed reasonable
to assume that the sizes of the matrices were the same as in the denominators,
and that the other index of the τ-function depended linearly on M and N.
Figuring out the shifts in n required some more trial and error, but eventually
we arrived at the result that worked:

For the $-2z$ hierarchy we got

$$x(n; \gamma + (M + \tfrac{1}{2})\delta, \gamma - (N + \tfrac{1}{2})\delta, \gamma) \propto$$
$$\frac{\cosh(\tfrac{1}{2}D_n)\, \tau_{N,M-N+1}(-\tfrac{1}{2}) \cdot \tau_{N+1,M-N+1}(-\tfrac{1}{2})}{\tau_{N,M-N+2}(-1)\, \tau_{N+1,M-N}(0)}, \qquad (27)$$

and the corresponding result for the $-1/z$ hierarchy was

$$x(n; \gamma + (M + \tfrac{1}{2})\delta, \gamma + (N + \tfrac{1}{2})\delta, \gamma) \propto$$
$$\frac{\sinh(\tfrac{1}{2}D_n)\, \tau_{N,M-2N+1}(N - \tfrac{1}{2}) \cdot \tau_{N+1,M-2N}(N + \tfrac{1}{2})}{\tau_{N,M-2N+1}(N)\, \tau_{N+1,M-2N}(N)}. \qquad (28)$$

In expressing the n-shifts we have used the usual Hirota bilinear derivative operator D_n. These results look a bit more symmetrical if we define
$\bar{\tau}_{N,A+B+N-1}(2B - N + 1) := \tau_{N,A}(B)$, then we get

$$x(n; \gamma + (M + \tfrac{1}{2})\delta, \gamma - (N + \tfrac{1}{2})\delta, \gamma) \propto$$
$$\frac{\cosh(\tfrac{1}{2}D_n)\, \bar{\tau}_{N,M-\tfrac{1}{2}}(-N) \cdot \bar{\tau}_{N+1,M+\tfrac{1}{2}}(-N - 1)}{\bar{\tau}_{N,M}(-N - 1)\, \bar{\tau}_{N+1,M}(-N)}, \qquad (29)$$

and

$$x(n; \gamma + (M + \tfrac{1}{2})\delta, \gamma + (N + \tfrac{1}{2})\delta, \gamma) \propto$$
$$\frac{\sinh(\tfrac{1}{2}D_n)\, \bar{\tau}_{N,M-\tfrac{1}{2}}(N) \cdot \bar{\tau}_{N+1,M+\tfrac{1}{2}}(N + 1)}{\bar{\tau}_{N,M}(N + 1)\, \bar{\tau}_{N+1,M}(N)}. \qquad (30)$$

7 Conclusions

We have shown here that the two hierarchies of rational solutions for d-P_{IV} (generated from $-2z$ and $-1/z$) can be expressed in terms of determinants, and that the matrix elements of these determinants are given by polynomials that can be regarded as discrete analogues of Hermite polynomials. It is interesting that in these expressions the parameter c, which vanishes in the continuous limit, plays an important role. However, the result is not yet complete, because the bilinear equation for the τ-function is still to be written, and the general proof must be given.

For the $-\frac{2}{3}z$ hierarchy things are an order of magnitude more difficult. There is no linear growth in any direction in the parameter space so there are no candidates for matrix elements. We have nevertheless found what the τ-functions should be, but no determinantal expression for them. Although no determinantal expression or basic polynomial is known even in the continuous case, we hope that similar structures in solutions of both continuous and discrete cases will be found.

Acknowledgments

One of the authors (K.K) was supported by the Grant-in-Aid for Encouragement of Young Scientists from The Ministry of Education, Science, Sports and Culture of Japan, No.08750090. This work was started when J.H. was visiting Doshisha University on an exchange grant between the Japanese Society for the Promotion of Science and the Academy of Finland.

References

[1] N. Lukashevich, Diff. Urav. **1**, 731 (1965). [Diff. Eqs. **1**, 561 (1965).]
N. Lukashevich, Diff. Urav. **3**, 771 (1967). [Diff. Eqs. **3**, 395 (1967).]
V.I. Gromak, Diff. Urav. **23**, 760 (1987). [Diff. Eqs. **23**, 506 (1987).]

[2] K. Okamoto, Math. Ann. **275**, 221 (1986).

[3] Y. Murata, Funkcial. Ekvac., **28**, 1 (1985).

[4] B. Grammaticos, A. Ramani and V. Papageorgiou, Phys. Rev. Lett. **67**, 1825 (1991).

[5] A. Ramani, B. Grammaticos and J. Hietarinta, Phys. Rev. Lett. **67**, 1829 (1991).

[6] A. Ramani and B. Grammaticos, Physica **A 228**, 160 (1996).

216   *5. Discrete Painlevé Equations*

[7] A. Ramani, B. Grammaticos and J. Satsuma: J. Phys. A: Math. Gen. **28**, 4655 (1995).

[8] A.P. Vorob'ev, Diff. Urav. **1**, 79 (1965). [Diff. Eqs. **1**, 58 (1965).]

[9] K. Kajiwara and Y. Ohta, J. Math. Phys. **37**, 4693 (1996).

[10] J. Satsuma, K. Kajiwara, B. Grammaticos, J. Hietarinta and A. Ramani, J. Phys. A: Math. Gen., **28** 3541 (1995).

[11] K. Kajiwara, K. Yamamoto and Y. Ohta, preprint `solv-int/9702001`.

[12] K. Kajiwara and Y. Ohta, unpublished.

[13] K.M. Tamizhmani, B. Grammaticos and A. Ramani, Lett. Math. Phys. **29**, 49 (1993).

[14] P. Clarkson and A. Bassom, preprint `solv-int/9409002`, `9412002`.

[15] A.C. Hearn, *REDUCE User's Manual*, v.3.6 Rand (1995).

The Discrete Painlevé II Equation and the Classical Special Functions

Kenji Kajiwara
Department of Electrical Engineering,
Doshisha University,
Tanabe, Kyoto 610-03, JAPAN
Email: kaji@elrond.doshisha.ac.jp

Abstract

Exact Solutions for the discrete Painlevé II Equation(dP_{II}) are constructed. It is shown that dP_{II} admits three kinds of exact solutions with determinant structure, namely, discrete Airy function solutions, rational solutions, and so-called "molecular type" solution. These solutions are expressed by classical special functions.

1 Introduction

The discrete Painlevé equations are now attracting much attention. One reason may be due to the importance of the six Painlevé equations in the continuous systems: their solutions play a role of special functions in the theory of nonlinear integrable systems.

We encounter various special functions when we reduce linear partial differential equations to ordinary differential equation by separation of variable. For example, when we solve the Helmholtz equation in polar coordinates, we separate the variable and we get the Bessel functions in radius. When we consider nonlinear integrable systems, simple separation of variables does not work indeed, but it is possible to reduce them to ordinary differential equations by considering traveling wave solutions or similarity solutions. In this case, it is believed that any reduced ordinary differential equations have so-called the "Painlevé property"[1] and usually we have one of the six types of Painlevé equations. We may expect that the discrete Painlevé equations plays a similar role in the discrete integrable systems. Here, natural questions arise: Is it possible to regard their solutions as the nonlinear versions of discrete analogue of special functions? What kind of solutions do we get for them?

The answer for the first question will be clear in some sense, if we look at the particular solutions of the discrete Painlevé equations. As for the second question, we have the class of exact solutions similar to the "classical solutions" for the continuous case, namely, rational solutions and special function type solutions. Moreover, we have another class of exact solutions which only discrete equations can have and collapse in the continuous limit. We call such solutions "molecular type".

In this contribution, we take the standard discrete Painlevé II equation (dP$_{II}$)[2],

$$x(n+1) + x(n-1) = \frac{(\alpha n + \beta)x(n) + \gamma}{1 - x(n)^2} \tag{1}$$

and discuss its exact solutions. It is shown that they are expressed by the special functions, which are, at the same time, regarded as the discrete analogue of other special functions.

2 Special Function Type Solutions: Hermite-Weber Functions

The Painlevé II equation(P$_{II}$),

$$\frac{d^2 w}{dt^2} = 2w^3 - 2tw + \alpha , \tag{2}$$

where α is a parameter, appears as the similarity reduction of the modified KdV equation[3],

$$v_{x_3} + \frac{3}{2}v^2 v_{x_1} - \frac{1}{4}v_{x_1 x_1 x_1} = 0 . \tag{3}$$

To obtain the simplest special function type solution to (2), it is convenient to consider the class of solutions which satisfies the Riccati equation,

$$\frac{dw}{dt} = a(x)w^2 + b(x)w + c(x) . \tag{4}$$

Substituting eq.(4) into eq.(2), one easily find that if w satisfies

$$\frac{dw}{dt} = -w^2 + t , \tag{5}$$

then w gives a particular solution to P$_{II}$ with $\alpha = 1$. Following to the standard procedure, eq.(5) is linearized by putting $w = (\log f)_t$ into

$$\frac{d^2 f}{dt^2} = tf , \tag{6}$$

which is nothing but the Airy equation. Thus, we obtain the simplest particular solution for $\alpha = 1$,

$$w = \frac{d}{dt} \log Ai \, , \qquad (7)$$

where Ai is the Airy function.

Crucial point is that we can extend this solution to "higher order" ones,
which are expressed by determinant as follows[4]:

Proposition 2.1 *Let* τ_N *be an* $N \times N$ *determinant defined by*

$$\tau_N = \begin{vmatrix} Ai & \frac{d}{dt}Ai & \cdots & \left(\frac{d}{dt}\right)^{N-1} Ai \\ \frac{d}{dt}Ai & \frac{d^2}{dt^2}Ai & \cdots & \left(\frac{d}{dt}\right)^{N} Ai \\ \vdots & \vdots & \ddots & \vdots \\ \left(\frac{d}{dt}\right)^{N-1} Ai & \left(\frac{d}{dt}\right)^{N} Ai & \cdots & \left(\frac{d}{dt}\right)^{2N-2} Ai \end{vmatrix} . \qquad (8)$$

Then,

$$u = \frac{d}{dt} \log \frac{\tau_{N+1}}{\tau_N}, \qquad (9)$$

satisfies P_{II} *(2) with* $\alpha = 2N + 1$.

The determinants which appear in the expression of the solution are called
the τ function.

Now let us construct the discrete analogue of the Airy-type solutions to
dP_{II}. We have the Hermite-Weber function for discrete analogue of the Airy
function as shown below.

Let us start from the discrete analogue of the Riccati equation,

$$x(n+1) = \frac{a(n)x(n) + b(n)}{c(n)x(n) + d(n)} \, , \qquad (10)$$

where $a(n), \cdots, d(n)$ are arbitrary functions in n. Substituting eq.(10) into
eq.(1), we find that if $x(n)$ satisfies,

$$x(n+1) = \frac{x(n) + (an + b)}{1 + x(n)}, \qquad (11)$$

then it gives a particular solution of dP_{II} with $\alpha = -2a$, $\beta = a - 2b + 2$,
$\gamma = a$. Putting $x(n) = F(n)/G(n)$, the above equation is linearized into

$$G(n+2) - 2G(n+1) + G(n) = (an + b)G(n) \, , \qquad (12)$$

$$x(n) = \frac{G(n+1)}{G(n)} - 1 . \tag{13}$$

It is clear that eqs.(12) and (13) can be regarded as the discrete analogue of the Airy equation and log derivative, respectively. Now compare eq.(12) with the contiguity relation of the Hermite-Weber function,

$$H_{n+2}(z) - zH_{n+1}(z) = -(n+1)H_n(z) . \tag{14}$$

Then we find that $G(n)$ can be expressed by the Hermite-Weber function as

$$G(n) = (-a)^{(n-1-(b-1)/a)/2} H_{n-1+(b-1)/a}\left(\frac{2}{(-a)^{1/2}}\right) . \tag{15}$$

Moreover, an easy calculation leads:

Proposition 2.2 *Let $H_n(z)$ be the Hermite-Weber function. Then,*

$$x(n) = \frac{2}{z}\frac{H_{n+1}(z)}{H_n(z)} - 1 , \tag{16}$$

satisfies dP_{II} (1) with $\alpha = \dfrac{8}{z^2}$, $\beta = \dfrac{4}{z^2}$, $\gamma = -\dfrac{4}{z^2}$.

It is possible to extend the above result to the "higher solutions"[7].

Theorem 2.3 *Let $\tau_N(n)$ be an $N \times N$ Casorati determinant defined by*

$$\tau_N(n) = \begin{vmatrix} H_n(z) & H_{n+1}(z) & \cdots & H_{n+N-1}(z) \\ H_{n+2}(z) & H_{n+3}(z) & \cdots & H_{n+N+1}(z) \\ \vdots & \vdots & \ddots & \vdots \\ H_{n+2N-2}(z) & H_{n+2N-1}(z) & \cdots & H_{n+3N-3}(z) \end{vmatrix} , \tag{17}$$

then

$$x(n) = \frac{2}{z}\frac{\tau_{N+1}(n+1)\tau_N(n)}{\tau_{N+1}(n)\tau_N(n+1)} - 1 , \tag{18}$$

satisfies dP_{II} with $\alpha = \dfrac{8}{z^2}$, $\beta = \dfrac{4(1+2N)}{z^2}$, $\gamma = -\dfrac{4(1+2N)}{z^2}$.

We remark that the simplest case(Proposition 2.2) is recovered by putting $N = 0$ and $\tau_0(n) = 1$ in Theorem 2.3.

Resemblance of continuous and discrete cases is remarkable, except that the τ function is not symmetric for discrete case while symmetric for continuous case. It is yet to understand the origin of such asymmetry in the discrete case.

Theorem 2.3 is derived from the following proposition:

Proposition 2.4 *The τ function satisfies the following bilinear difference equations.*

$$\tau_N(n+2)\tau_N(n-1) - \tau_N(n+1)\tau_N(n)$$
$$= \tau_{N+1}(n-1)\tau_{N-1}(n+2) , \tag{19}$$

$$\tau_{N+1}(n+2)\tau_N(n+1) - x\tau_{N+1}(n+1)\tau_N(n+1)$$
$$= -(n+1)\tau_{N+1}(n)\tau_N(n+3) , \tag{20}$$

$$\tau_{N+1}(n+1)\tau_{N-1}(n+2) + (n+2N-1)\tau_N(n+2)\tau_N(n+1)$$
$$= (n+1)\tau_N(n+3)\tau_N(n) . \tag{21}$$

The above bilinear difference equations are reduced to the Plücker relations, namely, quadratic identities of determinants. We refer [7] for proof.

It is interesting to point out that τ function which is similar to eq.(17) appears in the solution of continuous P_{IV}. In fact, we have the following solutions for P_{IV}[4].

Proposition 2.5 *Let τ_N^n be an $N \times N$ determinant defined by*

$$\tau_N^n = \begin{vmatrix} H_n(z) & H_{n+1}(z) & \cdots & H_{n+N-1}(z) \\ H_{n+1}(z) & H_{n+2}(z) & \cdots & H_{n+N}(z) \\ \vdots & \vdots & \ddots & \vdots \\ H_{n+N-1}(z) & H_{n+N}(z) & \cdots & H_{n+2N-2}(z) \end{vmatrix} . \tag{22}$$

Then,

$$q = \frac{\tau_{N+1}^{n+1}\tau_N^n}{\tau_{N+1}^n\tau_N^{n+1}} \tag{23}$$

satisfies P_{IV},

$$q_{zz} = \frac{1}{2q}q_z^2 + \frac{3}{2}q^3 - 2zq^2 + \frac{1}{2}(z^2 + 2(n-N))q - \frac{(n+N+1)^2}{2q} . \tag{24}$$

Compare this result with Theorem 2.3. The only difference with the solutions of dP_{II} is that the determinant structure. We do not know the meaning of this mysterious similarity between dP_{II} and P_{IV} yet.

Remark 2.6 *It is also possible to express q as log derivative of ratio of τ functions. In fact, Okamoto derived the expression in log derivative in ref.[4]. There are several equivalent expressions, which can be shown by using the bilinear equations satisfied by τ_N^n.*

3 Rational Solutions

It is known that P_{II} (2) admits another class of classical solutions, namely, rational solutions. It has been known that rational solutions are expressed by log derivative of ratio of some polynomials, which is called Yablonskii-Vorob'ev polynomials[8]. However, closer study reveals that rational solutions also have beautiful structure, as shown below.

Definition 3.1 *The Devisme polynomials of m variables $q_k(x_1, x_2, \cdots, x_m)$, $k = 0, 1, 2 \cdots$, are polynomials in $x_1, \cdots, x_m$ defined by*

$$\sum_{k=0}^{\infty} p_k(x_1, x_2, \cdots, x_m)\lambda^k = \exp\left[\sum_{n=1}^{m} x_n\lambda^n + \frac{1}{m+1}\lambda^{m+1}\right]. \tag{25}$$

We consider P_{II} in the following form:

$$\frac{d^2v}{dt^2} = 2v^3 - 4tv + 4\alpha, \tag{26}$$

Then we have the following representation for the rational solutions of P_{II}[6].

Theorem 3.2 *Let $p_k(t, s)$, $k = 0, 1, 2, \cdots$, be the Devisme polynomials and τ_N be an $N \times N$ determinant defined by*

$$\tau_N = \begin{vmatrix} p_N(t, s) & p_{N+1}(t, s) & \cdots & p_{2N-1}(t, s) \\ p_{N-2}(t, s) & p_{N-1}(t, s) & \cdots & p_{2N-3}(t, s) \\ \vdots & \vdots & \ddots & \vdots \\ p_{-N+2}(t, s) & p_{-N+3}(t, s) & \cdots & p_1(t, s) \end{vmatrix}, \quad p_k(t, s) = 0 \text{ for } k < 0. \tag{27}$$

Then

$$v = \frac{d}{dt} \log \frac{\tau_{N+1}}{\tau_N}, \tag{28}$$

satisfies P_{II} (26) with $\alpha = N + 1$.

Theorem 3.2 is a direct consequence of the following proposition.

Proposition 3.3 *The τ function satisfies the following bilinear equations.*

$$D_t^2\, \tau_N \cdot \tau_{N+1} = 0, \tag{29}$$

$$\left(D_t^3 + 4tD_t - 4(N+1)\right)\tau_N \cdot \tau_{N+1} = 0. \tag{30}$$

Remark 3.4

- *Bilinear equations (29) and (30) is reduced directly from the first two equations of the first modified KP hierarchy[6, 5].*

- *τ function itself does not depend on s.*

Now let us consider the discrete case. In this case, the generalized Laguerre polynomials appear as the entries of determinant instead of the Devisme polynomials.

Theorem 3.5 *Let $L_k^{(n)}(z)$, where $k = 0, 1, 2, \cdots$, be the generalized Laguerre polynomials defined by*

$$\sum_{k=0}^{\infty} L_k^{(n)}(z)\lambda^k = (1-\lambda)^{-1-n} \exp\frac{-z\lambda}{1-\lambda}, \quad L_k^{(n)}(z) = 0 \ (k < 0) , \quad (31)$$

and let $\tau_N(n)$ be an $N \times N$ determinant defined by

$$\tau_N(n) = \begin{vmatrix} L_N^{(n)} & L_{N+1}^{(n)} & \cdots & L_{2N-1}^{(n)} \\ L_{N-2}^{(n)} & L_{N-1}^{(n)} & \cdots & L_{2N-3}^{(n)} \\ \vdots & \vdots & \ddots & \vdots \\ L_{-N+2}^{(n)} & L_{-N+3}^{(n)} & \cdots & L_1^{(n)} \end{vmatrix} . \quad (32)$$

Then,

$$x(n) = \frac{\tau_{N+1}(n+1)\tau_N(n+1)}{\tau_{N+1}(n)\tau_N(n)} - 1 , \quad (33)$$

satisfies dP$_{\mathrm{II}}$ (1) with $\alpha = \dfrac{2}{z}$, $\beta = \dfrac{2}{z}$, $\gamma = -\dfrac{2}{z}(N+1)$.

Similar to the previous section, this result is derived from the following proposition.

Proposition 3.6 *The τ function satisfies the following bilinear difference equations.*

$$\tau_{N+1}(n+1)\tau_N(n-1) + \tau_{N+1}(n-1)\tau_N(n+1) = 2\tau_{N+1}(n)\tau_N(n) ,$$
$$\tau_N(n)\tau_N(n+1) - \tau_{N+1}(n+1)\tau_{N-1}(n) + \tau_{N+1}(n)\tau_{N-1}(n+1) = 0,$$
$$z\tau_N(n+2)\tau_N(n-1) - (n-N+1)\tau_N(n+1)\tau_N(n)$$
$$+(2N+1)\tau_{N+1}(n)\tau_{N-1}(n+1) = 0 .$$

The above three bilinear difference equations are reduced to the Plücker relations. However, proof is quite technical at present, which will be reported elsewhere.

Let us consider the continuous limit. We replace n by $n + z - 1$,

$$x(n+1) + x(n-1) = \frac{2}{z}\frac{(n+z)x(n) - (N+1)}{1 - x(n)^2} . \quad (34)$$

Putting

$$z = -\frac{1}{2\varepsilon^3}, \quad n = \frac{t}{\varepsilon}, \quad x(n) = \varepsilon v , \quad (35)$$

and take the limit $\varepsilon \to 0$, we get P$_{\text{II}}$,

$$\frac{d^2v}{dt^2} = 2v^3 - 4tv + (N + 1) \ . \tag{36}$$

However, the entries of τ function does not reduce to those of continuous case. To adjust them, we need some trick. We replace the entries of the τ function $\tau_N(n + z)$ by following polynomials,

$$\sum_{k=0}^{\infty} \hat{L}_k^{(n)}(z)\lambda^k = (1 - \lambda^2)^{-\frac{1}{2}z}(1 - \lambda)^{-n-z} \exp \frac{-z\lambda}{1 - \lambda},$$
$$\hat{L}_k^{(n)}(z) = 0 \ (k < 0). \tag{37}$$

This replacement does not change the τ function itself, since $\hat{L}_k^{(n)}(z)$ is a linear combination of $L_j^{(n+z)}(z)$, $j = k, k - 2, k - 4, \cdots$. Then, putting $z = -1/(2\varepsilon^3)$, $n = t/\varepsilon$, $\lambda = \varepsilon\eta$ and taking the limit $\varepsilon \to 0$ (note that $\hat{L}_k^{(n)}(z)$ is a polynomial of k-th degree in n), then the right hand side of eq.(37) reduces to $\exp(t\eta + \eta^3/3)$. Thus we have

$$\varepsilon^k \hat{L}_k^{(t/\varepsilon)}(-\frac{1}{2\varepsilon^3}) \to p_k(t, 0).$$

4   Molecular Type Solution: Modified Bessel Functions

There exists another type of solution, which we call "molecular type". This name comes from the semi-infinite or finite Toda lattice equation, which are sometimes called the "Toda molecule equation". The Toda molecule equation,

$$\frac{d^2}{dt^2} \log V_N = V_{N+1} + V_{N-1} - 2V_N, \quad n = 0, 1, 2, \cdots, \quad V_0 = 0 \ , \tag{38}$$

admits the solution given by

$$V_N = \frac{\tau_{N-1}\tau_{N+1}}{\tau_N^2} \ , \tag{39}$$

$$\tau_N = \begin{vmatrix} f & \frac{d}{dt}f & \cdots & \left(\frac{d}{dt}\right)^{N-1} f \\ \frac{d}{dt}f & \left(\frac{d}{dt}\right)^2 f & \cdots & \left(\frac{d}{dt}\right)^N f \\ \vdots & \vdots & \ddots & \vdots \\ \left(\frac{d}{dt}\right)^{N-1} f & \left(\frac{d}{dt}\right)^N f & \cdots & \left(\frac{d}{dt}\right)^{2N-2} f \end{vmatrix}, \tag{40}$$

where f is an arbitrary function in t. The crucial point is that the size of the determinant corresponds to the lattice site, while it corresponds to the number of solitons for normal soliton type solutions. Note that only the equation with discrete variables can have this type of solution, and it collapses in the continuous limit.

The molecular type solution for dP_{II} is given as follows:

Theorem 4.1 Let $I_n(z)$ be the modified Bessel function in z of n-th order satisfying

$$I_{n+1}(z) - I_{n-1}(z) = -\frac{2n}{z}I_n , \qquad (41)$$

and τ_N^n be a Toeplitz determinant given by

$$\tau_N^n = \begin{vmatrix} I_n & I_{n-1} & \cdots & I_{n-N+1} \\ I_{n+1} & I_n & \cdots & I_{n-N+2} \\ \vdots & \vdots & \ddots & \vdots \\ I_{n+N-1} & I_{n+N-2} & \cdots & I_n \end{vmatrix} , \qquad (42)$$

Then,

$$x(N) = \frac{\tau_N^1}{\tau_N^0} , \qquad (43)$$

satisfies dP_{II},

$$x(N+1) + x(N-1) = \frac{2}{z}\frac{Nx(N)}{1-x(N)^2} , \quad x(0) = 1 , \quad N \geq 0. \qquad (44)$$

Originally, dP_{II} was derived from the theory of matrix models by using the orthogonal polynomials technique [10]. This derivation immediately implies the above molecular type solution. It is also possible, however, to prove Theorem 4.1 by using the bilinear technique which was used in previous sections.

Proposition 4.2 $f_N = \tau_N^0$ and $g_N = \tau_N^1$ satisfy the following bilinear difference equations,

$$f_{N+1}g_{N-1} + f_{N-1}g_{N-1} - \frac{2N}{z}f_Ng_N = 0 , \qquad (45)$$

$$f_{N+1}f_{N-1} - f_N^2 = -g_N^2 . \qquad (46)$$

In ref.[11], it is shown that the τ function (42) satisfies the following bilinear difference equations,

$$\tau_{N+1}^{n+1}\tau_{N-1}^n + \tau_{N+1}^n\tau_{N-1}^{n-1} - \frac{2N}{z}\tau_N^n\tau_N^{n+1} = 0 , \qquad (47)$$

$$\tau_{N+1}^n\tau_{N-1}^n - (\tau_N^n)^2 = -\tau_N^{n+1}\tau_N^{n-1} . \qquad (48)$$

Putting $n = 0$ in eqs. (47) and (48) and noticing that

$$I_{-n} = I_n ,\qquad (49)$$

and hence

$$\tau_N^{-n} = \tau_N^n ,\qquad (50)$$

then we get eqs.(45) and (46) from eqs. (47) and (48), respectively.

We remark that the same τ function as eq.(42) appears as the that of P_{III}. In fact, we can show that[11, 12]

Proposition 4.3

$$u = -\frac{\tau_{N+1}^{n+1}\tau_N^{n}}{\tau_{N+1}^{n}\tau_N^{n+1}}\qquad (51)$$

satisfies P_{III},

$$\frac{d^2u}{dz^2} = \frac{1}{u}\left(\frac{du}{dz}\right)^2 + \frac{1}{z}\frac{du}{dz} - \frac{1}{z}\left((2N-2n)u^2 - (2n+2N+2)\right) - u^3 + \frac{1}{u} .\qquad (52)$$

Remark 4.4 *Similar to the case of P_{IV}, it is also possible to express u by the log derivative of ratio of the τ functions.*

This result immediately implies that

$$u = -\frac{x(N+1)}{x(N)} ,\qquad (53)$$

satisfies P_{III},

$$\frac{d^2u}{dz^2} = \frac{1}{u}\left(\frac{du}{dz}\right)^2 + \frac{1}{z}\frac{du}{dz} - \frac{1}{z}\left(2Nu^2 - (2N+2)\right) - u^3 + \frac{1}{u} .\qquad (54)$$

5   Concluding Remarks

In this contribution, we have presented three classes of exact solutions for dP_{II}. Namely, special function type solutions, rational solutions and molecular type solution. All of them admits determinant representation with good structures, and the first two solutions reduce to "classical solutions" of P_{II} in the continuous limit.

It is interesting that all of the above solutions are expressed in terms of special functions. This implies that there might be several connections between the discrete and continuous Painlevé equations. Moreover, we might expect that there is some object unifying them, and we are just looking at it from various aspects.

There are so many discrete analogue for each continuous Painlevé equations, and their solutions are not known for most of them except for the simplest ones. Studies from various aspects reveals that they have abundant mathematical structure. Why are there so many varieties and why do they have so rich mathematical structures? At present, nobody can answer this question. At least, these results tells us that the discrete Painlevé equations are worth further study. There are several tools and keys to understand the discrete Painlevé equations. The structure of the solutions would be one of the most useful keys for better understandings.

Acknowledgement

This work was supported by the Grant-in-Aid for Encouragement of Young Scientists from The Ministry of Education, Science, Sports and Culture of Japan, No.08750090.

References

[1] M. J. Ablowitz, A. Ramani and H. Segur, J. Math. Phys. **21** (1980) 715.

[2] A. Ramani, B. Grammaticos and J. Hietarinta, Phys. Rev. Lett. **67** (1991) 1829.

[3] M. J. Ablowitz and H. Segur, *Solitons and Inverse Scattering Transform* (SIAM, Philadelphia, 1981).

[4] K. Okamoto, Math. Ann. **275**(1986), 221.

[5] M. Jimbo and T. Miwa, Publ. RIMS, Kyoto Univ., **19**(1983) 943.

[6] K. Kajiwara and Y. Ohta, J. Math. Phys. **37**(1996) 4693.

[7] K. Kajiwara, Y. Ohta, J. Satsuma, B. Grammaticos and A. Ramani, J. Phys. A **27**(1994) 915.

[8] A. P. Vorob'ev, Diff. Eq. **1**(1965) 58.

[9] K. Kajiwara and Y. Ohta, unpublished.

[10] V. Periwal and D. Shevitz, Phys. Rev. Lett. **64**(1990) 135.

[11] F. Nijhoff, J. Satsuma, K. Kajiwara, B. Grammaticos, and A. Ramani, Inverse Problems **12**(1996) 697.

[12] K. Okamoto, Funkcialaj Ekvacioj, **30**(1987) 305.

Freud's equations for orthogonal polynomials as discrete Painlevé equations

Alphonse P. Magnus

Institut de Mathématique Pure et Appliquée,
Université Catholique de Louvain,
Chemin du Cyclotron,2,
B-1348 Louvain-la-Neuve, Belgium

Abstract

We consider orthogonal polynomials p_n with respect to an exponential weight function $w(x) = \exp(-P(x))$. The related equations for the recurrence coefficients have been explored by many people, starting essentially with Laguerre [49], in order to study special continued fractions, recurrence relations, and various asymptotic expansions (G. Freud's contribution [28, 56]).

Most striking example is $n = 2tw_n + w_n(w_{n+1} + w_n + w_{n-1})$ for the recurrence coefficients $p_{n+1} = xp_n - w_n p_{n-1}$ of the orthogonal polynomials related to the weight $w(x) = \exp(-4(tx^2 + x^4))$ (notations of [26, pp.34–36]). This example appears in practically all the references below. The connection with discrete Painlevé equations is described here.

1 Construction of orthogonal polynomials recurrence coefficients

Consider the set $\{p_n\}_0^\infty$ of orthonormal polynomials with respect to a weight function w on (a part of) the real line:

$$\int_{-\infty}^{\infty} p_n(x)p_m(x)w(x)\,dx = \delta_{n,m}, \qquad n,m = 0,1,\ldots \qquad (1)$$

The very useful *recurrence formula* has the form $a_1 p_1(x) = (x - b_0)p_0$,

$$a_{n+1}p_{n+1}(x) = (x - b_n)p_n(x) - a_n p_{n-1}(x), \quad n = 1,2,\ldots \qquad (2)$$

The connection between the nonnegative integrable function w and the real sequences $\{a_n\}_1^\infty, \{b_n\}_0^\infty$ is of the widest interest. It is investigated in combinatorics [9, 10, 52], asymptotic analysis [11, 12, 28, 30, 53, 54, 55, 61, 77, 78, 79], numerical analysis [15, 21, 31, 32, 51, 57], and, of course, Lax-Painlevé-Toda theory (all the other references, excepting *perhaps* Chaucer).

However, the connection may seem quite elementary and explicit:

1. From w to a_n and b_n: let $\mu_k = \int_{-\infty}^{\infty} x^k w(x)\,dx$, $k = 0, 1, \ldots$ be the moments of w, then

$$a_n^2 = \frac{H_{n-1}H_{n+1}}{(H_n)^2}, n = 1, 2, \ldots \qquad b_n = \frac{H_{n+1}}{H_{n+1}} - \frac{H_n}{H_n}, n = 0, 1, \ldots \quad (3)$$

where H_n and H_n are the determinants

$$H_n = \begin{vmatrix} \mu_0 & \cdots & \mu_{n-1} \\ \vdots & & \vdots \\ \mu_{n-1} & \cdots & \mu_{2n-2} \end{vmatrix}; H_n = \begin{vmatrix} \mu_0 & \cdots & \mu_{n-2} & \mu_n \\ \vdots & & \vdots & \vdots \\ \mu_{n-1} & \cdots & \mu_{2n-3} & \mu_{2n-1} \end{vmatrix}, \quad (4)$$

$n = 0, 1, \ldots, H_0 = 1, H_0 = 0$.

2. From the a_n's and b_n's to w: consider the Jacobi matrix

$$J = \begin{bmatrix} b_0 & a_1 & & \\ a_1 & b_1 & a_2 & \\ & a_2 & b_2 & a_3 \\ & & \ddots & \ddots & \ddots \end{bmatrix}$$

then, remarking that (2) may be written $xp_n(x) = a_n p_{n-1}(x) + b_n p_n(x) + a_{n+1}p_{n+1}(x)$, (Remark that the coefficients appear now with positive signs, a very elementary fact, but always able to puzzle weak souls), or $x\,p(x) = J p(x)$, where $p(x)$ is the column (infinite) vector of $p_0(x)$, $p_1(x), \ldots$, so that the expansion of $x^k p_n(x)$ in the basis $\{p_0, p_1, \ldots\}$ is $x^k p_n(x) = \sum_\ell (J^k)_{n,\ell} p_\ell(x)$. As we are dealing with a basis of orthonormal polynomials, this means that $(J^k)_{n,\ell}$ is the ℓth Fourier coefficient of $x^k p_n(x)$, (where row and column indexes start at zero), i.e.,

$$(J^k)_{n,\ell} = \int_{-\infty}^{\infty} x^k p_n(x)p_\ell(x)w(x)\,dx.$$

For any polynomial p, one has

$$(p(J))_{n,\ell} = \int_{-\infty}^{\infty} p(x)p_n(x)p_\ell(x)w(x)\,dx. \quad (5)$$

In particular, $(J^k)_{0,0} = \mu_k/\mu_0$.

Direct numerical use of (3) is almost always unsatisfactory, for stability reasons. Together with various ways to cope with the numerical stability problem [31, 32], compact formulas or equations for the recurrence coefficients a_n and b_n in special cases have been sought. Of course, quite a number of exact solutions are known (appendix of [14]) and new ones are steadily discovered (Askey-Wilson polynomials and other instances of the new "q–calculus" [4, 35, 36, 47]). The trend explored here is essentially related to weight functions satisfying simple differential equations. Main innovators were Laguerre in 1885 [49], Shohat in 1939 [77], and Freud in 1976 [28]. Their contributions are now described in reverse order. On Géza Freud (1922-1979), see [29, 68, 69].

2   Definition of Freud's equations

In order to reduce the technical contents of what will follow, only *even* weights: $\forall$ real x, $w(-x) = w(x)$ will be used. Then, the orthogonal polynomials are even or odd according to their degrees; equivalently, all the b_n's in (2), in (3), and in the Jacobi matrix J vanish. Let a be the sequence of coefficients $\{a_1, a_2, \dots\}$.

Theorem 1. *Let P be the real even polynomial $P(x) = c_0 x^{2m} + c_1 x^{2m-2} + \cdots + c_m$, with $c_0 > 0$, and let $a_1, a_2, \dots$ be the recurrence coefficients (2) of the orthonormal polynomials with respect to the weight $w(x) = \exp(-P(x))$ on the whole real line. Then, the* **Freud's equations**

$$F_n(a) := a_n \left(P'(J) \right)_{n,n-1} = n, \quad n = 1, 2, \dots \qquad (6)$$

hold, where $(\ )_{n,n-1}$ means the element at row n and column $n-1$ of the (infinite) matrix $P'(J)$. A finite number of computations is involved in each of these elements. The row and column indexes start at 0. We could as well take the indexes $(n-1, n)$, as J and any polynomial function of J are symmetric matrices.

   Remark. One has

$$F_1(a) = (JP'(J))_{0,0} . \qquad (7)$$

Indeed, the first element of the first row of $JP'(J)$ is a_1 times the element $(1, 0)$ of $P'(J)$, i.e., $F_1(a)$.

   Examples.

$$w(x) = \exp(-x^2) \quad F_n(a) = 2a_n^2, \qquad (8)$$
$$w(x) = \exp(-\alpha x^4 - \beta x^2) \quad F_n(a) = 4\alpha a_n^2(a_{n-1}^2 + a_n^2 + a_{n+1}^2) + 2\beta a_n^2, \qquad (9)$$
$$w(x) = \exp(-x^6) \quad F_n(a) = 6a_n^2(a_{n-2}^2 a_{n-1}^2 + a_{n-1}^4 + 2a_{n-1}^2 a_n^2 + a_n^4$$
$$+ a_{n-1}^2 a_{n+1}^2 + 2a_n^2 a_{n+1}^2 + a_{n+1}^4 + a_{n+1}^2 a_{n+2}^2), \qquad (10)$$

The elementary Hermite polynomials case is of course immediately recovered in (8). (9) and (10) were used by Freud [28] in investigations on asymptotic behaviour, see [11, 53, 56, 57, 67, 69] for more. The rich connection of (9) with discrete and continuous Painlevé theory will be recalled later on. For a much more general setting, see [1, Theor. 4.1].

Proof of (6). We consider two different ways to write the integral $I_n = \int_{-\infty}^{\infty} p_n'(x) p_{n-1}(x) \exp(-P(x)) \, dx$. First, let $p_{n-1}(x) = \pi_{n-1} x^{n-1} + \dots$ and $p_n(x) = \pi_n x^n + \dots$. From (2) (with $n-1$ instead of n), $\pi_n = \pi_{n-1}/a_n$; so, $p_n'(x) = n\pi_n x^{n-1} + \dots = \dfrac{n}{a_n} p_{n-1}(x) +$ a polynomial of degree $\leqslant n-2$. As p_{n-1} is orthogonal to all polynomials of degree $\leqslant n - 2$, what remains is $I_n = \dfrac{n}{a_n}$, as the orthonormal polynomials have a unit square integral. Next, we perform an integration by parts on $I_n = -\int_{-\infty}^{\infty} p_n(x) \left[p_{n-1}(x) \exp(-P(x)) \right]' \, dx = \int_{-\infty}^{\infty} P'(x) p_n(x) p_{n-1}(x) \exp(-P(x)) \, dx$, using the orthogonality of p_n and p_{n-1}' of degree $< n - 1 < n$, and the latter integral is $(P'(J))_{n,n-1}$, according to the spectral representation (5). $\qquad \square$

3 Freud's equations software

The interested reader (if there is still one left) will find in the "software" part of http://www.math.ucl.ac.be/~magnus/ three FORTRAN programs related to Freud's equations. **freud1.f** asks for an exponent m and puts data associated to the Freud's equations for $x^2, x^4, \dots, x^{2m}$ in the file **freud00m.dat**. **freud2.f** produces a readable (sort of) listing out of such a data file. Finally, **freud3.f** reads the coefficients of a polynomial P and computes in a stable way a sequence of positive recurrence coefficients $a_1, a_2, \dots$ associated to the weight $\exp(-P(x))$ on the whole real line. Actually, weight functions $|x|^\alpha \exp(-P(x))$, with P even, on the whole real line, and $x^\beta \exp(-Q(x))$ on the *positive* real line are considered simultaneously ($P(x) = Q(x^2)$).

4 Solution of Freud's equations

We started from the *solution* (3), and built the *equation* (6) afterwards, when the a_n's are recurrence coefficients of orthogonal polynomials associated to the weight $\exp(-P(x))$.

We try now to discuss the general solution of (6). With P even of degree $2m$, the nth equation of (6) involves $a_{n-m+1}, \dots, a_{n+m-1}$ (see [56, 57] or the examples (8) (9), (10)), so that the general solution depends on $2m - 2$ arbitrary constants, for instance $a_{n_0-m+1}, \ldots, a_{n_0+m-2}$. I shall only solve the case $a_{-m+1} = \cdots = a_0 = 0$, and investigate how the solution depends on the $m - 1$ initial data $a_1, \ldots, a_{m-1}$:

Theorem 2. *With $P(x) = c_0 x^{2m} + c_1 x^{2m-2} + \cdots + c_m$, the solution of (6) with $a_{-m+1} = \cdots = a_0 = 0$ is (3), with (4), where the μ_j's of (4) satisfy the linear recurrence relation of order $2m$*

$$2mc_0\mu_{2n+2m} + (2m - 2)c_1\mu_{2n+2m-2} + \cdots + 2c_{m-1}\mu_{2n+2} = (2n + 1)\mu_{2n}, \quad (11)$$

for $n = 0, 1, \ldots$, and $\mu_{2j+1} = 0$. $\mu_2, \mu_4, \ldots, \mu_{2m-2}$ are given by $\mu_{2j}/\mu_0 = (J^{2j})_{0,0}$.

The following lemma will be used:

Lemma 1. *Let $a^{(k)} = \{a_{n,k}\}_{n=1}^{\infty}$, $k = 0, 1, \ldots$, be the solutions of the* **quotient-difference** *equations*

$$\begin{aligned} a_{n-1,k+1}^2 + a_{n,k+1}^2 &= a_{n,k}^2 + a_{n+1,k}^2, && \text{if } n \text{ is odd,} \\ a_{n-1,k+1}a_{n,k+1} &= a_{n,k}a_{n+1,k}, && \text{if } n \text{ is even,} \end{aligned} \quad (12)$$

then, $F_n(a^{(k)})$ satisfies

$$\frac{a_{n-1,k+1}}{a_{n,k}}\left(F_n(a^{(k+1)}) - F_n(a^{(k)})\right) = \frac{a_{n,k}}{a_{n-1,k+1}}\left(F_{n+1}(a^{(k)}) - F_{n-1}(a^{(k+1)})\right),$$

$$n \text{ even,}$$

$$F_n(a^{(k+1)}) - F_n(a^{(k)}) = F_{n+1}(a^{(k)}) - F_{n-1}(a^{(k+1)}), \quad n \text{ odd.}$$
$$(13)$$

where $F_n(a^{(k)}) = a_{n,k}\left(P'(J_k)\right)_{n,n-1}$, and where J_k is the Jacobi matrix

$$J_k = \begin{bmatrix} 0 & a_{1,k} & & \\ a_{1,k} & 0 & a_{2,k} & \\ & a_{2,k} & 0 & a_{3,k} \\ & & \ddots & \ddots & \ddots \end{bmatrix}$$

The quotient-difference equations are well known in orthogonal polynomials identities investigations, see [74, 75] for recent examples. The notation chosen here is motivated by the interpretation of $a_{n,k}$ as recurrence coefficient of orthogonal polynomials related to the weight $x^{2k}w(x)$, but we do not need this interpretation in order to establish the theorem and the lemma. Actually, the theorem 2 is meant to recover an interpretation of the solutions of (6) in terms of orthogonal polynomials, but arbitrary initial data $a_1, a_2, \ldots, a_{m-1}$ will yield formal orthogonal polynomials through formal moments μ_j which, from (11), can still be written $\mu_j = \int_S x^j \exp(-P(x))\,dx$, but where the support S is now a system of arcs in the complex plane, see [66, "Laplace's method"], [62, 63].

Remark that, with $a_{n,0} = a_n$ and $a_{0,k} = 0$, all the $a_{n,k}$'s are completely determined by (12) as functions of $a_1, a_2, \ldots$

Proof of Lemma 1. The quotient-difference equations (12) can be written $J_k L_k = L_k J_{k+1}$, where L_k is the infinite lower triangular matrix

$$
L_k = \begin{bmatrix}
a_{1,k} & & & & \\
0 & a_{1,k+1} & & & \\
a_{2,k} & 0 & a_{3,k} & & \\
& a_{2,k+1} & 0 & a_{3,k+1} & \\
& & a_{4,k} & 0 & a_{5,k} \\
& & & \ddots & \ddots & \ddots
\end{bmatrix}
$$

One finds also $J_k^2 = L_k R_k$, $J_{k+1}^2 = R_k L_k$, where R_k is the transposed of L_k (the famous LR relations). So,

$$J_k^3 L_k = L_k R_k L_k J_{k+1} = L_k J_{k+1}^3, \ldots,$$
$$J_k^{2p+1} L_k = L_k R_k L_k J_{k+1}^{2p-1} = L_k J_{k+1}^{2p+1},$$

for any odd power, and so $\quad P'(J_k) L_k = L_k P'(J_{k+1})$, and we look at the element of indexes $(n, n-1)$:

$$
(P'(J_k))_{n,n-1}(L_k)_{n-1,n-1} + (P'(J_k))_{n,n+1}(L_k)_{n+1,n-1} = \\
(L_k)_{n,n-2}(P'(J_{k+1}))_{n-2,n-1} + (L_k)_{n,n}(P'(J_{k+1}))_{n,n-1},
$$

or $\quad F_n(a^k)\dfrac{(L_k)_{n-1,n-1}}{a_{n,k}} + F_{n+1}(a^k)\dfrac{(L_k)_{n+1,n-1}}{a_{n+1,k}} =$

$$
\dfrac{(L_k)_{n,n-2}}{a_{n-1,k+1}} F_{n-1}(a^{k+1}) + \dfrac{(L_k)_{n,n}}{a_{n,k+1}} F_n(a^{k+1}),
$$

leading to (13). $\qquad\square$

Proof of Theorem 2. From (13), with (6), $(a^{(0)} = a)$, and $F_0(a^{(k)}) = 0$ for all k, one finds by induction $F_n(a^{(k)}) = n + k(1 - (-1)^n)$, so, using (7), $F_1(a^{(k)}) = (J_k P'(J_k))_{0,0} = 2k + 1$, $k = 0, 1, \ldots$ Let $\mu_{2k} = $ constant $\times a_1^2 a_{1,1}^2 a_{1,2}^2 \cdots a_{1,k-1}^2$ and $\mu_{2k+1} = 0$. From the LR relations, $(J_k^{2p})_{0,0} = (L_k L_{k+1} \cdots L_{k+p} R_{k+p} \cdots R_{k+1} R_k)_{0,0} = a_{1,k}^2 a_{1,k+1}^2 \cdots a_{1,k+p-1}^2 = \mu_{2k+2p}/\mu_{2k}$. Finally, $2k + 1 = (J_k P'(J_k))_{0,0} = 2mc_0 (J_k^{2m})_{0,0} + (2m - 2)c_1 (J_k^{2m-2})_{0,0} +$

$\cdots + 2c_{m-1}(\boldsymbol{J}_k^2)_{0,0} = 2mc_0\mu_{2k+2m}/\mu_{2k} + (2m - 2)c_1\mu_{2k+2m-2}/\mu_{2k} + \cdots + 2c_{m-1}\mu_{2k+2}/\mu_{2k}$, whence (11), and the determinant ratios (3) follow from (12): $a_{n,k}^2 = H_{n-1,k}H_{n+1,k}/(H_{n,k})^2$ where $H_{n,k}$ is the Hankel determinant of rows $\mu_{2k}, \mu_{2k+1}(= 0), \ldots, \mu_{2k+n-1}; \mu_{2k+1}, \ldots, \mu_{2k+n}$, etc., up to $\mu_{2k+n-1}, \ldots, \mu_{2k+2n-2}$. $\qquad\square$

The *isomonodromy approach* looks at the linear differential equation satisfied by an orthogonal polynomial p_n [5, 16, 17, 37, 38] and produces isomonodromy identities of interest here. For instance, working (9) leads to $a_n^2 = (4\alpha)^{-1/2}\mathrm{P}_{IV}((4\alpha)^{-1/2}\beta; -n/2, -n^2/4)$ (Kitaev, [23, 24, 25, 26, 43]), where $\mathrm{P}_{IV}(t; A, B)$ is the solution of the fourth Painlevé equation

$$\ddot{y} = \frac{\dot{y}^2}{2y} + \frac{3y^3}{2} + 4ty^2 + 2(t^2 - A)y + \frac{B}{y}$$

which remains $O(t^{-1})$ when $t \to +\infty$ [60, pp. 231-232] (the importance and relevance of this latter condition is not yet quite clear). See [1, 2, 3, 16, 17, 18, 19, 20, 22, 23, 24, 25, 26, 39, 43, 44, 46] for more on such connections.

5 Freud's equations as discrete Painlevé equations

As far as I understand the matter, discrete Painlevé equations were first designed as clever discretisations of the genuine Painlevé equations, so to keep interesting features of these equations, especially integrability. For instance, (9) when $\beta \neq 0$ is a discretisation of the Painlevé-I equation [23, 24, 25, 26, 33, 34, 43]. Then, these features are examined on various discrete equations not necessarily associated to Painlevé equations. So, for instance, [33, p.350] encounter (9) with $\beta = 0$ as a discrete equation which is no more linked to the discretisation of a continuous Painlevé equation (and the authors of [33] call (9) a discrete Painlevé-0 equation when $\beta = 0$).

If we want to check that the Freud's equations are a valuable instance of discrete Painlevé equation, the following features are of interest, according to experts [48]:

5.1 Analyticity

Each a_{n+m-1}^2 is a rational function of preceding elements, so a meromorphic function of initial conditions.

5.2 Reversibility

There is no way to distinguish past and future, (6) is left unchanged when $a_{n+1}, a_{n+2}, \ldots$ are permuted with $a_{n-1}, a_{n-2}, \ldots$ Indeed ([56, p.369]), (6) is a sum of terms $a_n^2 a_{n+i_1}^2 a_{n+i_2}^2 \cdots a_{n+i_p}^2$, provided $0 \leqslant i_1 + 1 \leqslant i_2 + 2 \leqslant \cdots \leqslant i_p + p \leqslant p + 1$. Then, $\{-i_1, -i_2, \ldots, -i_p\}$ satisfies the same conditions and is therefore present too: $0 \leqslant -i_p + 1 \leqslant -i_{p-1} + 2 \leqslant \cdots \leqslant -i_1 + p \leqslant p + 1$.

5.3 Symmetry

An even stronger property is the following:

Theorem 3. *The matrix of derivatives* $\left(\dfrac{\partial F_n(a)}{\partial \log a_m} \right)_{n,m=1,2,\ldots}$ *is symmetric.*

Proof: see [56, 57].

5.4 Integrability

The existence of the formula (3) according to Theorem 2 shows how (6) can be solved. The similarity of (3) with known solutions of discrete Painlevé equations is also striking [42, 45]. The structure of the solution leads also to

5.5 Singularity confinement

The movable poles property of continuous equations is replaced by the following [34]: if initial values are such that some a_n, say a_{n_0} becomes very large, only a finite number of neighbours $a_{n_0+1}, \ldots, a_{n_0+p}$ should be liable to be very large too; moreover, a_{n_0+p+1}, etc. should be continuous functions of a_{n_0-1}, a_{n_0-2} etc. The formula (3) shows that the singularities of a_n are the zeros of H_n which are determinants built with solutions of (11), therefore continuous functions of the initial data.

6 Generalizations of Freud's equations: semi-classical orthogonal polynomials, etc.

Let the weight function satisfy the differential equation $w'/w = 2V/W$, where V and W are polynomials. The related orthogonal polynomials on a support S are then called *semi-classical* [6, 7, 8, 40, 41, 64, 65, 72] (classical: degrees of V and $W \leqslant 1$ and 2). One can still build equations for the recurrence coefficients by working out the integrals $I_n = \displaystyle\int_S W(x) p_n'(x) p_{n-1}(x) w(x)\, dx$, and $J_n = \int_S W(x)p'_n(x)\, p_n(x)w(x)\, dx$. Many equivalent forms can be found, and it not yet clear to know which one is the most convenient. If we manage to have $W(x)w(x) = 0$ at the endpoints of the support S, explicit formulas follow [77] (see also formula (4.5) of [1]). For instance, with $w(x) = \exp(-Ax^2)$ on $S = [-a, a]$ (see [15] for other equations), one takes $W(x) = x^2 - a^2$ and $V(x) = -Ax(x^2 - a^2)$, uses $p'_n = np_{n-1}/a_n + [2(a_1^2 + \cdots + a_{n-1}^2) - na_{n-1}^2]p_{n-3}/(a_{n-2}a_{n-1}a_n) + \cdots$ from (2), to find $2(a_1^2 + \cdots + a_{n-1}^2) + (2n + 1)a_n^2 - na^2 - 2Aa_n^2(a_{n-1}^2 + a_n^2 + a_{n+1}^2 - a) = 0$, or $(2n+1)x_{n+1} - (2n-1)x_n - na^2 - 2A(x_{n+1} - x_n)(x_{n+2} - x_{n-1} - a) = 0$, where $x_n = a_1^2 + \cdots + a_{n-1}^2$.

It is not clear how to recover reversibility and symmetry. For an even much more nasty case, see [61].

But the root of the matter lies even deeper, as already recognized by Laguerre in 1885 [49]! The point is to derive equations for the coefficients in the Jacobi continued fraction $f(z) = \cfrac{1}{z - b_0 - \cfrac{a_1^2}{z - b_1 - \cdots}}$ from a differential equation $Wf' = 2Vf + U$ with polynomial coefficients (same W and V as above: orthogonal polynomials lead to this continued fraction through $f(z) = \int_S (z-x)^{-1}w(x)\, dx$). See [30, 72] for this technique.

The continued fraction approach leads readily to the *Riccati* extension $Wf' = Rf^2 + 2Vf + U$ [20, 55], as each continued fraction $f_k(z) = \cfrac{1}{z - b_k - \cfrac{a_{k+1}^2}{z - b_{k+1} - \cdots}}$ is found to satisfy a Riccati equation $Wf'_k = -a_k^2\Theta_{k-1}f_k^2 + 2\Omega_k f_k - \Theta_k$ too, where Θ_k and Ω_k are polynomials of fixed degrees (i.e., independent. of k). Equations for the recurrence coefficients a_k's and b_k's follow from recurrence relations for the Θ_k's and Ω_k's (use $f_0(z) = f(z)$ and $f_k(z) = 1/[z - b_k - a_{k+1}^2 f_{k+1}(z)]$).

Finally, the differential operator can be replaced by a suitable difference operator, amounting to $W(z)\dfrac{f(\varphi_2(z)) - f(\varphi_1(z))}{\varphi_2(z) - \varphi_1(z)} = R(z)f(\varphi_1(z))f(\varphi_2(z)) + V(z)[f(\varphi_1(z)) + f(\varphi_2(z))] + U(z)$, where $\varphi_1(z)$ and $\varphi_2(z)$ are the two roots of a quadratic equation $A\varphi^2(z) + 2Bz\varphi(z) + Cz^2 + 2D\varphi(z) + 2Ez + F = 0$, see [36, 58].

Acknowledgements

Thanks to the Conference organizers, and to Darwin's College hospitality.

References

[1] M. Adler, P. van Moerbeke, Matrix integrals, Toda symmetries, Virasoro constraints, and orthogonal polynomials, *Duke Math. J.* **80** (1995) 863-911.

[2] A.I. Aptekarev, Asymptotic properties of polynomials orthogonal on a system of contours and periodic motions of Toda lattices, *Mat. Sb.* **125** (1984) 231-258, = *Math. USSR Sbornik* **53** (1986) 233-260.

[3] A.I. Aptekarev, A. Branquinho, F. Marcellán, Toda-type differential equations for the recurrence coefficients of orthogonal polynomials and Freud transformation, Pré-Publicações Dep. Matemática Univ. Coimbra, February 1996.

[4] N.M. Atakishiyev, These Proceedings.

[5] F.V. Atkinson, W.N. Everitt, Orthogonal polynomials which satisfy second order differential equations, pp. 173-181 *in E.B. Christoffel* (P.L. Butzer and F. Fehér, editors), Birkhäuser, Basel, 1981.

[6] S. Belmehdi, On semi-classical linear functionals of class $s = 1$. Classification and integral representations. *Indag. Mathem.* N.S. **3** (3) (1992), 253-275.

[7] S. Belmehdi, A. Ronveaux, Laguerre-Freud's equations for the recurrence coefficients of semi-classical orthogonal polynomials, *J. Approx. Theory* **76** (1994) 351-368.

[8] S. Belmehdi, A. Ronveaux, On the coefficients of the three-term recurrence relation satisfied by some orthogonal polynomials. *Innovative Methods in Numerical Analysis*, Bressanone, Sept. 7-11th, 1992.

[9] D. Bessis, A new method in the combinatorics of the topological expansion, *Comm. Math. Phys.* **69** (1979), 147-163.

[10] D. Bessis, C. Itzykson, J.B. Zuber, Quantum field theory techniques in graphical enumeration, *Adv. in Appl. Math.* **1** (1980), 109-157.

[11] S. Bonan, P. Nevai, Orthogonal polynomials and their derivatives,I, *J. Approx. Theory* **40** (1984), 134-147.

[12] S.S. Bonan, D.S. Lubinsky, P. Nevai, Orthogonal polynomials and their derivatives,II, *SIAM J. Math. An.* **18** (1987), 1163-1176.

[13] G. Chaucer, *The works of Geoffrey Chaucer* 2nd. ed. editor: F.N. Robinson. : Houghton Mifflin. Boston, Mass. 1957

[14] T. Chihara, *An Introduction to Orthogonal Polynomials*, Gordon & Breach, 1978.

[15] R.C.Y. Chin, A domain decomposition method for generating orthogonal polynomials for a Gaussian weight on a finite interval, *J. Comp. Phys.* **99** (1992) 321-336.

[16] D.V. Chudnovsky, Riemann monodromy problem, isomonodromy deformation eqations and completely integrable systems, pp.385-447 *in Bifurcation Phenomena in Mathematical Physics and Related Topics, Proceedings Cargèse, 1979* (C. Bardos & D. Bessis, editors), NATO ASI series C, vol. **54**, D.Reidel, Dordrecht, 1980.

[17] G.V. Chudnovsky, Padé approximation and the Riemann monodromy problem, pp.449-510 *in Bifurcation Phenomena in Mathematical Physics and Related Topics, Proceedings Cargèse, 1979* (C. Bardos & D. Bessis, editors), NATO ASI series C, vol. **54**, D.Reidel, Dordrecht, 1980.

[18] D.V. Chudnovsky, G.V. Chudnovsky, Laws of composition of Bäcklund transformations and the universal form of completely integrable systems in dimensions two and three, *Proc. Nat. Acad. Sci. USA* **80** (1983) 1774-1777.

[19] D.V. Chudnovsky, G.V. Chudnovsky, High precision computation of special function in different domains, talk given at Symbolic Mathematical Computation conference, Oberlech, July 1991.

[20] D.V. Chudnovsky, G.V. Chudnovsky, Explicit continued fractions and quantum gravity, *Acta Applicandæ Math.* **36** (1994) 167–185.

[21] A.S. Clarke, B. Shizgal, On the generation of orthogonal polynomials using asymptotic methods for recurrence coefficients. *J. Comp. Phys.* **104** (1993) 140-149.

[22] P. Deift, K. T-R McLaughlin, *A Continuum Limit of the Toda Lattice*, to appear in *Memoirs of the AMS*.

[23] A.S. Fokas, A.R. Its and A.V. Kitaev, Isomonodromic approach in the theory of two-dimensional quantum gravity, *Uspekhi Mat. Nauk* **45** (6) (276) (1990) 135–136 (in Russian).

[24] A.S. Fokas, A.R. Its, A.V. Kitaev, Discrete Painlevé equations and their appearance in quantum gravity, *Commun. Math. Phys.* **142** (1991) 313-344.

[25] A.S. Fokas, A.R. Its and A.V. Kitaev, The isomonodromy approach to matrix models in 2D quantum gravity, *Commun. Math. Phys.* **147** (1992) 395–430.

[26] A.S. Fokas, A.R. Its and Xin Zhou, Continuous and discrete Painlevé equations, pp. 33-47 *in* [50].

[27] G. Freud, *Orthogonale Polynome*, Birkhäuser, 1969 = *Orthogonal Polynomials*, Akadémiai Kiadó/Pergamon Press, Budapest/Oxford, 1971.

[28] G. Freud, On the coefficients in the recursion formulæ of orthogonal polynomials, *Proc. Royal Irish Acad. Sect. A* **76** (1976), 1-6.

[29] G. Freud, Publications of Géza Freud in approximation theory, *J. Approx. Theory* **46** (1986) 9-15.

[30] J.L. Gammel, J. Nuttall, Note on generalized Jacobi polynomials, *in* "The Riemann Problem, Complete Integrability and Arithmetic Applications" (D. Chudnovsky and G. Chudnovski, Eds.), pp.258-270, Springer-Verlag (Lecture Notes Math. **925**), Berlin, 1982.

[31] W. Gautschi, Computational aspects of orthogonal polynomials, pp.181-216 *in Orthogonal Polynomials: Theory and Practice* (P. Nevai, editor) *NATO ASI Series C* **294**, Kluwer, Dordrecht, 1990.

[32] W. Gautschi, Algorithm 726: ORTHPOL– A package of routines for generating orthogonal polynomials and Gauss-type quadratures, *ACM Trans. Math. Soft.* **20** (1994) 21-62.

[33] B. Grammaticos, B. Dorizzi, Integrable discrete systems and numerical integrators, *Math. and Computers in Simul.* **37** (1994) 341-352.

[34] B. Grammaticos, A. Ramani, V. Papageorgiou, Do integrable mappings have the Painlevé property? *Phys. Rev. Lett.* **67** (1991) 1825-1828.

[35] F.A. Grünbaum, L. Haine, The q−version of a theorem of Bochner, *J. Comp. Appl. Math.* **68** (1996) 103-114.

[36] F.A. Grünbaum, L. Haine, Some functions that generalize the Askey-Wilson polynomials, to appear in *Commun. Math. Phys.*

[37] W. Hahn, On differential equations for orthogonal polynomials, *Funk. Ekvacioj*, **21** (1978) 1-9.

[38] W. Hahn, Über Orthogonalpolynome, die linearen funktionalgleichungen genügen, pp. 16-35 in *Polynômes Orthogonaux et Applications, Proceedings, Bar-le-Duc 1984*, (C. Brezinski *et al.*, editors), *Lecture Notes Math.* **1171**, Springer, Berlin 1985.

[39] L. Haine, E. Horozov, Toda orbits of Laguerre polynomials and representations of the Virasoro algebra. *Bull. Sci. Math.* (2) **117** (1993) 485-518.

[40] E. Hendriksen, H. van Rossum, A Padé-type approach to non-classical orthogonal polynomials, *J. Math. An. Appl.* **106** (1985) 237-248.

[41] E. Hendriksen, H. van Rossum, Semi-classical orthogonal polynomials, pp. 354-361 in *Polynômes Orthogonaux et Applications, Proceedings, Bar-le-Duc 1984*, (C. Brezinski *et al.*, editors), *Lecture Notes Math.* **1171**, Springer, Berlin 1985.

[42] J. Hietarinta, Rational solutions to d$-$P$_{IV}$, these Proceedings.

[43] A.R. Its, these Proceedings.

[44] M. Kac, P. van Moerbeke, On an explicitly soluble system on nonlinear differential equations related to certain Toda lattices, *Adv. Math.* **16** (1975) 160-169.

[45] K. Kajiwara, Y. Ohta, J. Satsuma, B. Grammaticos, A. Ramani, Casorati determinant solutions for the discrete Painlevé-II equation, *J. Phys. A: Math. Gen.* **27** (1994) 915-922.

[46] Y. Kodama, K. T-R McLaughlin, Explicit integration of the full symmetric Toda hierarchy and the sorting property, *Letters in Math. Phys.* **37** (1996) 37-47.

[47] R. Koekoek, R.F. Swarttouw, The Askey-scheme of hypergeometric orthogonal polynomials and its q-analogue, Report 94-05, Delft Univ. of Technology, Faculty TWI, 1994. See also the "Askey-Wilson scheme project" entry in http://www.cs.vu.nl/~rene/ for errata and electronic access.

[48] M. Kruskal, Remarks on trying to unify the discrete with the continuous, these Proceedings.

[49] E. Laguerre, Sur la réduction en fractions continues d'une fraction qui satisfait à une équation différentielle linéaire du premier ordre dont les coefficients sont rationnels, *J. Math. Pures Appl. (4)* **1** (1885), 135-165 = pp. 685-711 in *Oeuvres*, Vol.II, Chelsea, New-York 1972.

[50] D. Levi, P. Winternitz, editors: *Painlevé Transcendents. Their Asymptotics and Physical Applications. NATO ASI Series: Series B: Physics* **278**, Plenum Press, N.Y., 1992.

[51] J.S. Lew, D.A. Quarles, Nonnegative solutions of a nonlinear recurrence, *J. Approx. Th.* **38** (1983), 357-379.

[52] J.-M. Liu, G. Müller, Infinite-temperature dynamics of the equivalent-neighbor XYZ model, *Phys. Rev. A* **42** (1990) 5854-5864.

[53] D.S. Lubinsky, A survey of general orthogonal polynomials for weights on finite and infinite intervals, *Acta Applicandæ Mathematicæ* **10** (1987) 237-296.

[54] D.S. Lubinsky, An update on general orthogonal polynomials and weighted approximation on the real line, *Acta Applicandæ Math.* **33** (1993) 121-164.

[55] A.P. Magnus, Riccati acceleration of Jacobi continued fractions and Laguerre-Hahn orthogonal polynomials, pp. 213-230 *in Padé Approximation and its Applications, Proceedings, Bad Honnef 1983*, Lecture Notes Math. **1071** (H. Werner & H.T. Bünger, editors), Springer-Verlag, Berlin, 1984.

[56] A.P. Magnus, A proof of Freud's conjecture about orthogonal polynomials related to $|x|^\rho \exp(-x^{2m})$ for integer m, pp. 362-372 *in Polynômes Orthogonaux et Applications, Proceedings Bar-le-Duc 1984*. (C. Brezinski *et al.*, editors), *Lecture Notes Math.* **1171**, Springer-Verlag, Berlin 1985.

[57] A.P. Magnus, On Freud's equations for exponential weights, *J. Approx. Th.* **46** (1986) 65-99.

[58] A.P. Magnus, Associated Askey-Wilson polynomials as Laguerre-Hahn orthogonal polynomials, pp. 261-278 *in Orthogonal Polynomials and their Applications, Proceedings Segovia 1986*. (M. Alfaro *et al.*, editors), *Lecture Notes Math.* **1329**, Springer-Verlag, Berlin 1988.

[59] A.P. Magnus, Refined asymptotics for Freud's recurrence coefficients, pp. 196-200 *in* V.P. Havin, N.K. Nikolski (Eds.): *Linear and Complex Analysis Book 3, Part II*, Springer Lecture Notes Math. **1574**, Springer, 1994.

[60] A.P. Magnus: Painlevé-type differential equations for the recurrence coefficients of semi-classical orthogonal polynomials. *J. Comp. Appl. Math.* **57** (1995) 215-237.

[61] A.P. Magnus Asymptotics for the simplest generalized Jacobi polynomials recurrence coefficients from Freud's equations: numerical explorations, *Annals of Numerical Mathematics* **2** (1995) 311-325.

[62] F. Marcellán, I.A. Rocha, On semiclassical linear functionals: integral representations, *J. Comp. Appl. Math.* **57** (1995) 239-249.

[63] F. Marcellán, I.A. Rocha, Complex path integral representation for semiclassical linear functionals, preprint.

[64] P. Maroni, Le calcul des formes linéaires et les polynômes orthogonaux semi-classiques, pp. 279-290 *in Orthogonal Polynomials and their Applications. Proceedings, Segovia 1986* (M. Alfaro *et al.*, editors), *Lecture Notes Math.* **1329**, Springer, Berlin 1988.

[65] P. Maroni, Une théorie algébrique des polynômes orthogonaux. Application aux polynômes orthogonaux semi-classiques. pp. 95-130 *in Orthogonal Polynomials and their Applications* (C. Brezinski *et al.*, editors), *IMACS Annals on Computing and Applied Mathematics* **9** (1991), Baltzer AG, Basel.

[66] L.M. Milne-Thomson, *The Calculus of Finite Differences*, Macmillan, London, 1933.

[67] P. Nevai, Two of my favorite ways of obtaining asymptotics for orthogonal polynomials, pp.417-436 *in Anniversary Volume on Approximation Theory and Functional Analysis*, (P.L. Butzer, R.L. Stens and B.Sz.-Nagy, editors), **ISNM 65**, Birkhäuser Verlag, Basel, 1984.

[68] P. Nevai, Letter to a friend, *J. Approx. Theory*, **46** (1986) 5-8.

[69] P. Nevai, Géza Freud, orthogonal polynomials and Christoffel functions. A case study, *J. Approx. Theory* **48** (1986), 3-167.

[70] P. Nevai, Research problems in orthogonal polynomials, pp. 449-489 *in Approximation Theory VI*, vol. **2** (C.K. Chui, L.L. Schumaker & J.D. Ward, editors), Academic Press, 1989.

[71] P. Nevai, Introduction to chapter 13 (Orthogonal polynomials) of V.P. Havin, N.K. Nikolski (Eds.), *Linear and Complex Analysis Problem Book 3, Part II, Springer Lecture Notes Math.* **1574**, Springer, 1994.

[72] J. Nuttall, Asymptotics of diagonal Hermite-Padé polynomials, *J. Approx. Th.* **42** (1984) 299-386.

[73] E.P. O'Reilly, D. Weaire, On the asymptotic form of the recursion basis vectors for periodic Hamiltonians, *J. Phys. A: Math. Gen.* **17** (1984) 2389-2397.

[74] V. Papageorgiou, these Proceedings

[75] V. Papageorgiou, B. Grammaticos, A. Ramani, Orthogonal polynomial approach to discrete Lax pairs for initial boundary-value problems of the qd algorithm, *Lett. Math. Phys.* **34** (1995) 91-101.

[76] O. Perron, *Die Lehre von den Kettenbrüchen*, 2nd edition, Teubner, Leipzig, 1929 = Chelsea,

[77] J.A. Shohat, A differential equation for orthogonal polynomials, *Duke Math. J.* **5** (1939),401-417.

[78] W. Van Assche, *Asymptotics for Orthogonal Polynomials, Springer Lecture Notes Math.* **1265**, Springer, 1987.

[79] J. Wimp, Current trends in asymptotics: some problems and some solutions, *J. Comp. Appl. Math.* **35** (1991) 53-79.

Chapter 6

Symmetries of Difference Equations

An Approach to Master Symmetries of Lattice Equations

Benno Fuchssteiner[†] and Wen-Xiu Ma[‡]
[†] Universität-GH Paderborn,
D-33098 Paderborn, Germany
Email: benno@uni-paderborn.de
[‡] Department of Mathematics,
University of Manchester Institute of Science and Technology,
Manchester M60 1QD, UK
Email: wenxiuma@uni-paderborn.de

Abstract

An approach to master symmetries of lattice equations is proposed by the use of discrete zero curvature equation. Its key is to generate isospectral flows from the discrete spectral problem associated with a given lattice equation. A Volterra-type lattice hierarchy and the Toda lattice hierarchy are analyzed as two illustrative examples.

1 Introduction

Symmetries are one of important aspects of soliton theory. When any integrable character hasn't been found for a given equation, among the most efficient ways is to consider its symmetries. It is through symmetries that Russian scientists et al. classified many integrable equations including lattice equations [1] [2]. They gave some specific description for the integrability of nonlinear equations in terms of symmetries, and showed that if an equation possesses higher differential-difference degree symmetries, then it is subject to certain conditions, for example, the degree of its nonlinearity mustn't be too large, compared with its differential-difference degree. Usually an integrable equation in soliton theory is referred as to an equation possessing infinitely many symmetries [3] [4]. Moreover these symmetries form beautiful algebraic structures [3] [4].

The appearance of master symmetries [5] gives a common character for integrable differential equations both in $1+1$ dimensions and in $1+2$ dimensions, for example, the KdV equation and the KP equation. The resulting symmetries are sometimes called τ-symmetries [6] and constitute centreless Virasoro algebras together with time-independent symmetries [7]. Moreover this kind of τ-symmetries may be generated by use of Lax equations [8] or zero curvature equations [9]. In the case of lattice equations, there also exist some similar results. For instance, a lot of lattice equations have τ-symmetries and centreless Virasoro symmetry algebras [10]. So far, however, there has not been a systematic theory to construct this kind of τ-symmetries for lattice equations.

The purpose of this paper is to provide a procedure to generate those master symmetries for a given lattice hierarchy. The discrete zero curvature equation is our basic tool to give such a procedure. A Volterra-type lattice hierarchy and the Toda lattice hierarchy are chosen and analyzed as two illustrative examples, which have one dependent variable and two dependent variables, respectively.

The paper begins by discussing discrete zero curvature equations. It will then go on to establish an approach to master symmetries by using discrete zero curvature equations. The fourth section will give rise to applications of our approach to two concrete examples of lattice hierarchies. Finally, the fifth section provides some concluding remarks.

2 Discrete zero curvature equations

Let $u(n, t)$ be a function defined over $Z \times \mathbf{R}$, E be a shift operator: $(Eu)(n) = u(n+1)$, and $K^{(m)} = E^m K$, $m \in Z$, K being a vector function. Consider the discrete spectral problem

$$\begin{cases} E\phi = U\phi = U(u, \lambda)\phi, \\ \phi_t = V\phi = V(u, \lambda)\phi, \end{cases} \tag{2.1}$$

where U, V are the same order square matrix operators and λ is a spectral parameter. Its integrability condition is the following discrete zero curvature equation

$$\begin{aligned} U_t &= (EV)U - UV = ((E-1)V)U - UV + VU \\ &= ((E-1)V)U - [U, V] = (\Delta_+ V)U - [U, V]. \end{aligned} \tag{2.2}$$

Recall that the continuous zero curvature equation reads as

$$U_t = V_x - [U, V].$$

Therefore we see that there is a slight difference between two zero curvature equations. The Gateaux derivative of $X(u)$ at a direction $S(u)$ is defined by

$$X'[S] = \frac{d}{d\varepsilon}\bigg|_{\varepsilon=0} X(u + \varepsilon S). \tag{2.3}$$

Throughout the paper, we assume that the spectral operator U has the injective property, that is, if $U'[K] = 0$, then $K = 0$. Therefore the Gateaux derivative U' is an injective linear map. For example, in the case of Toda lattices, the spectral operator U reads as [11]

$$U = \begin{pmatrix} 0 & 1 \\ -v & \lambda - p \end{pmatrix}, \quad u = \begin{pmatrix} p \\ v \end{pmatrix}, \tag{2.4}$$

and thus we have

$$U'[K] = \begin{pmatrix} 0 & 0 \\ -K_2 & -K_1 \end{pmatrix} = 0 \Rightarrow K = \begin{pmatrix} K_1 \\ K_2 \end{pmatrix} = 0,$$

which means that the spectral operator of the Toda lattice hierarchy has the injective property. We also need another property of the spectral operator

$$\text{if } (EV)U - UV = U'[K] \text{ and } V|_{u=0} = 0, \text{ then } V = 0, \tag{2.5}$$

which is called the uniqueness property. Further we obtain $K = 0$ from (2.5) by the injective property. It may be shown that (2.4) share such a property, when V is Laurent-polynomial dependent on λ.

Note that $U_t = U'[u_t] + \lambda_t U_\lambda = U'[u_t] + f(\lambda)U_\lambda$ when $\lambda_t = f(\lambda)$, where $U_\lambda = \frac{\partial U}{\partial \lambda}$. If (V, K, f) satisfies a so-called key discrete zero curvature equation

$$(EV)U - UV = U'[K] + fU_\lambda, \tag{2.6}$$

then we can say that when $\lambda_t = f(\lambda)$,

$$u_t = K(u) \qquad \Longleftrightarrow \qquad U_t = (EV)U - UV.$$
$$\text{a discrete evolution equation} \qquad \text{a discrete zero curvature equation}$$

This result may be proved as follows.
Proof: $(\Rightarrow)$

$$U_t = U'[u_t] + \lambda_t U_\lambda = U'[K] + fU_\lambda = (EV)U - UV.$$

$(\Leftarrow)$

$$U'[K] + fU_\lambda = (EV)U - UV = U_t = U'[u_t] + \lambda_t U_\lambda$$
$$= U'[u_t] + fU_\lambda \Rightarrow U'[u_t - K] = 0 \Rightarrow u_t = K,$$

where the linearity of U' and the injective property of U are used in the last two steps. ∎

Definition 2.1 *A matrix operator V is called a Lax operator corresponding to the spectral operator U with a spectral evolution law $\lambda_t = f(\lambda)$ if a key discrete zero curvature equation (2.6) holds. Moreover V is called an isospectral Lax operator if $f = 0$ or a nonisospectral Lax operator if $f \neq 0$.*

The equation (2.6) exposes an essential relation between a discrete equation and its discrete zero curvature representation. It will play an important role in the context of our construction of master symmetries.

3   An approach to master symmetries

What are master symmetries? For a given evolution equation $u_t = K(u)$, where $K(u)$ does not depend explicitly on t, the definition of master symmetries is the following [5].

Definition 3.1 *A vector field $\rho(u)$ is called a master symmetry of $u_t = K(u)$ if $[K,[K,\rho]] = 0$, where the commutator of two vector fields is defined by*

$$[K,\rho] = K'[\rho] - \rho'[K]. \tag{3.1}$$

If $\rho(u)$ is a master symmetry of $u_t = K(u)$, then the vector field $\tau = t[K,\rho] + \rho$, depending explicitly on the time t, is a symmetry of $u_t = K(u)$, namely, to satisfy the linearized equation of $u_t = K(u)$:

$$\frac{d\tau}{dt} = K'[\tau], \quad i.e. \quad \frac{\partial \tau}{\partial t} = [K,\rho]. \tag{3.2}$$

Main Observation:

Nonisospectral flows with $\lambda_t = \lambda^{n+1} \Rightarrow$ master symmetries.

So for (K,V,f) and (S,W,g), we introduce a new product

$$[V,W] = V'[S] - W'[K] + [V,W] + gV_\lambda - fW_\lambda, \tag{3.3}$$

in order to discuss master symmetries. An algebraic structure for the key discrete zero curvature equation is shown in the following theorem.

Theorem 3.1 *If two discrete zero curvature equations*

$$(EV)U - UV = U'[K] + fU_\lambda, \tag{3.4}$$

$$(EW)U - UW = U'[S] + gU_\lambda, \tag{3.5}$$

hold, then we have the third discrete zero curvature equation

$$(E[V,W])U - U[V,W] = U'[T] + [f,g]U_\lambda, \tag{3.6}$$

where $T = [K,S]$ and $[f,g]$ is defined by

$$[f,g](\lambda) = f'(\lambda)g(\lambda) - f(\lambda)g'(\lambda). \tag{3.7}$$

Proof: The proof is an application of equalities (3.4), (3.5) and

$$(U'[K])'[S] - (U'[S])'[K] = U'[T], \ T = [K, S],$$

which is a similar result to one in the continuous case in [12] and may be immediately checked. We first observe that

$$(\text{Equ. } (3.4))'[S] - (\text{Equ. } (3.5))'[K] + g(\text{Equ. } (3.4))_\lambda - f(\text{Equ. } (3.5))_\lambda,$$

and then a direct calculation may give the equality (3.6). ∎

This theorem shows that a product equation $u_t = [K, S]$ has a discrete zero curvature representation

$$U_t = (E[V, W])U - U[V, W] \ \text{ with } \ \lambda_t = [f, g],$$

when two evolution equations $u_t = K(u)$ and $u_t = S(u)$ have the discrete zero curvature representations:

$$U_t = (EV)U - UV \ \text{ with } \ \lambda_t = f(\lambda), \ U_t = (EW)U - UW \ \text{ with } \ \lambda_t = g(\lambda), \tag{3.8}$$

receptively. According to the above theorem, we can also easily find that if an equation $u_t = K(u)$ is isospectral ($f = 0$), then the product equation $u_t = [K, S]$ for any vector field $S(u)$ is still isospectral because we have $[f, g] = [0, g] = 0$, where g is the evolution law corresponding $u_t = S(u)$ (see [13] for the continuous case).

Let us now assume that we already have a hierarchy of isospectral equations of the form

$$u_t = K_k = \Phi^k K_0, \ k \geq 0, \tag{3.9}$$

associated with a discrete spectral problem

$$E\phi = U\phi, \ \phi = (\phi_1, \cdots, \phi_r)^T. \tag{3.10}$$

Usually a discrete spectral problem $E\phi = U\phi$ yields a hereditary operator Φ (see [14] for instance), i.e. a square matrix operator to satisfy

$$\Phi^2[K, S] + \Phi[\Phi K, \Phi S] - \{\Phi[K, \Phi S] + \Phi[\Phi K, S]\} = 0$$

for any vector fields K, S.

In order to generate nonisospectral flows, we further introduce an operator equation of $\Omega(X)$:

$$(E\Omega(X))U - U\Omega(X) = U'[\Phi X] - \lambda U'[X]. \tag{3.11}$$

We call it a characteristic operator equation of $E\phi = U\phi$.

Theorem 3.2 *Let two matrices V_0, W_0 and two vector fields K_0, ρ_0 satisfy*

$$(EV_0)U - UV_0 = U'[K_0], \tag{3.12}$$

$$(EW_0)U - UW_0 = U'[\rho_0] + \lambda U_\lambda, \tag{3.13}$$

and $\Omega(X)$ be a solution to (3.11). If we define a hierarchy of new vector fields and two hierarchies of square matrix operators as follows

$$\rho_l = \Phi^l \rho_0, \ l \geq 1, \tag{3.14}$$

$$V_k = \lambda^k V_0 + \sum_{i=1}^{k} \lambda^{k-i} \Omega(K_{i-1}) , \ k \geq 1, \tag{3.15}$$

$$W_l = \lambda^l W_0 + \sum_{j=1}^{l} \lambda^{l-j} \Omega(\rho_{j-1}) , \ l \geq 1, \tag{3.16}$$

then the square matrix operators $V_k, W_l, k, l \geq 0$, satisfy

$$(EV_k)U - UV_k = U'[K_k], \ k \geq 0, \tag{3.17}$$

$$(EW_l)U - UW_l = U'[\rho_l] + \lambda^{l+1} U_\lambda, \ l \geq 0, \tag{3.18}$$

respectively. Therefore $u_t = K_k$ and $u_t = \rho_l$ possess the isospectral $(\lambda_t = 0)$ and nonisospectral $(\lambda_t = \lambda^{l+1})$ discrete zero curvature representations

$$U_t = (EV_k)U - UV_k, \ U_t = (EW_l)U - UW_l,$$

respectively.

Proof: We prove two equalities (3.17) and (3.18). We can compute that

$$(EV_k)U - UV_k$$

$$= \lambda^k[(EV_0)U - UV_0] + \sum_{i=1}^{k} \lambda^{k-i}[E\Omega(K_{i-1})U - U\Omega(K_{i-1})]$$

$$= \lambda^k U'[K_0] + \sum_{i=1}^{k} \lambda^{k-i}\{U'[\Phi K_{i-1}] - \lambda U'[K_{i-1}]\}$$

$$= \lambda^k U'[K_0] + \sum_{i=1}^{k} \lambda^{k-i}\{U'[K_i] - \lambda U'[K_{i-1}]\}$$

$$= U'[K_k], \ k \geq 1;$$

$$(EW_l)U - UW_l$$

$$= \lambda^l[(EW_0)U - UW_0] + \sum_{j=1}^{l} \lambda^{l-j}[E\Omega(\rho_{j-1})U - U\Omega(\rho_{j-1})]$$

$$= \lambda^l\{U'[\rho_0] + \lambda U_\lambda\} + \sum_{j=1}^{l} \lambda^{l-j}\{U'[\Phi \rho_{j-1}] - \lambda U'[\rho_{j-1}]\}$$

$$= \lambda^l \{U'[\rho_0] + \lambda U_\lambda\} + \sum_{j=1}^{l} \lambda^{l-j} \{U'[\rho_j] - \lambda U'[\rho_{j-1}]\}$$
$$= U'[\rho_l] + \lambda^{l+1} U_\lambda, \; l \geq 1.$$

Note that we have used the characteristic operator equation (3.11). The rest is obviously and the proof is therefore finished. ∎

This theorem gives rise to the structure of Lax operators associated with the isospectral and nonisospectral hierarchies. In fact, the theorem provides us with a method to construct an isospectral hierarchy and a nonisospectral hierarchy associated with a discrete spectral problem (3.10) by solving two initial key discrete zero curvature equations (3.12) and (3.13) and by solving a characteristic operator equation (3.11), if a hereditary operator is known. However, the hereditary operator Φ and the isospectral hierarchy (3.9) are often determined from the spectral problem (3.10) simultaneously. Therefore we obtain just a new nonisospectral hierarchy (3.14). A operator solution to (3.11) may be generated by changing the element K_k (or G_k) in the following equality

$$\Omega(K_k) = V_{k+1} - \lambda V_k, \qquad (3.19)$$

which may be checked through (3.15). The Lax operator matrices V_{k+1} and V_k are known, when the isospectral hierarchy has already been given. Therefore the whole process of construction of nonisospectral hierarchies becomes an easy task: finding ρ_0, W_0 to satisfy (3.13) and computing $V_{k+1} - \lambda V_k$.

The nonisospectral hierarchy (3.14) is exactly the required master symmetries. The reasons are that the product equations of the isospectral hierarchy and the nonisospectral hierarchy are still isospectral by Theorem 3.1, and the isospectral equations often commute with each other. Therefore it is because there exists a nonisospectral hierarchy that there exist master symmetries for lattice equations derived from a given spectral problem. In the next section, we will in detail explain our construction process by two concrete examples and establish their corresponding centreless Virasoro symmetry algebras.

4 Application to lattice hierarchies

We explain by two lattice hierarchies how to apply the method in the last section to construct master symmetries.

Example 4.1. *A Volterra-type lattice hierarchy.* Let us first consider the discrete spectral problem [15]:

$$E\phi = U\phi, \; U = \begin{pmatrix} 1 & u \\ \lambda^{-1} & 0 \end{pmatrix}, \; \phi = \begin{pmatrix} \phi_1 \\ \phi_2 \end{pmatrix}. \qquad (4.1)$$

The corresponding isospectral lattice hierarchy:

$$u_t = K_k = \Phi^k K_0 = u(a_{k+1}^{(1)} - a_{k+1}^{(-1)}), \ \ K_0 = u(u^{(-1)} - u^{(1)}), \ k \geq 0, \quad (4.2)$$

where the hereditary operator Φ is defined by

$$\Phi = u(1 + E^{-1})(-u^{(1)}E^2 + u)(E - 1)^{-1}u^{-1}.$$

The associated Lax operators are as follows

$$V_k = (\lambda^{k+1}V)_{\geq 1} + \begin{pmatrix} a_{k+1} & 0 \\ c_{k+1} & a_{k+1}^{(-1)} \end{pmatrix}, \ k \geq 0, \quad (4.3)$$

where $(P)_{\geq i}$ denotes the selection of the terms with degrees of λ no less than i. The matrix $V = \sum_{i \geq 0} V_{(i)}\lambda^{-i} = \sum_{i \geq 0} \begin{pmatrix} a_i & b_i \\ c_i & -a_i \end{pmatrix}\lambda^{-i}$ solves the stationary $(U_t = 0)$ zero curvature equation $(EV)U - UV = 0$, and we choose $a_0 = \frac{1}{2}, b_0 = u, c_0 = 0, a_1 = -u, b_1 = -(u^2 + uu^{(-1)}), c_1 = 1$ and require that $a_{i+1}|_{u=0} = b_{i+1}|_{u=0} = c_{i+1}|_{u=0} = 0, i \geq 1$. Actually we have

$$\begin{cases} a_{i+1}^{(1)} - a_{i+1} = -u^{(1)}c_{i+1}^{(2)} + uc_{i+1}, \ i \geq 1, \\ b_{i+1} = uc_{i+2}^{(1)}, \ i \geq 1 \\ c_{i+1} = a_i + a_i^{(-1)}, \ i \geq 1. \end{cases} \quad (4.4)$$

In particular, we can obtain

$$V_0 = \begin{pmatrix} \frac{1}{2}\lambda - u & \lambda u \\ 1 & -\frac{1}{2}\lambda - u^{(-1)} \end{pmatrix}.$$

Nonisospectral Hierarchy:
Step 1: To solve the nonisospectral ($\lambda_t = \lambda$) initial key discrete zero curvature equation (3.13) yields a pair of solutions:

$$\rho_0 = u, \ W_0 = \begin{pmatrix} \frac{1}{2} & 0 \\ 0 & -\frac{1}{2} \end{pmatrix}. \quad (4.5)$$

Step 2: We compute that

$$V_{k+1} - \lambda V_k$$

$$= (\lambda^{k+2}V)_{\geq 1} + \begin{pmatrix} a_{k+2} & 0 \\ c_{k+2} & a_{k+2}^{(-1)} \end{pmatrix} - \lambda(\lambda^{k+1}V)_{\geq 1} - \lambda\begin{pmatrix} a_{k+1} & 0 \\ c_{k+1} & a_{k+1}^{(-1)} \end{pmatrix}$$

$$= \begin{pmatrix} a_{k+2} & \lambda b_{k+1} \\ c_{k+2} & -\lambda(a_{k+1} + a_{k+1}^{(-1)}) + a_{k+2}^{(-1)} \end{pmatrix}.$$

On the other hand, by (4.2) and (4.4), we have

$$\begin{cases} a_{k+1} = (E - E^{-1})^{-1}u^{-1}X_k, \\ a_{k+2} = (E - E^{-1})^{-1}u^{-1}\Phi X_k, \\ c_{k+2} = a_{k+1} + a_{k+1}^{(-1)} = (1 + E^{-1})a_{k+1}, \\ b_{k+1} = uc_{k+2}^{(1)} = u(E + 1)a_{k+1}. \end{cases}$$

Now by interchanging the element X_k into X in the quantity $V_{k+1} - \lambda V_k$, we obtain a solution to the corresponding characteristic operator equation:

$$\Omega(X) = \begin{pmatrix} \Omega_{11}(X) & \Omega_{12}(X) \\ \Omega_{21}(X) & \Omega_{22}(X) \end{pmatrix}, \tag{4.6}$$

where $\Omega_{ij}(X)$, $i, j = 1, 2$, are given by

$$\Omega_{11}(X) = (E - 1)^{-1}(-u^{(1)}E^2 + u)(E - 1)^{-1}u^{-1}X$$
$$\Omega_{12}(X) = \lambda uE(E - 1)^{-1}u^{-1}X$$
$$\Omega_{21}(X) = (E - 1)^{-1}u^{-1}X$$
$$\Omega_{22}(X) = [-\lambda + E^{-1}(E - 1)^{-1}(-u^{(1)}E^2 + u)](E - 1)^{-1}u^{-1}X.$$

Therefore we obtain a hierarchy of nonisospectral discrete equations $u_t = \rho_l = \Phi^l \rho_0$, $l \geq 0$, by Theorem 3.2.

Symmetry Algebra:

Step 3: We make the following computation at $u = 0$:

$$K_k|_{u=0} = 0, \ \rho_l|_{u=0} = \Phi^l \rho_0|_{u=0} = 0, \ k, l \geq 0,$$

$$V_k|_{u=0} = \lambda^k \begin{pmatrix} \frac{1}{2}\lambda & 0 \\ 1 & -\frac{1}{2}\lambda \end{pmatrix}, \ k \geq 0,$$

$$W_l|_{u=0} = \lambda^l \begin{pmatrix} \frac{1}{2} & 0 \\ 0 & -\frac{1}{2} \end{pmatrix} + (1 - \delta_{l0})\lambda^{l-1} \begin{pmatrix} 0 & 0 \\ [n] & -\lambda[n] \end{pmatrix}, \ l \geq 0,$$

where δ_{l0} represents the Kronecker symbol, and $[n]$ denotes a multiplication operator $[n] : u \mapsto [n]u$, $([n]u)(n) = nu(n)$, involved in the construction of nonisospetral hierarchies. While computing $W_l|_{u=0}$, we need to note that $\Omega(\rho_0)|_{u=0} \neq 0$, but $\Omega(\rho_l)|_{u=0} = 0$, $l \geq 1$. Now we may find by the definition (3.3) of the product of two Lax operators that

$$\begin{cases} [V_k, V_l]|_{u=0} = 0, \ k, l \geq 0, \\ [V_k, W_l]|_{u=0} = (k + 1)V_{k+l}|_{u=0}, \ k, l \geq 0, \\ [W_k, W_l]|_{u=0} = (k - l)W_{k+l}|_{u=0}, \ k, l \geq 0. \end{cases} \tag{4.7}$$

Step 4: It is easy to prove that

$$[V_k, V_l], [V_k, W_l] - (k+1)V_{k+l}, [W_k, W_l] - (k-l)W_{k+l}, \ k, l \geq 0,$$

are all isospectral ($\lambda_t = 0$) Lax operators. For example, the spectral evolution laws of the Lax operators of the third kind are

$$[\lambda^{k+1}, \lambda^{l+1}] - (k-l)\lambda^{k+l+1} = (k+1)\lambda^k \lambda^{l+1} - (l+1)\lambda^{k+1}\lambda^l - (k-l)\lambda^{k+l+1} = 0.$$

Then by the uniqueness property of the spectral problem (4.1), we obtain a Lax operator algebra

$$\begin{cases} [V_k, V_l] = 0, \ k, l \geq 0, \\ [V_k, W_l] = (k+1)V_{k+l}, \ k, l \geq 0, \\ [W_k, W_l] = (k-l)W_{k+l}, \ k, l \geq 0. \end{cases} \tag{4.8}$$

Further, due to the injective property of U', we finally obtain a vector field algebra of the isospectral hierarchy and the nonisospectral hierarchy

$$\begin{cases} [K_k, K_l] = 0, \ k, l \geq 0, \\ [K_k, \rho_l] = (k+1)K_{k+l}, \ k, l \geq 0, \\ [\rho_k, \rho_l] = (k-l)\rho_{k+l}, \ k, l \geq 0. \end{cases} \tag{4.9}$$

This implies that ρ_l, $l \geq 0$, are all master symmetries of each equation $u_t = K_{k_0}$ of the isospectral hierarchy, and the symmetries

$$K_k, \ k \geq 0, \ \tau_l^{(k_0)} = t[K_{k_0}, \rho_l] + \rho_l = (k_0 + 1)tK_{k_0+l} + \rho_l, \ l \geq 0, \tag{4.10}$$

constitute a symmetry algebra of Virasoro type possessing the same commutator relations as (4.9).

Example 4.2: *The Toda lattice hierarchy.* Let us second consider the discrete spectral problem [11]:

$$E\phi = U\phi, \ U = \begin{pmatrix} 0 & 1 \\ -v & \lambda - p \end{pmatrix}, \ u = \begin{pmatrix} p \\ v \end{pmatrix}, \ \phi = \begin{pmatrix} \phi_1 \\ \phi_2 \end{pmatrix}, \tag{4.11}$$

which is equivalent to $(E + vE^{-1} + p)\psi = \lambda\psi$. The corresponding isospectral ($\lambda_t = 0$) integrable Toda lattice hierarchy reads as

$$u_t = K_k = \Phi^k K_0, \ K_0 = \begin{pmatrix} v - v^{(1)} \\ v(p - p^{(-1)}) \end{pmatrix}, \ k \geq 0, \tag{4.12}$$

where the hereditary operator Φ is given by

$$\Phi = \begin{pmatrix} p & (v^{(1)}E^2 - v)(E-1)^{-1}v^{-1} \\ v(E^{-1} + 1) & v(pE - p^{(-1)})(E-1)^{-1}v^{-1} \end{pmatrix}. \tag{4.13}$$

The first nonlinear discrete equation is exactly the Toda lattice [16]

$$\begin{cases} p_t(n) = v(n) - v(n+1), \\ v_t(n) = v(n)(p(n) - p(n-1)), \end{cases} \tag{4.14}$$

up to a transform of dependent variables. The corresponding Lax operators read as

$$V_k = (\lambda^{k+1}V)_{\geq 0} + \begin{pmatrix} b_{k+2} & 0 \\ 0 & 0 \end{pmatrix}, \quad k \geq 0, \tag{4.15}$$

Here $V = \sum_{i \geq 0} V_{(i)} \lambda^i = \sum_{i \geq 0} \begin{pmatrix} a_i & b_i \\ c_i & -a_i \end{pmatrix} \lambda^{-i}$ solves the stationary discrete zero curvature equation $(EV)U - UV = 0$, and we choose $a_0 = \frac{1}{2}$, $b_0 = 0$, $c_0 = 0$, $a_1 = 0$, $b_1 = -1$, $c_1 = -v$ and require that $a_{i+1}|_{u=0} = b_{i+1}|_{u=0} = c_{i+1}|u = 0 = 0$, $i \geq 1$. More precisely, we have

$$\begin{cases} a_{i+1}^{(1)} - a_{i+1} = p(a_i^{(1)} - a_i) + (vb_i - v^{(1)}b_i^{(2)}), \ i \geq 1, \\ b_{i+1}^{(1)} = pb_i^{(1)} - (a_i^{(1)} + a_i), \ i \geq 1, \\ c_{i+1} = -vb_{i+1}^{(1)}, \ i \geq 1. \end{cases} \tag{4.16}$$

Now it is easy to find an isospectral Lax operator

$$V_0 = \begin{pmatrix} \frac{1}{2}\lambda - p^{(-1)} & -1 \\ v & -\frac{1}{2}\lambda \end{pmatrix}.$$

Nonisospectral Hierarchy:
Step 1: To solve find the nonisospectral $(\lambda_t = \lambda)$ initial key discrete zero curvature equation (3.2) leads to a pair of solutions

$$\rho_0 = \begin{pmatrix} p \\ 2v \end{pmatrix}, \ W_0 = \begin{pmatrix} [n] - 1 & 0 \\ 0 & [n] \end{pmatrix},$$

$[n]$ still being a multiplication operator $[n] : u \mapsto [n]u$, $([n]u)(n) = nu(n)$.
Step 2: To compute $V_{k+1} - \lambda V_k$ leads to a solution to the corresponding characteristic operator equation:

$$\Omega(X) = \begin{pmatrix} \Omega_{11}(X) & \Omega_{12}(X) \\ \Omega_{21}(X) & \Omega_{22}(X) \end{pmatrix}, \ X = \begin{pmatrix} X_1 \\ X_2 \end{pmatrix},$$

where $\Omega_{ij}(X)$, $i, j = 1, 2$, are given by

$$\Omega_{11} = E^{-1}(E-1)^{-1}X_1 + (p^{(-1)} - \lambda)(E-1)^{-1}v^{-1}X_2,$$
$$\Omega_{12} = (E-1)^{-1}v^{-1}X_2,$$
$$\Omega_{21} = vE(E-1)^{-1}v^{-1}X_2,$$
$$\Omega_{22} = (E-1)^{-1}X_1.$$

At this stage, we obtain a hierarchy of nonisospectral discrete equations $u_t = \rho_l = \Phi^l \rho_0$, $l \geq 0$, by Theorem 3.2.

Symmetry Algebra:

Step 3: We make the following computation at $u = 0$:

$$K_k|_{u=0} = 0, \quad \rho_l|_{u=0} = \Phi^l \rho_0|_{u=0} = 0, \quad k, l \geq 0,$$

$$V_k|_{u=0} = \lambda^k \begin{pmatrix} \frac{1}{2}\lambda & -1 \\ 0 & -\frac{1}{2}\lambda \end{pmatrix}, \quad k \geq 0,$$

$$W_l|_{u=0} = \lambda^l \begin{pmatrix} [n] - 1 & 0 \\ 0 & [n] \end{pmatrix} + (1 - \delta_{l0}) \begin{pmatrix} -2\lambda[n] & 2[n] \\ 0 & 0 \end{pmatrix}, \quad l \geq 0.$$

Now we may similarly find by the product definition (3.3) of the product of two Lax operators that

$$\begin{cases} [V_k, V_l]|_{u=0} = 0, \quad k, l \geq 0, \\ [V_k, W_l]|_{u=0} = (k+1)V_{k+l}|_{u=0}, \quad k, l \geq 0, \\ [W_k, W_l]|_{u=0} = (k-l)W_{k+l}|_{u=0}, \quad k, l \geq 0. \end{cases}$$

Step 4: Because $[V_k, V_l]$, $[V_k, W_l] - (k+1)V_{k+l}$, $[W_k, W_l] - (k-l)W_{k+l}$, $k, l \geq 0$, are still isospectral ($\lambda_t = 0$) Lax operators, a Lax operator algebra may be similarly obtained by using the uniqueness property:

$$\begin{cases} [V_k, V_l] = 0, \quad k, l \geq 0, \\ [V_k, W_l] = (k+1)V_{k+l}, \quad k, l \geq 0, \qquad\qquad (4.17) \\ [W_k, W_l] = (k-l)W_{k+l}, \quad k, l \geq 0. \end{cases}$$

Further, through the injective property of U', a vector field algebra is yielded

$$\begin{cases} [K_k, K_l] = 0, \quad k, l \geq 0, \\ [K_k, \rho_l] = (k+1)K_{k+l}, \quad k, l \geq 0, \qquad\qquad (4.18) \\ [\rho_k, \rho_l] = (k-l)\rho_{k+l}, \quad k, l \geq 0. \end{cases}$$

This shows that the symmetries for $u_t = K_{k_0}$:

$$K_k, \ k \geq 0, \ \tau_l^{(k_0)} = t[K_{k_0}, \rho_l] + \rho_l = (k_0 + 1)tK_{k_0+l} + \rho_l, \ l \geq 0, \qquad (4.19)$$

constitute the same centreless Virasoro algebra as (4.18).

5   Concluding remarks

We introduced a simple procedure to construct master symmetries for isospectral lattice hierarchies associated with discrete spectral problems. The crucial points are an algebraic structure related to discrete zero curvature equations and the structure of Lax operators of isospectral and nonisospectral lattice hierarchies. The procedure may be divided into four steps. The first two steps yields the required nonisospectral lattice hierarchy, by solving an initial key nonisospectral discrete zero curvature equation, and by computing a difference $V_{k+1} - \lambda V_k$ between two isospectral Lax operators V_{k+1} and V_k. The second two steps yields the required centreless Virasoro symmetry algebra, by using an algebraic relation between two discrete zero curvature equations, and the uniqueness property and the injective property of the corresponding spectral operators. Two lattice hierarchies are shown as illustrative examples.

There is a common Virasoro algebraic structure for symmetry algebras of a Volterra-type lattice hierarchy and the Toda lattice hierarchy. In general, we have

$$[K_k, \rho_l] = (k + \gamma)K_{k+l}, \ \gamma = const., \quad (semi - direct \ product), \qquad (5.1)$$

but the other two equalities do not change:

$$\begin{cases} [K_k, K_l] = 0 & (Abelian \ algebra), \\ [\rho_k, \rho_l] = (k - l)\rho_{k+l} & (centreless \ Virasoro \ algebra). \end{cases} \qquad (5.2)$$

This is also a common characteristic for continuous integrable hierarchies [17]. Interestingly, this kind of centreless Virasoro algebra itself may also be used to generate variable-coefficient integrable equations which possess higher-degree polynomial time-dependent symmetries [18].

We point out that there are some other methods to construct master symmetries by directly searching for some suitable vector fields ρ_0 [10] [19] [20]. However the approach above takes full advantage of discrete zero curvature equations, and therefore it is more systematic and may be easily applied to other lattice hierarchies.

Acknowledgments

One of the authors (W. X. Ma) would like to thank the Alexander von Humboldt Foundation for financial support. He is also grateful to Prof. R. K. Bullough and Dr. P. J. Caudrey for their warm hospitality at UMIST and to Prof. W. I. Fushchych for his kind and helpful conversations.

References

[1] A. V. Mikhailov, A. B. Shabat and V. V. Sokolov, in: *What is Integrability?*, ed. V. E. Zakhalov (Springer Verlag, Berlin, 1991) p115–184.

[2] D. Levi and R. I. Yamilov, "Classification of evolutionary equations on the lattice. I. The general theory", solv-int/9511006.

[3] B. Fuchssteiner and A. S. Fokas, *Physica D*, 4(1981), 47–66.

[4] A. S. Fokas, *Studies in Appl. Math.*, 77(1987), 253–299.

[5] B. Fuchssteiner, *Prog. Theor. Phys.*, 70(1983), 1508–1522.

[6] W. X. Ma, *J. Phys. A: Math. Gen.*, 23(1990), 2707–2716.

[7] H. H. Chen and Y. C. Lee, in: *Advances in Nonlinear Waves* Vol. II, ed. L. Debnath, Research Notes in Mathematics 111 (Pitman, Boston, 1985) p233-239.

[8] W. X. Ma, *J. Phys. A: Math. Gen.*, 25(1992), L719-L726.

[9] W. X. Ma, *Phys. Lett. A*, 179(1993), 179–185; *Proceedings of the 21st International Conference on the Differential Geometry Methods in Theoretical Physics*, ed M. L. Ge (World Scientific, Singapore, 1993) p535-538.

[10] W. Oevel, B. Fuchssteiner and H. W. Zhang, *Prog. Theor. Phys.*, 81(1989), 294–308.

[11] G. Z. Tu, *J. Phys. A: Math. Gen.*, 23(1990), 3903–3922.

[12] W. X. Ma, *J. Phys. A: Math. Gen.*, 25(1992), 5329–5343.

[13] W. X. Ma, *J. Phys. A: Math. Gen.*, 26(1993), 2573–2582.

[14] A. S. Fokas and R. L. Anderson, *J. Math. Phys.*, 23(1982), 1066–1073.

[15] H. W. Zhang, G. Z. Tu, W. Oevel and B. Fuchssteiner, *J. Math. Phys.*, 32(1991), 1908–1918.

[16] M. Toda, *Theory of Nonlinear Lattice* (Springer-Verlag, New York, 1981).

[17] W. X. Ma, *J. Math. Phys.*, 33(1992), 2464–2476.

[18] W. X. Ma, R. K. Bullough and P. J. Caudrey, preprint, 1996.

[19] I. Y. Cherdantsev and R. I. Yamilov, Physica D 87, 140–144 (1995).

[20] B. Fuchssteiner, S. Ivanov and W. Wiwianka, Algorithmic determination of infinite dimensional symmetry groups for integrable systems in 1 + 1 dimensions, preprint, 1995.

Symmetries and Generalized Symmetries for Discrete Dynamical Systems

D. Levi

Dipartimento di Fisica – Universitá di Roma 3
and INFN-Sezione di Roma I,
Via della Vasca Navale 84,
00146 Roma, Italy
Email: levi@roma1.infn.it, levifis.uniroma3.it

1   Introduction

Nonlinear differential difference equations are always more important in applications. They enter as models for many biological chains, are encountered frequently in queuing problems and as discretisations of field theories. So, both as themselves and as approximations of continuous problems, they play a very important role in many fields of mathematics, physics, biology and engineering.

Not many tools are available to solve such kind of problems. If one is interested in obtaining just a few exact solutions a very important and sufficiently simple technique is provided by the group theoretical approach to the costruction of point simmetries. This approach is algorithmic and can be carried out by computer. In the case of differential difference equations it has been applied with success to a few classes of equations [1-5].

However, apart for a few exceptional cases the complete solution of nonlinear differential difference equations can be obtained only by numerical calculations or by going to the continuous limit when the lattice spacing vanishes and the system is approximated by a continuous nonlinear partial differential equation. Exceptional cases are those equations which, in a way or another, are either linearizable or integrable via the solution of an associated spectral problem on the lattice. In all such cases we can write down for those equations a denumerable set of exact solutions corresponding to symmetries of the nonlinear differential-difference equations. Such symmetries can be either depending just on the dependent field and independent variable and are denoted as point symmetries or can depend on the dependent field in various positions of the lattice and in this case we speak of generalized symmetries. Any differential difference equation can have point symmetries, but the existence of generalized symmetries is usually associated only to the integrable ones.

Few classes of integrable nonlinear differential difference equations are known [6-9] and are important for all kind of applications both as themselves and as starting point for perturbation analysis [10]. However not all cases of physical interest are covered and so it would be nice to be able to recognize if a given nonlinear differential-difference equation is integrable or not, so as it can be used as model of nonlinear systems on the lattice or as starting point of perturbation theory.

Integrable equations, be they differential, or differential-difference ones, tend to have large Lie point symmetry groups. For instance, the Korteweg-de Vries equation with variable coefficients was shown to have at most a four-dimensional Lie point symmetry group [11]. More important, it has a four-dimensional symmetry algebra precisely if it is equivalent, under point transformations, to the KdV equation itself.

In the same context, we mention that integrable equations involving 3 independent variables (like the Kadomtsev-Petviashvili, Davey-Stewartson, 3-wave equations, and others) have infinite dimensional Lie point symmetry algebras with a specific Kac-Moody-Virasoro structure [12-15]. Interestingly enough, this is also true when one of the 3 variables is discrete, as in the two-dimensional Toda lattice [1,2].

In short, nonlinear differential and differential-difference equations with a large group of point symmetries are prime candidates for being integrable, or having some of the attributes of integrability [16]. Generically the equations which have the largest group of point symmetries are the integrable ones.

In this paper we want to present results which show that the situation may be different in the discrete case: the classification of dynamical systems on the lattice seems to show that there are many nonlinear differential difference equations which possess a large group of point symmetries. Among them, for the sake of concreteness we will focus our attention on a class of equations which have a group of point symmetries of dimension four and include the Toda lattice among them. For this class we will apply the formal symmetry approach introduced by A.B. Shabat and collaborators in Ufa (see e.g. articles [7-9; 17-19]) to classify partial differential equations and differential difference equations which possess few generalized symmetries of a certain kind and prove that the request of existence of higher symmetries in this case, restrict the class of equations to the Toda lattice equation and its integrable generalizations. The class of nonlinear differential-difference equations we will consider in the following is given by

$$\Delta_n = u_{n,tt}(t) - F(n, t, u_{n-1}(t), \ u_n(t), \ u_{n+1}(t)) = 0, \tag{1.1}$$

where $u_n(t)$ is a complex dependent field expressed in terms of its independent variables, t varying over the real numbers and n varying over the integers. Eq.(1.1) is a differential functional relation which correlates the "time" evolution of a function calculated at the point n to its values in its nearest neighboring points $(n+1, \ n-1)$.

In Section 2 and 3 we will present a brief review of the classification of equations of the form (1.1) into conjugacy classes and of the Lie point symmetries for each conjugacy class. Conjugacy is considered under a group of "allowed transformations", preserving the form of (1.1), while possibly changing the function F. We restrict the allowed transformations to be fiber preserving, i.e. to have the form

$$u_n(t) = \Omega_n(\tilde{u}_n(\tilde{t}), t, g), \quad \tilde{t} = \tilde{t}(t, g), \quad \tilde{n} = n, \tag{1.2}$$

where Ω_n and $\tilde{t}$ are some locally smooth and monotonous (invertible) functions, and g represents the group parameters. These functions are such that $\tilde{u}_n(\tilde{t})$ satisfies an equation of the form (1.1) with F replaced by some function $\tilde{F}(n, \tilde{t}, \tilde{u}_k(\tilde{t}))$ $(k = n-1, n, n+1)$.

With this formulation, Lie symmetries of (1.1) are special cases of allowed transformations, namely those for which we have

$$\tilde{F}(n, \tilde{t}, \tilde{u}_k(\tilde{t})) = F(n, \tilde{t}, \tilde{u}_k(\tilde{t})). \tag{1.3}$$

We assume that the Lie algebra of the symmetry group is realized by vector fields of the form

$$\widehat{X} = \tau(t, u_n)\partial_t + \phi_n(t, u_n)\partial_{u_n}, \tag{1.4}$$

and request that the prolongation $pr\widehat{X}$ of $\widehat{X}$ should annihilate the equation on its solution set

$$pr\widehat{X}\Delta_n \big|_{\Delta_n=0} = 0. \tag{1.5}$$

The prolongation formula for differential difference equations was given earlier [1].

To decide on the integrability of the obtained equations we investigate the existance of generalized symmetries for the following class of nonlinear differential-difference equations:

$$u_{n,t}(t) = f_n(u_{n-1}(t), \ u_n(t), \ u_{n+1}(t)), \tag{1.6}$$

A peculiarity of the choice of (1.6) is the fact that the right hand side is not just a function, i.e. it is not the same for all points in the lattice, but for each point of the lattice one has an a priori different right hand side. By proper choices of the functions f_n (1.6) can be reduced to a system of k coupled differential difference equations for the k unknown u_n^k or to a system of dynamical equations on the lattice. In fact, for example, by imposing periodicity conditions on the dependent field in the lattice variables one is able to rewrite (1.6) as a coupled system of nonlinear differential difference equations. Let us assume that f_n and u_n are periodic functions of n of period k, i.e.

$$f_n(u_{n-1}(t),\ u_n(t),\ u_{n+1}(t)) = \sum_{j=0}^{k-1} P_{n-j}^k f^j(u_{n-1}(t),\ u_n(t),\ u_{n+1}(t))$$

$$u_n = \sum_{j=0}^{k-1} P_{n-j}^k u_m^j$$

where we have defined the projection operator P_n^k such that for any integer m such that $n = km + j$ with $0 \le j \le k - 1$ we have:

$$P_{km}^k = 1, \qquad P_{km+j}^k = 0, \qquad (j = 1, 2, ..., k - 1) \tag{1.7}$$

than (1.6) becomes the system:

$$u_{m,t}^0 = f^0(u_{m-1}^{k-1}(t), u_m^0(t), u_m^1(t))$$

$$u_{m,t}^1 = f^1(u_m^0(t), u_m^1(t), u_m^2(t))$$

$$\cdots \qquad \cdots\cdots$$

$$u_{m,t}^{k-1} = f^{k-1}(u_m^{k-2}(t), u_m^{k-1}(t), u_{m+1}^0(t)) \tag{1.8}$$

Of particular interest is the case of periodicity $k = 2$ when we have:

$$u_{n,t}^0 = f^0(u_{n-1}^1, u_n^0, u_n^1)$$

$$u_{n,t}^1 = f^1(u_n^0, u_n^1, u_{n+1}^0) \tag{1.9}$$

A subclass of (1.9) of particular relevance for its physical applications is given by dynamical systems on the lattice, i.e. equations of the type

$$\chi_{n,tt} = g(\chi_{n+1} - \chi_n, \chi_n - \chi_{n-1}) \tag{1.10}$$

Eq.(1.10) is obtained from (1.9) for $u_n^0 = \chi_{n,t}$, $u_n^1 = \chi_{n+1} - \chi_n$ by choosing $f^0 = g(u_n^1, u_{n-1}^1)$ and $f^1 = u_n^0 - u_{n+1}^0$. Than, by choosing

$$g(z, z') = e^z - e^{z'}$$

(1.6) reduces to the Toda lattice equation

$$\chi_{n,tt} = e^{\chi_{n+1}-\chi_n} - e^{\chi_n-\chi_{n-1}}. \tag{1.11}$$

In term of the projection operator (1.7), (1.11) can obviously also be written in polynomial form as:

$$u_{n,t} = (P_{n+1}^2 u_n + P_n^2)(u_{n+1} - u_{n-1}) \tag{1.12}$$

the polynomial Toda Lattice.

In Section 4 we will provide the general theorems on the conditions for the integrability of (1.6) and than we will present results on the formal symmetry approach for the class of equations which has a four dimensional symmetry group and which include the Toda lattice.

2 Formulation of the problem and equations with one-dimensional symmetry algebras

By implement the symmetry algorithm (1.5) for equation (1.1) [1,2] we get that the coefficients in the vector field (1.4) are given by:

$$\phi_n(t, u_n) = (\tfrac{1}{2}\dot{\tau}(t) + a_n)u_n + \beta_n(t), \quad \tau(t, u_n) = \tau(t), \quad \dot{a}_n = 0. \tag{2.1}$$

The constants a_n and functions $\tau(t)$, $\beta_n(t)$ satisfy the remaining determining equation

$$\tfrac{1}{2}\ddot{\tau}u_n + \ddot{\beta}_n + (a_n - \tfrac{3}{2}\dot{\tau})F - \tau F_t - \sum_\alpha [(\tfrac{1}{2}\dot{\tau} + a_\alpha)u_\alpha + \beta_\alpha]F_{u_\alpha} = 0. \tag{2.2}$$

Let us now determine the allowed transformation (1.2). Substituting (1.2) into (1.1) and requiring that the terms $(\tilde{u}_{n,\tilde{t}})^2$ and $\tilde{u}_{n,\tilde{t}}$ be absent we find that the allowed transformation (1.2) must be linear and satisfy

$$u_n(t) = \frac{A_n}{\sqrt{\tilde{t}_t}}\tilde{u}_n(\tilde{t}) + B_n(t), \quad \tilde{t} = \tilde{t}(t), \quad A_{n,t} = 0, \quad \tilde{t}_t \neq 0, \quad A_n \neq 0, \quad \tilde{n} = n. \tag{2.3}$$

Let us now assume that the interaction F is given and that it is invariant under a one parameter symmetry group, generated by the vector field (2.1) with coefficients satisfying (2.2). Let us now use the allowed transformation (2.3) to simplify the vector field $\widehat{X}$, i.e. transform it into a convenient "canonical" form. Once this is done, we insert the coefficients of the canonical vector field into the determining equation (2.2) and solve this equation for $F(n, t, u_k)$. This is easy to do, since we have a first order linear partial differential equation and we simply apply the method of characteristics.

We see that three different possibilities occur.

1. $\tau(t) \not\equiv 0$.

If we have $\tau(t) \neq 0$ (in some open neighbourhood), we choose $\tilde{t}(t)$ and $B_n(t)$ to satisfy:

$$\frac{d\tilde{t}}{dt} = [\tau(t)]^{-1}, \quad \tau\frac{dB_n}{dt} - (\tfrac{1}{2}\dot{\tau} + a_n)B_n - \beta_n = 0.$$

We obtain

$$A_{1,1}: \qquad \hat{X} = \partial_t + a_n u_n \partial_{u_n}, \tag{2.4}$$

and using (2.2) with $\tau = 1$, $\beta_n = 0$ we find

$$F(n, t, u_k) = f(n, \xi_k)e^{a_n t}, \quad \xi_k = u_k e^{-a_k t}, \qquad k = n-1, n, n+1. \tag{2.5}$$

In particular, for $a_n = 0$ we have invariance with respect to time translations: F does not depend on t.

2. $\tau(t) = 0$, $a_n \not\equiv 0$.

We choose $B_n = -\beta_n(t)/a_n$ and obtain

$$A_{1,2}: \qquad \hat{X} = a_n u_n \partial_n, \tag{2.6}$$

$$F(n, t, v_k) = u_n f(n, t, \xi_k), \quad \xi_k = u_k^{a_n} u_n^{-a_k}, \qquad k = n \pm 1. \tag{2.7}$$

The vector field (2.6) can be interpreted as generating site dependent (n dependent) dilations of the function u_n.

3. $\tau(t) = 0$, $a_n = 0$, $\beta_n(t) \neq 0$.

In this case we already have:

$$A_{1,3}: \qquad \hat{X} = \beta_n(t)\partial_{u_n}, \tag{2.8}$$

and obtain

$$F(n, t, u_k) = \frac{\ddot{\beta}_n}{\beta_n}u_n + f(n, t, \xi_k), \quad \xi_k = \beta_n(t)u_k - \beta_k(t)u_n, \qquad k = n \pm 1. \tag{2.9}$$

In this case allowed transformations provide the equivalence

$$\beta_n(t) \sim \beta_n(t)(\tilde{t}_t)^{-1/2}A_n^{-1}. \tag{2.10}$$

In particular, if $\beta_n(t)$ factorizes as a function of n and t, i.e. $\beta_n(t) = \mu_n h(t)$, $\dot{\mu}_n = 0$, we can transform $\beta_n(t)$ into $\beta_n(t) = 1$. In this case the vector field (2.8) corresponds to a translation of the dependent variable: $u_n \longrightarrow u_n + c$.

We see that the existence of a one-dimensional symmetry algebra imposes a certain restriction on the form of F. Instead of being an arbitrary function of four variables, it will involve a function of only three "symmetry" variables.

By allowed transformations it can be taken into one of three "standard" form (2.5), (2.7), or (2.9).

Starting from these results, assuming that F and one of the symmetry generators is already in standard form, one further restrict F by requiring the existence of a higher dimensional symmetry algebra.

In [20] one has proceeded into higher dimensional symmetry algebras "structurally". Thus at first one has found all interactions that allow abelian symmetry algebras (of any dimension N, but it turns out that in this case we have $N \leq 4$). Then proceeded to classify nilpotent (nonabelian) symmetry algebras. For these one has found that their dimensions satisfy $3 \leq N \leq 5$. The classification of abelian and nilpotent symmetry algebras is then used to find all solvable (nonabelian) Lie algebras. To do this one uses a known result, namely that a solvable Lie algebra L of dimension $\dim L = d$, has a (unique) nilradical $NR(L)$ of dimension $\dim NR(L) \geq d/2$. Finally, one construct all nonsolvable Lie algebras. These are either simple, or they have a nontrivial Levi decomposition [21,22] into a semidirect sum of a simple Lie algebra and a solvable one (the radical, i.e. the maximal solvable ideal, unique up to equivalence).

3   Summary of the classification

1. The results of the symmetry classification of the nonlinear differential difference equation (1.1) are summed up in the following Table.

$dimL$	A	N	SN	SA	NS	T
7	0	0	0	0	1	1
6	0	0	3	0	0	3
5	0	1	0	7	2	10
4	2	1	4	8	1	16
3	4	2	0	12	1	19
2	5	0	0	4	0	9
1	3	0	0	0	0	3

In the first row A, N, SN, SA, NS and T mean abelian, nilpotent, solvable with nonabelian nilradical, solvable with abelian nilradical, nonsolvable, and total respectively. In the second to fifth column we give the number of each type of symmetry algebra for each dimension $dimL$. The total number for each dimension is given in the last column.

We see that we have $dimL \leq 7$. For $dimL = 6$, or 7 the interactions F_n are completely specified (up to 1 or more functions of n). Several types of "symmetry variables" occur for the higher dimensional symmetry algebras. Let us denote them as follows

$$\xi = (\gamma_n - \gamma_{n+1})u_{n-1} + (\gamma_{n+1} - \gamma_{n-1})u_n + (\gamma_{n-1} - \gamma_n)u_{n+1},$$

$$\eta = \xi h(t),$$

$$\zeta = p_{n-1}(t)u_{n-1} + p_n(t)u_n + p_{n+1}u_{n+1},$$

$$\rho = \frac{u_{n+1} - u_n}{u_{n-1} - u_n},$$

$$\sigma = u_{n-1}^{q_{n-1}} u_n^{q_n} u_{n+1}^{q_{n+1}}, \tag{3.1}$$

where γ_n and q_n depend only on n, $h(t)$ is some (specific in each case) function of t and $p_n(t)$ is a (specific) function of n and t.

One writes down here just the interactions having symmetry algebras of dimension $5 \leq dimL \leq 7$; they read:

$$NS_{7,1} \quad F = c_n \xi^{-3} \tag{3.2}$$

$$SN_{6,1} \quad F = c_n \xi^p \tag{3.3a}$$

$$SN_{6,2} \quad F = c_n + (a + b\gamma_n)\ln\xi \tag{3.3b}$$

$$SN_{6,3} \quad F = c_n + e^{b_n \xi} \tag{3.4a}$$

$$SN_{6,4} \quad F = c_n \xi^{-1} \tag{3.4b}$$

$$SN_{6,5} \quad F = \frac{2k}{\gamma_{n+1} - \gamma_n} + f_n(\xi) \tag{3.4c}$$

$$SA_{5,1}, \ldots, SA_{5,7}: \quad F = \frac{\alpha(t) + \beta(t)\gamma_n}{\gamma_{n+1} - \gamma_n}(u_{n+1} - u_n) + g(t)f_n(\eta), \tag{3.5}$$

where the functions of α, β, g and h are different in each case [20]

$$NS_{5,1} \quad F = c_n \sigma \tag{3.6a}$$

$$SN_{5,2} \quad F = (u_{n+1} - u_n)^{-3} f_n(\rho). \tag{3.6b}$$

2. Among the interactions with four-dimensional symmetry algebras a case of particular interest is $SN_{4,3}$

$$F = \exp\left(\frac{u_{n+1} - u_n}{\varepsilon_{n+1}}\right) G_n \left(\frac{u_{n+1} - u_n}{\varepsilon_{n+1}} - \frac{u_n - u_{n-1}}{\varepsilon_n}\right). \tag{3.7}$$

which can be rewritten in the form (1.6)

$$v_{k,t} = P_{k+1}^2 e^{v_{k+1}} g_k(v_{k+1} - v_{k-1}) + P_k^2 \lambda_k(v_{k+1} - v_{k-1}) \qquad (3.8)$$

by setting:

$$v_{2n} = \frac{u_{n+1} - u_n}{\varepsilon_{n+1}}, \quad v_{2n-1} = u_{n,t}, \quad g_{2n-1} = G_n, \quad \lambda_{2n} = \varepsilon_{n+1}^{-1}. \qquad (3.9)$$

This case contains the well-known and integrable Toda lattice (1.11). Indeed, if we choose

$$\gamma_n = 1/\varepsilon_n = 2n, \quad G_n(\xi) = -1 + e^{1/2\xi}, \qquad (3.10)$$

we obtain for $u_n = \chi_n$ the equation (1.11).

Thus, the Toda lattice is not singled out by its Lie point symmetry group, at least not by symmetries generated by vector fields of the type (1.4). Instead, it comes in a family involving two arbitrary functions, γ_n and $G_n(\xi)$. Nor is the Toda lattice the one with the largest symmetry group of the considered type. This distinction goes to the interaction (3.2).

4   Construction of the classifying conditions

Given a nonlinear chain (1.6) we will say that the restricted function

$$g_n(u_{n+i}, \ldots, u_{n+j})$$

is a generalized (or higher) local symmetry of *order* i of our equation iff

$$u_{n,\tau} = g_n(u_{n+i}, \ldots, u_{n+j}) \qquad (4.1)$$

is compatible with (1.6), i.e. iff

$$\partial_t \partial_\tau(u_n) = \partial_\tau \partial_t(u_n). \qquad (4.2)$$

Explicitating condition (4.1) we get:

$$\partial_t g_n = \partial_\tau f_n = \frac{\partial f_n}{\partial u_{n+1}} u_{n+1,\tau} + \frac{\partial f_n}{\partial u_n} u_{n,\tau} + \frac{\partial f_n}{\partial u_{n-1}} u_{n-1,\tau}$$

$$= \left[\frac{\partial f_n}{\partial u_{n+1}} D + \frac{\partial f_n}{\partial u_n} + \frac{\partial f_n}{\partial u_{n-1}} D^{-1} \right] g_n = f_n^* g_n,$$

i.e.

$$(\partial_t - f_n^*) g_n = 0, \qquad (4.3)$$

where by f_n^* we mean the Frechet derivative of the function f_n.

Eq.(4.3) is an equation for g_n once the function f_n is given, an equation for the symmetries.

Given a symmetry we can construct a new symmetry by applying a recursive operator, i.e. an operator which transforms symmetries into symmetries. Given a symmetry g_n of (1.6) the operator

$$L_n = \sum_{j=-\infty}^{m} l_n^{(j)}(t) D^j \qquad (4.4)$$

will be a recursive operator for (1.6) if

$$\tilde{g}_n = L_n g_n \qquad (4.5)$$

is a new generalized symmetry associated to (1.6). Eq.(4.3) and Eq.(4.5)imply that

$$A(L_n) = L_{n,t} - [f_n^*, L_n] = 0. \qquad (4.6)$$

Moreover from (4.3) it follows that

$$A(g_n^*) = \partial_\tau(f_n^*). \qquad (4.7)$$

Eq.(4.7) implies that, as its rhs is an operator of the order 1, the highest terms in the lhs must be zero. We can define an *approximate* symmetry of *order i* and *length m*, i.e. an operator

$$G_n = \sum_{k=i-m+1}^{i} g_n^{(k)} D^k$$

such that the highest m terms of the operator

$$A(G_n) = \sum_{k=i-m}^{i+1} a_n^{(k)} D^k$$

are zeroes. Taking into account (4.7), we find that we must have $i - m + 2 > 1$ if the equation

$$A(G_n) = 0 \qquad (4.8)$$

is to be satisfied and G_n is to be an approximation to the Frechet derivative of g_n, the local generalized symmetry of our equation.

From this results we can derive the integrability conditions which can be stated in the following theorems:

Theorem 1. If (1.6) has a local generalized symmetry of order $i \geq 2$, then it must have a conservation law given by

$$\partial_t \log f_n^{(1)} = (D - 1) q_n^{(1)}, \qquad (4.9)$$

where $q_n^{(1)}$ is an RF.

Theorem 2. If the equation (1.6) has two generalized local symmetries of order i and $i+1$, with $i \geq 4$, then the following conservation laws must be true:

$$\partial_t p_n^{(k)} = (D-1)q_n^{(k)} \qquad (k = 1, 2, 3),$$

$$p_n^{(1)} = \log \frac{\partial f_n}{\partial u_{n+1}}, \qquad p_n^{(2)} = q_n^{(1)} + \frac{\partial f_n}{\partial u_n},$$

$$p_n^{(3)} = q_n^{(2)} + \tfrac{1}{2}(p_n^{(2)})^2 + \frac{\partial f_n}{\partial u_{n+1}} \frac{\partial f_{n+1}}{\partial u_n}, \qquad (4.10)$$

where $q_n^{(k)}$ ($k = 1, 2, 3$) are some RFs.

So, if (1.6) has local generalized symmetries of high enough order, we can construct a few conservation laws, which we will call *canonical conservation laws*, depending on the function at the rhs of (1.6). Than we can state the following theorem:

Theorem 3. If the chain (1.6) has a conservation law of order $N \geq 3$, and the condition (4.9) is satisfied, then the following conditions must take place:

$$r_n^{(k)} = (D-1)s_n^{(k)} \qquad (k = 1, 2) \qquad (4.11a)$$

with

$$r_n^{(1)} = \log[-f_n^{(1)}/f_n^{(-1)}], \qquad r_n^{(2)} = s_{n,t}^{(1)} + 2f_n^{(0)}, \qquad (4.11b)$$

where $s_n^{(k)}$ are RFs.

Now, using (4.9-4.11) we can classify chains of the form (3.8). We can than formulate the following theorem:

Theorem. A chain of the form (3.8) satisfies the classifying conditions (4.9-4.11) iff it is related by a point transformation of the form $\tilde{u}_n = \alpha_n u_n + \beta_n$ to one of two following chains:

$$u_{nt} = P_{n+1}^2(\exp u_{n+1} - \exp u_{n-1}) + P_n^2(u_{n+1} - u_{n-1}), \qquad (4.12A)$$

$$u_{nt} = P_{n+1}^2 \exp\left(\frac{u_{n+1} - u_{n-1}}{a_n}\right) + P_n^2 a_{n+1} a_{n-1}(u_{n+1} - u_{n-1}), \qquad (4.12B)$$

with $a_n = \alpha n + \beta$ where α and β are arbitrary constants.

For the prove, see [23]. In the case (4.12A)

$$\varepsilon_n = 1, \qquad G_n = 1 - \exp((u_n - u_{n-1}) - (u_{n+1} - u_n)),$$

and this is nothing but the Toda model (1.11) for the function u_n. The chain (4.12B) is a new example of integrable (and n-dependent) equation. In this case, the chain equation can be rewritten as, (setting, for semplicity, $c_n = a_{2n-1}$)

$$u_{n,tt} = \exp[c_{n+1}(u_{n+1} - u_n) - c_{n-1}(u_n - u_{n-1})].\tag{4.13}$$

It belongs to the class (3.7), as

$$G_n(\zeta_n) = \exp(\delta_n\zeta_n),\quad \delta_n = c_{n-1}\varepsilon_n,\quad c_{n+1}\varepsilon_{n+1} - c_{n-1}\varepsilon_n = 1.$$

As c_n is linear in n, (4.13) can be written as

$$u_{n,tt} = \exp(c_{n+1}u_{n+1} - 2c_nu_n + c_{n-1}u_{n-1}).$$

and by an obvious point transformation, we can remove the c_n and will have the potential Toda equation:

$$u_{n,tt} = \exp(u_{n+1} - 2u_n + u_{n-1})\tag{4.14}$$

which reduces to the Toda by the following transformation:

$$\tilde{u}_n = u_{n+1} - u_n.$$

This implies that (4.13) is completely integrable.

References

[1] D. Levi and P. Winternitz, Phys. Lett. A **152**, 335 (1991).

[2] D. Levi and P. Winternitz, J. Math. Phys. **34**, 3713 (l993).

[3] S. Maeda, Math. Jap. **25**, 405 (1980), **26**, 85 (1981), IMA J. Appl. Math. **38**, 129 (1987).

[4] G. Quispel, H. W. Capel, and R. Sahadevan, Phys. Lett. A **170**, 379 (1992).

[5] D. Levi, L. Vinet, and P. Winternitz, *Lie group formalism for difference equations*, J. Phys. A: Math. Gen. 30 (1997) 633-649.

[6] M.J. Ablowitz and J. Ladik, *Nonlinear differential-difference equations*, J. Math. Phys. 16 (1975) 598-603; *Nonlinear differential-difference equations and Fourier transform*, J. Math. Phys. 17 (1976) 1011-1018; *On the solution of a class of nonlinear partial difference equations*, Stud. Appl. Math. 55 (1976) 213; *On the solution of a class of nonlinear partial difference equations*, Stud. Appl. Math. 57 (1976) 1.

[7] R.I. Yamilov, *Classification of discrete evolution equations*, Uspekhi Mat. Nauk 38:6 (1983) 155-156 [in Russian].

[8] R.I. Yamilov, *Generalizations of the Toda model, and conservation laws*, Preprint, Institute of Mathematics, Ufa, 1989 [in Russian];

English version: R.I. Yamilov, *Classification of Toda type scalar lattices*, Proc. Int. Conf. NEEDS'92, World Scientific Publishing, 1993, p. 423-431.

[9] A.B. Shabat and R.I. Yamilov, *Symmetries of nonlinear chains*, Algebra i Analiz 2:2 (1990) 183-208 [in Russian]; English transl. in Leningrad Math. J. 2:2 (1991) 377-400.

[10] C. Claude, Yu. S. Kivshar, O. Kluth and K.H. Spatschek, *Moving localized modes in nonlinear lattices*, Phys. Rev. B47 (1993) 14228

[11] J. P. Gazeau and P. Winternitz, Phys. Lett. A **167**, 246 (1992), J. Math. Phys. 33, 4087 (1992).

[12] D. David, N. Kamran, D. Levi, and P. Winternitz, Phys. Rev. Lett. **55**, 2111 (1985), J. Math. Phys. **27**, 1225 (1986).

[13] B. Champagne and P. Winternitz, J. Math. Phys. **29**, 1 (1988).

[14] L. Martina and P. Winternitz, Ann. Phys. **196**, 231 (1989).

[15] A. Yu. Orlov and P. Winternitz, Preprint CRM-1936 (1994), and CRM-2149 (1994).

[16] F. Güngör, M. Sanielevici, and P. Winternitz, *On the integrability properties of variable coefficient Korteweg-de Vries equations*, Can. J. Phys. 74(1996) 676-684).

[17] V.V. Sokolov and A.B. Shabat, *Classification of integrable evolution equations*, Soviet Scientific Rev., Section C, Math. Phys. Rev. 4 (1984) 221-280.

[18] A.V. Mikhailov, A.B. Shabat, and R.I. Yamilov, *The symmetry approach to the classification of nonlinear equations. Complete lists of integrable systems*, Uspekhi Mat. Nauk 42:4 (1987) 3-53 [in Russian]; English transl. in Russian Math. Surveys 42:4 (1987) 1-63.

[19] A.V. Mikhailov, A.B. Shabat, and V.V. Sokolov, *The symmetry approach to classification of integrable equations*, in "What is Integrability?", Springer-Verlag (Springer Series in Nonlinear Dynamics), 1991, p. 115-184.

[20] D. Levi and P. Winternitz, *Symmetries of discrete dynamical systems*, J. Math. Phys. 37(1996) 5551-5576

[21] N. Jacobson, *Lie algebras*, Dover, New York, 1962.

[22] E. E. Levi, *Opere*, volume 1, Edizioni Cremonese, Roma, 1959.

[23] D. Levi and R. Yamilov, *Conditions for the existance of higher symmetries of evolutionary equations on the lattice*, submitted to J.M.P. (1996)

Nonlinear Difference Equations with Superposition Formulas

P. Winternitz

Centre de Recherches Mathématiques et
Département de Mathématiques et Statistique,
Université de Montréal, CP 6128,
succ. Centre-ville, Montréal, QC, Canada, H3C 3J7 *

Abstract

Lie group theory can be used to construct systems of nonlinear equations with superposition formulas. The superposition formula expresses the general solution in terms of a finite number of particular ones. We show how these equations can be discretized, while preserving the group structure and its consequences (the superposition formula, linearizability and the Painlevé property). The method is then applied to a second order equation, namely the Pinney equation, arising in many applications.

1   Introduction

Lie theory plays an important role in the theory of differential equations in many different guises. The most obvious one is that of invariance groups, i.e. groups of transformations leaving the solution set of a system invariant. This aspect of the application of group theory is currently being adapted to the treatment of difference, and differential-difference equations. [1],$\cdots$, [6].

A different application in that of Lie algebras underlying the existence of superposition formulas for certain systems of ordinary differential equations (ODEs). Indeed, S. Lie proved the following theorem [7].

Theorem 1.1 (Lie):
The general solution $\overrightarrow{y}(t, c_1, \cdots, c_n)$ *of a system of ODEs*

$$\dot{\overrightarrow{y}} = \overrightarrow{\eta}(\overrightarrow{y}, t), \qquad \overrightarrow{y} \in K^n, \ t \in K \tag{1.1}$$

---

*currently visiting The University of New South Wales, Sydney 2052, Australia

(with $K = \mathbb{C}$, or $K = \mathbb{R}$) can be expressed as a (nonlinear) superposition of m particular solutions

$$\vec{y}(t) = \vec{F}(\vec{y}_1, \cdots, \vec{y}_m, c_1, \cdots c_n) \qquad (1.2)$$

if and only if the system has the form

$$\dot{\vec{y}}(t) = \sum_{k=1}^{r} Z_k(t) \vec{\eta}_k(\vec{y}), \qquad (1.3)$$

where the functions $\vec{\eta}_k(\vec{y})$ are such that the vector fields

$$X_k = \sum_{\mu=1}^{n} \eta_k^\mu(\vec{y}) \frac{\partial}{\partial y^\mu} \quad 1 \le k \le r \qquad (1.4)$$

generate a finite dimensional Lie algebra.

This application of Lie theory has been developed in a recent series of publications. [8]-[13] In particular, it was shown that a classification of finite transitive primitive Lie subalgebras of diff(n) (the infinite dimensional Lie algebra of all local diffeomorphisms of $\mathbb{C}^n$ or $\mathbb{R}^n$) provides a classification of all indecomposable systems of n ODEs with a superposition formula.

For $n = 1$ the only finite-dimensional subalgebras of diff(1) are sl(2, K) and its subalgebras. Up to local changes of variables, sl(2, K) can be realized in one way only, namely

$$X_1 = \frac{d}{dy}, \qquad X_2 = y\frac{d}{dy}, \qquad X_3 = y^2\frac{d}{dy} \qquad (1.5)$$

The corresponding ODE is the Riccati equation:

$$\dot{y}(t) = Z_1(t) + Z_2(t)y + Z_3(t)y^2. \qquad (1.6)$$

Its superposition formula is well-known. Indeed, the anharmonic ratio of any 4 solutions is constant:

$$\frac{(y - y_1)}{(y - y_2)}\frac{(y_2 - y_3)}{(y_1 - y_3)} = c. \qquad (1.7)$$

If y_1, y_2 and y_3 are any three distinct solutions, then the general solution $y(t)$ is obtained from (1.7) algebraicly. Further crucial properties of the Riccati equation are that it is linearizable and that it has the Painlevé property.

These properties are shared by large classes of systems of ODE's, constructed in previous articles. [8], $\cdots$, [13] The equations are associated with Lie algebras pairs $\{L, L_0\}$, where $L_0 \subset L$ is a maximal subalgebra of L (the algebra of Lie's theorem) and is realized by vector fields of the form (1.4) that vanish at the coordinate origin. Group theoretically, we are considering a homogeneous space $M \sim G/G_0$, with $G \sim \exp L$, $G_0 \sim \exp L_0$. The group G acts locally transitively on the neighbourhood of the origin of M, G_0 is the isotropy group of the origin.

The Riccati equation (1.6) has been discretized [14] in such a manner as to preserve its essential properties: linearizability, the superposition formula, and the Painlevé property (expressing itself as singularity confienment [15]). The obtained "discrete Riccati equation" is a homographic mapping.

The purpose of this article is to present discretizations of large classes of ODEs with superposition formulas [16] (Section 2) and to apply the corresponding ideas to discretize a second order ODE with many applications in physics, namely the Pinney equation [17],[18],[19] (Section 3).

2 Discretization of Matrix Riccati Equations

Let us consider the Lie group $G \sim SL(N,K)$ and its maximal parabolic subgroup $G_0 \sim P(r,s)$. An element $g \in G$ will be parametrized as

$$g = \begin{pmatrix} G_{11} & G_{12} \\ G_{21} & G_{22} \end{pmatrix} \quad \begin{matrix} G_{11} \in K^{r \times r}, \; G_{22} \in K^{s \times s} \\ G_{12} \in K^{r \times s}, \; G_{21} \in K^{s \times r} \end{matrix} \qquad (2.1)$$

$$r + s = N, \qquad r \geq s \geq 1, \qquad \det g = 1$$

Elements of the subgroup G_0 have $G_{12} = 0$ in (2.1).

The homogeneous space $M \sim G/G_0$ in this case satisfies

$$M \sim SL(N,K)/P(r,s), \qquad \dim M = n = rs. \qquad (2.2)$$

This space can be realized as a grassmannian of r-planes in K^N. We can introduce homogeneous coordinates $\xi \in K^{(r+s) \times s}$ in this space, as well as affine ones $W \in K^{r \times s}$. The group action is linear in the first ones, however they are redundant, ξ and ξ' define the same point with:

$$\xi = \begin{pmatrix} X \\ Y \end{pmatrix} \sim \xi' = \begin{pmatrix} X & H \\ Y & H \end{pmatrix}, \qquad W = XY^{-1}, \qquad (2.3)$$

$$W, \; X \in K^{r \times s}, \qquad Y, H \in GL(s, K).$$

The differential equations in homogeneous coordinates are

$$\begin{pmatrix} \dot{X} \\ \dot{Y} \end{pmatrix} = \begin{pmatrix} C & A \\ -D & -B \end{pmatrix} \begin{pmatrix} X \\ Y \end{pmatrix}, \qquad \dot{\xi} = Z\xi, \qquad trC = TrB, \quad (2.4)$$

where $A, \cdots D$ are given functions of t. In affine coordinate (2.4) implies the matrix Riccati equation

$$\dot{W} = A + WB + CW + WDW. \qquad (2.5)$$

In other words the matrix Riccati equation is a system of $n = r \cdot s$ nonlinear equations with a superposition formula and the system (2.4) is its linearization (obtained by an embedding into a higher dimensional space). The superposition formula for $s = 1$ ("projective Riccati equations") was obtained in [8],[9], for $r = s$ ("square matrix Riccati equation") in [10], for other cases in [11].

The discretization, preserving all group properties, and hence also the linearizability, the superposition formula and the Painlevé property is immediate. We replace the continuous variable t by a discrete one k, we denote e.g. $W \equiv W(k)$, $\overline{W} = W(k + \delta)$ where δ is the lattice spacing. Instead of the linear ODE's (2.4) we write a linear system

$$\begin{pmatrix} \overline{X} \\ \overline{Y} \end{pmatrix} = \begin{pmatrix} G_{11} & G_{12} \\ G_{21} & G_{22} \end{pmatrix} \begin{pmatrix} X \\ Y \end{pmatrix} \qquad \begin{matrix} \bar{\xi} = g\xi, \ \det g = 1 \\ G_{\mu\nu} = G_{\mu\nu}(k), \ \mu, \nu = 1, 2 \end{matrix} \qquad (2.6)$$

Thus, $Z \in sl(N, K)$ in (2.4) is replaced by $g \in SL(N, K)$ in (2.6) (the Lie algebra is replaced by the Lie group). The "discrete matrix Riccati equations" are obtained by rewriting (2.6) in affine coordinates

$$\overline{W} = (G_{11}W + G_{12})(G_{21}W + G_{22})^{-1}. \qquad (2.7)$$

We see that this is a matrix homographic mapping. It is, by construction, linearisable to (2.6), and will satisfy singularity confinement and allow a superposition formula, representing the general solution.

For further details and discrete equations based on other group-subgroup pairs, we refer to a forthcoming article [16].

3   Group Theory of the Pinney Equation and its Discretization

The Pinney equation

$$\ddot{u} + \omega(t)u = \frac{\alpha}{u^3} \qquad (3.1)$$

arises in numerous applications, including nonlinear optics, nonlinear elasticity and the description of waves in rotating shallow water (see e.g. [17] for a review of the literature). Its general solution can be expressed as a "superposition" of solutions of a linear equation, namely (3.1) with $\alpha = 0$. It has recently been discretized in a manner preserving its linearizability [18]. Here we shall show how the Pinney equation fits into the framework of nonlinear equations with superposition formulas and use this to generalize the equation and discretize it in a very natural manner.

First of all, we rewrite it as a first order system

$$\begin{aligned} \dot{u} &= v \\ \dot{v} &= -\omega(t)u + \frac{\alpha}{u^3}. \end{aligned} \qquad (3.2)$$

Following Lie's theorem we read off two vector fields $\{X_1,\ X_2\}$ from (3.2) and add their commutator X_3. We have:

$$X_1 = v\partial_u + \frac{\alpha}{u^3}\partial_v, \qquad X_2 = u\partial_v, \qquad X_3 = v\partial_v - u\partial_u \qquad (3.3)$$

$$[X_1, X_2] = X_3, \quad [X_1, X_3] = -2X_1, \quad [X_2, X_3] = 2X_2.$$

Thus, $L \sim \{X_1, X_2, X_3\}$ is the Lie algebra $sl(2, K)$, with $K = \mathbb{R}$ if we assume that u and v must be real, $K = \mathbb{C}$, otherwise. The algebra $L_0 \subset L$ of vector fields vanishing at the origin, which we can choose to be $(u, v) = (1, 0)$ is one-dimensional and given by

$$X = X_1 - \alpha X_2 = v\partial_u + \alpha \left(\frac{1}{u^3} - u\right)\partial_v, \qquad \tilde{X} = \begin{pmatrix} 0 & 1 \\ -\alpha & 0 \end{pmatrix} \qquad (3.4)$$

where $\tilde{X}$ is the $sl(2, K)$ matrix representation of X. Over the field of real numbers $(K = \mathbb{R})$ $\alpha > 0$ corresponds to rotations $O(2)$, $\alpha < 0$ to Lorentz transformations $O(1, 1)$. Over the complex numbers $(K = \mathbb{C})$ the two are equivalent.

An important fact is that the algebra $O(2)$ is maximal in $sl(2, \mathbb{R})$, whereas $O(1, 1)$ is not maximal in $sl(r, \mathbb{R})$, nor is $O(2, \mathbb{C})$ maximal in $sl(2, \mathbb{C})$. Quite generally, when L_0 is not maximal in L, the system of ODEs is decomposable. In the case under consideration, that means that we can transfer to new variable $(u, v) \to (x, y)$ such that at least one of the new variables satisfies a Riccati equation.

Indeed, if we put

$$x = \frac{v}{u} + \frac{\sqrt{-\alpha}}{u^2}, \qquad y = \frac{v}{u} - \frac{\sqrt{-\alpha}}{u^2} \qquad (3.5)$$

we transform the vector fields (3.3) into

$$X_1 = -(x^2\partial_x + y^2\partial_y), \qquad X_2 = \partial_x + \partial_y, \qquad X_3 = 2(x\partial_x + y\partial_y) \qquad (3.6)$$

and the Pinney system (3.2) is transformed into two identical Riccati equations

$$\dot{x} = -x^2 - \omega(t), \qquad \dot{y} = -y^2 - \omega(t) \qquad (3.7)$$

This has three immediate consquences:

1. We obtain a superposition formula for the Pinney system itself. Indeed the transformation (3.5) is invertible:

$$u = 2^{1/2}(-\alpha)^{1/4}(x - y)^{-1/2}, \qquad v = 2^{-1/2}(-\alpha)^{1/4}(x - y)^{-1/2}(x + y). \qquad (3.8)$$

Hence we can express the general solution x and y in terms of 3 solutions of (3.7) and use (3.8) to obtain the general solution u. More practically, we can use the Cole-Hopf transformation

$$x = \frac{\dot{f}}{f}, \qquad \ddot{f} + \omega(t)f = 0 \qquad (3.9)$$

to linearize the Riccati equation. If f_1 and f_2 are two linearly independent solutions of (3.9) and A, B, C and D are constants, we obtain

$$u = \sqrt{2}(-\alpha)^{1/4}\frac{[ACf_1^2 + (AD + BC)f_1f_2 + BDf_2^2]^{1/2}}{[-(AD - BC)W_0]^{1/2}}, \qquad (3.10)$$

where W_0 is the Wronskian of the two solutions. We can use this formula to obtain real solutions $u(t)$, even when we have $\alpha > 0$, by appropriately choosing complex constants $A, \cdots, D$.

2. The Lie algebra (3.6) is not maximal in the algebra diff(2). It can be imbedded into $sl(2, K) \oplus sl(2, K)$:

$$\begin{array}{lll} A_1 = -x^2\partial_x, & A_2 = x\partial_x, & A_3 = \partial_x, \\ B_1 = -y^1\partial_y, & B_2 = y\partial_y, & B_3 = \partial_y. \end{array} \qquad (3.11)$$

Instead of (3.7) we then write two different completely general Riccati equations. The transformation (3.8) then takes them into a linearizable generalization of the Pinney system (3.2). The implications of this fact will be investigated elsewhere [19].

3. The Pinney equation can be discretized while preserving its linearizability, using the approach of Section 2 and [16]. The appropriate discretization of the Riccati equations (3.7) is

$$x_{n+\delta} = \frac{x_n - \delta\omega(n)}{\delta x_n + 1}, \qquad y_{n+\delta} = \frac{y_n - \delta\omega(n)}{\delta y_n + 1} \qquad (3.12)$$

where δ is the lattice spacing. Using the transformation (3.8), and putting $\bar{u} = u_{n+\delta}, \qquad \bar{v} = v_{n+\delta}$, we obtain

$$\bar{u}^2 = \frac{u^2(\delta v + u)^2 + \delta^2\alpha}{(\delta\omega + 1)u^2}$$

$$\bar{v} = \frac{1}{u}\{(\delta\omega + 1)[u^2(\delta v + u)^2 + \delta^2\alpha]\}^{-1/2}\{u^2v^2 + \alpha$$

$$+ vu^3(1 - \delta^2\omega) - \epsilon\omega u^4\}. \qquad (3.13)$$

Calculating $\bar{\bar{u}}$ and eliminating v and $\bar{v}$, we obtain the discrete Pinney equation

$$[u^2(1 + \omega) = \bar{\bar{u}}^2(1 + \bar{\omega}) + 4\bar{u}^2]^2 = 16[\bar{\bar{u}}^2u^2(1 + \omega) - \alpha] \qquad (3.14)$$

A different linearization and discretization may be more appropriate for $\alpha > 0$. It can also be used for $\alpha < 0$. The underlying algebraic structure is $L \sim sl(2, \mathbb{R})$, $L_0 \sim o(2)$. We use the following diffeomorphism between two real homogeneous spaces

$$M \sim SO(p, q)/SO(p) \times SO(q) \sim SL(p, q)/P(p, q), \qquad (3.15)$$

(in our case $p = 2$, $q = 1$). A metric K on M is introduced and we have

$$K = \begin{pmatrix} I_2 & \\ & -1 \end{pmatrix}, \qquad \xi K + K\xi^T = 0, \qquad gKg^T = K$$

$$\xi \in o(2, 1) \sim sl(2, \mathbb{R}), \qquad g \in SO(2, 1). \qquad (3.16)$$

A system of linear ODE's is given by

$$\begin{pmatrix} \dot{u}_1 \\ \dot{u}_2 \\ \dot{u}_0 \end{pmatrix} = \begin{pmatrix} 0 & c & a \\ -c & 0 & b \\ a & b & 0 \end{pmatrix} \begin{pmatrix} u_1 \\ u_2 \\ u_0 \end{pmatrix}, \qquad u_0^2 - u_1^2 - u_2^2 = 1. \qquad (3.17)$$

We introduce affine coordinates (x, y) in the neighbourhood of the origin $(0, 1)$ and obtain a system of $O(2, 1)$ Riccati equations

$$\dot{x} = a(1 - x^2) - bxy + cy, \qquad x = \frac{u_1}{u_0}, \qquad y = \frac{u_2}{u_0},$$

$$\dot{y} = -axy + b(1 - y^2) - cx. \qquad (3.18)$$

The corresponding vector fields are

$$\hat{A} = (1-x^2)\partial_x - xy\partial_y, \qquad \hat{B} = -xy\partial_x + (1-y^2)\partial_y, \qquad \hat{C} = y\partial_x - x\partial_y. \quad (3.19)$$

Thus, we can transform the vector fields (3.3) into (3.19), more precisely $\hat{X}_1 \to \hat{A} + \hat{C}$, $\hat{X}_2 \to \hat{A} - \hat{C}$, $X_3 \to 2\hat{B}$.

The appropriate transformation is

$$x = \frac{2u^3v}{\alpha + u^2v^2 + u^4}, \qquad y = \frac{\alpha + u^2v^2 - u^4}{\alpha + u^2v^2 + u^4}, \qquad (3.20)$$

and its inverse is

$$u = \left(\frac{\alpha(1 - y^2)}{1 - (x^2 + y^2)} \right)^{1/4}, \qquad v = \frac{x}{1 - y} \left(\frac{\alpha(1 - y^2)}{1 - (x^2 + y^2)} \right)^{1/4}. \qquad (3.21)$$

$\alpha > 0$, $x^2 + y^2 < 1$, or $\alpha < 0$, $x^2 + y^2 > 1$.

The discretization of the Pinney equation, preserving the $O(2, 1)/O(2)$ structure, and hence its consequences: linearizability, a superposition formula and the Painlevé property, is now obvious. We replace the Lie algebra element in the linear system (3.17) by a Lie group element:

$$\bar{u} = g(n)u \quad u^T = (u_1(n),\ u_2(n),\ u_0(n))$$
$$\bar{u}^T = (u_1(n+\delta), u_2(n+\delta),\ u_0(n+\delta))$$
$$gKg^T = K, \quad u^TKu = 0. \tag{3.22}$$

The $O(2,1)$ Riccati equations are replaced by

$$\bar{x} = \frac{g_{11}x + g_{12}y + g_{10}}{g_{01}x + g_{02}y + g_{00}}, \qquad \bar{y} = \frac{g_{21}x + g_{22}y + g_{20}}{g_{01}x + g_{02}y + g_{00}} \tag{3.23}$$

The $O(2,1)/O(2)$ discrete Pinney equations are obtained from (3.23) using the transformations (3.20) and (3.21). For details and applications we refer to a forthcoming article [19].

The results of this section can be used also for $\alpha < 0$. The homogeneous manifold then is $O(2,1)/O(1,1)$ and in (3.17) we have $u_0^2 - u_1^2 - u_2^2 = -1$. (3.18),$\cdots$, (3.21) are still valid, but in (3.21) we have $\alpha < 0$, $x^2 + y^2 > 1$. We mention that $x^2 + y^2 < 1$ and $x^2 + y^2 > 1$ correspond to the Beltrami models of real and imaginary Lobashevsky spaces, respectively.

Acknowledgements

The author's research was partly supported by research grants from NSERC of Canada and FCAR du Quebec. This report was written during his visit to the Department of Mathematics of the University of New South Wales. He thanks this Department and specially C. Rogers for hospitality. Helpful discussions with Colin Rogers, W. Schief, B. Grammaticos and A. Ramani are much appreciated.

References

[1] S. Maeda, IMA J. Appl. Math. **38**, 129(1987).

[2] D. Levi and P. Winternitz, Phys. Lett. **A152**, 335 (1991), J. Math. Phys. **34**, 3713 (1993), J. Math. Phys. **37**, 5551 (1996).

[3] D. Levi, L. Vinet and P. Winternitz, J. Phys. A. Math. Gen. **30**, xxxx (1997).

[4] R. Floreanini and L. Vinet, J. Math. Phys. **36**, 7024 (1995).

[5] V.A. Dorodnitryn, in Symmetries and Integrability of Difference Equations, Providence, R.I., 1995, AMS.

[6] G.R.W. Quispel, H.W. Capel and R. Sahadevan, Phys. Lett. **A170**, 379 (1992).

[7] S. Lie, Vorlesungen über continuirliche Gruppen mit geometrischen und anderen Anwendungen. (Bearbeitet und herausgegeben von Dr. G. Scheffers). Teubner, Leipzig, 1893.

[8] R.L. Anderson, Lett. Math. Phys. **4**, 1 (1980).

[9] R.L. Anderson, J. Harnad and P. Winternitz, Physica **D4**, 164 (1982).

[10] J. Harnad, P. Winternitz and R.L. Anderson, J. Math. Phys. **24**, 1062, (1983).

[11] M.A. del Olmo, M.A. Rodriguez and P. Winternitz, J. Math. Phys. **28**, 530, (1987).

[12] S. Shnider and P. Winternitz, J. Math. Phys. **25**, 3155 (1984), Lett. Math. Phys. **5**, 154 (1984).

[13] L. Michel and P. Winternitz, Families of transitive primitive maximal simple Lie subagebras of diff (n). In "Advances in Mathematical Sciences, CRM's 25 years", AMS, Providence, R.I., (1997), to appear.

[14] B. Grammaticos, A. Ramani, K.M. Tamizhmani, Jour. Phys. **A27**, 559 (1994).

[15] B. Grammaticos, A. Ramani and V. Papageorgiou, Phys. Rev. Lett. **67**, 1825 (1991).

[16] B. Grammaticos, A. Ramani and P. Winternitz, "Discretizing families of linearizable equations", Preprint CRM, Montréal, 1997.

[17] C. Rogers and W.K. Schief, J. Math. Anal. Appl. **198**, 194 (1996).

[18] W.K. Schief, A discrete Pinney equation, Appl. Math. Letters, to appear, 1997.

[19] C. Rogers, W.K. Schief and P. Winternitz, Lie theory and generalizations fo the Pinney equation, to be published.

Chapter 7

Numerical Methods and Miscellaneous

Generalising Painlevé truncation: expansions in Riccati pseudopotentials

A.P. Fordy[†] and A. Pickering[‡]

[†] Department of Applied Mathematical Studies
and Centre for Nonlinear Studies
University of Leeds
Leeds LS2 9JT, U.K.

[‡] Institute of Mathematics and Statistics
University of Kent at Canterbury
Canterbury, Kent
CT2 7NF, U.K.

Abstract

In the search for the Lax pair and Darboux transformation of a (completely integrable) partial differential equation, a natural generalisation of the constant-level truncation in Painlevé analysis is to seek an expansion as a polynomial in the components of some (to be determined) Riccati pseudopotential. In this direct approach the relationship with Painlevé analysis becomes a secondary consideration. Here we give two generalisations of this method. The first is to finding Lax pairs of higher order, which is relatively straightforward. The second concerns the assumption of the form of the Darboux transformation within the context of a Lax pair having some gauge freedom: this allows us to deal easily with a modification of the Boussinesq equation.

1 Introduction

Given a partial differential equation (PDE) suspected of being completely integrable, for example because it passes the Weiss-Tabor-Carnevale (WTC) Painlevé test [1], there are several techniques that might be used to find its Lax pair. One could use Walquhist-Estabrook prolongation, or one could put the PDE into Hirota bilinear form and then try to find the Bäcklund transformation (BT), or one could use the information provided by truncating the

WTC Painlevé expansion. Overviews of these approaches can be found in [2]. Here we will be concerned with an alternative method, which involves seeking a solution to the PDE as an expansion in some Riccati pseudopotential. As we will see, this provides a natural extension of the constant-level truncation in Painlevé analysis.

From the truncated Painlevé expansion there are several ways to proceed in order to find the Lax pair. Perhaps the method most commonly referred to is the "singular manifold method" of Weiss [3]. This identifies the singular manifold of the truncated Painlevé expansion with the quotient of two linearly independent solutions of the sought-after Lax pair. This approach has been rewritten by Musette and Conte [4] using the language of the homographic invariant Painlevé analysis [5], which has the advantage of simplifying the expressions involved. These authors also remarked that for many PDEs the truncated Painlevé expansion U_T provides the Darboux transformation (DT):

$$U_T = \mathcal{D} \log \xi + u, \tag{1}$$

where $\mathcal{D}$ is the "singular part operator" obtained from Painlevé analysis i.e. $\mathcal{D} \log \varphi$ is the principal part of the WTC Painlevé expansion, and u is the solution of the PDE appearing in the sought-after Lax pair — assumed in [4] to be second or third order — having eigenfunction ξ. That is, the form of the DT is given by the BT between U_T and the constant-level coefficient u' of the truncation in WTC form (this coefficient being a third solution of the PDE). Using this as an additional *assumption* can prove extremely useful [4].

The connection between the invariant Painlevé analysis (see Appendix A) and the sought-after spectral problem is made by expressing the quantity $\chi^{-1} - \xi_x/\xi$ and the two homographic invariants S and C in terms of the ratios of certain Wronskians Y_1, Y_2, Y_3 [3, 4] of two linearly independent solutions ξ_1 and ξ_2 of the scattering problem:

$$\chi^{-1} - \frac{\xi_x}{\xi} = -\frac{1}{2}Y_1, \qquad S = Y_{1,x} - \frac{1}{2}Y_1^2 + 2Y_2, \qquad C = -Y_3. \tag{2}$$

In the case of a second order scattering problem $Y_1 = 0$ and from (2) we see that Y_2 and Y_3 are just the coefficients of the Lax pair obtained by the linearisation of the Riccati system in χ via $\chi^{-1} = \xi_x/\xi$. For a third order scattering problem, the relations (2) are used to replace S and C in the determining equations resulting from the truncation in χ. Only two of the three functions Y_1, Y_2, Y_3 are independent and (Y_1, Y_2) are found to satisfy a projective Riccati system equivalent to the third order Lax pair; this Riccati system is used to express the determining equations as polynomials in (Y_1, Y_2). Further details of this method can be found in [4].

When the expansion variable satisfies a system of Riccati equations, truncation at constant level is not the only possibility [6]; this observation has been used in [7, 8, 9] to obtain DTs which cannot be obtained from the constant-level truncation. For example, for the classical Boussinesq system [10, 11, 12] the DT is written in terms of the Riccati pseudopotential Z as [9]

$$U = 2\kappa(\log Z)_x + u, \qquad V = 2\kappa^2(Z^{-1} + BZ)_x + v - \kappa u_x, \qquad (3)$$

where $B = -(v - \kappa u_x)/(4\kappa^2)$. The DT is expressed in terms of the eigenfunctions ψ_1, ψ_2 of the corresponding matrix spectral problem in $sl(2)$ using the identities $Z = \psi_1/\psi_2$, and $Z^{-1} + BZ = (\log(\psi_1\psi_2))_x$. We note that since Z is invariant under transformations $\psi_i = \alpha\sigma_i$, there is a trivial gauge freedom in the choice of spectral problem, corresponding to the addition of a multiple of the identity matrix, which is fixed to be zero if we insist that the spectral problem is in $sl(2)$. However if we no longer insisit on this, then this gauge freedom can be used, with $\alpha = \exp[(1/(4\kappa)) \int u dx]$, to rewrite (3) as

$$U = 2\kappa(\log(\sigma_1/\sigma_2))_x + u, \qquad V = 2\kappa^2(\log(\sigma_1\sigma_2))_{xx} + v, \qquad (4)$$

which can be compared to the BT given in [13], expressed in terms of "two singular manifolds" by a double iteration of the WTC BT. However, since Z is invariant under the above gauge transformation, the representation of the DT in terms of Z (3) would be difficult to derive in advance.

In [9] we showed how corresponding to the DT (3) we have infinite rather than truncated Painlevé expansions for U and V. This extension of the singular manifold method, which allows the possibility that the DT might correspond to an infinite Painlevé expansion for some choice of the arbitrary data therein, makes use of only one singular manifold. In the present work it is an extension of the constant-level truncation that we will be dealing with. This does however have applications to "two singular manifold" examples.

We note that the information we are interested in is the Lax pair and the DT, or equivalently the Riccati pseudopotential and the DT. This means that we might as well expand directly in terms of our pseudopotential, rather than first as a truncated Painlevé expansion and then expanding the resulting determining equations in terms of the pseudopotential. This remark has in fact already been made [14], although under the assumption that the expansion of U in terms of its DT is given by (1). The natural extension of a constant-level truncation in χ — which is a Riccati variable for second order Lax pairs — is to expand U as a polynomial in the components $Y_1, Y_2, Y_3, \ldots$ of a Riccati system corresponding to a higher order Lax pair. Once we no longer insist on maintaining a connection with Painlevé analysis the extension to Lax pairs of order higher than three is immediate. We also extend this approach further in order to obtain the Lax pair of a modified Boussinesq equation.

2   A direct approach

In this direct approach to finding the Lax pair and DT for a PDE we seek an expansion for U as a polynomial in some Riccati pseudopotential. The form of polynomial sought need not be of the form dictated by (1), although of course this can be a useful ansatz. In the case of second and third order Lax pairs this amounts to stripping the language of Painlevé analysis away from the method outlined in [4]. The extension to finding higher order Lax pairs can then be made without any reference to Painlevé analysis.

Not restricting the DT to be of the form (1), and a further extension to be detailed later, means that the class of equations to which this approach is applicable is larger than that to which the methods in [4, 14] can be applied. In order that we be as algorithmic as possible we make the following remarks:

1. We do not derive the DT in advance but we do allow an assumption of its form (1). This may not always be valid, but for some equations it can be used to fix a gauge freedom in the assumed Lax pair.

2. In searching for a second order Lax pair we use an expansion in a single Riccati variable; for a third order Lax pair we use an expansion in two Riccati variables, and so on. After substitution of the expansion for U, the coefficients of the resulting expansion of the PDE in terms of $Y_1, Y_2, Y_3, \ldots$ are to be set equal to zero independently. This means that we prefer to avoid the step of adding and subtracting terms at different levels in φ, advocated within the context of the WTC approach in [16].

3. The spectral parameter should be introduced as a result of solving this set of determining equations.

4. We should be able to analyse the PDE in isolation, i.e. a knowledge of the relationship of the PDE being studied to other PDEs should not be necessary.

5. Since our aim is to obtain both the Lax pair and the DT, a necessary condition is that we should be able to obtain the one-soliton solution of the PDE. However this is not of course a sufficient condition. For the modified Korteweg-de Vries equation it is possible to obtain the one-soliton solution from the constant-level truncation in χ (though not with S and C constant, which gives only the kink solution [6]). Recovery of the Lax pair requires a more general approach [7, 15, 9].

These assertions certainly hold for the examples dealt with in this paper, although as always further extensions may be necessary. We note that in these examples, for reasons of brevity, we proceed directly to an assumption of a Lax pair of correct order. Full details will be presented elsewhere [17].

3 Examples

3.1 Cylindrical KdV equation

As a simple example we consider the cylindrical Korteweg-de Vries (KdV) equation [18],

$$E[U] \equiv U_t + U_{xxx} + 6UU_x + \frac{1}{2(t+\tau)}U = 0. \tag{5}$$

This equation can of course be mapped back onto KdV itself [19]; we consider it here in order to illustrate some of the points made above. We will see that derivation of the Lax pair is straightforward once we regard this as a search for the pseudopotential and DT, as opposed to truncating a Painlevé expansion.

Seeking a second order Lax pair

$$\eta_{xx} = a\eta, \tag{6}$$
$$\eta_t = b\eta_x + c\eta, \tag{7}$$

we seek an expansion

$$U = U_0 Y^2 + U_1 Y + U_2, \tag{8}$$

as a polynomial in the Riccati pseudopotential $Y = Y_1 = \eta_x/\eta$:

$$Y_x = -Y^2 + a, \tag{9}$$
$$Y_t = -bY^2 + b_x Y + (ab + c_x). \tag{10}$$

The cross-derivative conditions on (6), (7) are

$$(\eta_{xx})_t - (\eta_t)_{xx} \equiv X_1 \eta_x + X_0 \eta = 0, \tag{11}$$

where

$$X_1 \equiv -(2c + b_x)_x, \tag{12}$$
$$X_0 \equiv a_t - a_x b - 2ab_x - c_{xx}, \tag{13}$$

these being equivalent of course to those on (9), (10). The Riccati system (9), (10) is of course equivalent to that used in the invariant analysis (see Appendix A) with $Y = \chi^{-1}$, $a = -S/2$, and $b = -C$. However, from the point of view of finding the Lax pair and DT these identifications are irrelevant.

Substitution of (8) into (5) gives

$$E \equiv \sum_{i=0}^{5} E_i Y^i = 0, \tag{14}$$

and the solution of the system of equations $X_0 = 0$, $X_1 = 0$, $E_i = 0$ is easily obtained as

$$U_0 = -2, \tag{15}$$

$$U_1 = 0, \tag{16}$$

$$U_2 = 2\frac{\Lambda}{t+\tau} + \frac{x}{6(t+\tau)} - u, \tag{17}$$

$$a = \frac{\Lambda}{t+\tau} + \frac{x}{12(t+\tau)} - u, \tag{18}$$

$$b = -4\frac{\Lambda}{t+\tau} - \frac{x}{3(t+\tau)} - 2u, \tag{19}$$

$$c = \frac{1}{6(t+\tau)} + u_x, \tag{20}$$

where Λ is a constant of integration and u is a second solution of the cylindrical KdV equation: $E[u] = 0$. The compatibility conditions (12), (13) give

$$X_1 \equiv 0, \qquad X_0 \equiv -E[u] = 0. \tag{21}$$

Thus we obtain both the Lax pair (6), (7) and the DT

$$U = -2Y^2 + 2\frac{\Lambda}{t+\tau} + \frac{x}{6(t+\tau)} - u = 2Y_x + u = 2\left(\log\eta\right)_{xx} + u. \tag{22}$$

Setting $U = W_x$ and $u = w_x$ it is then a simple matter to solve the DT for $Y = (W - w)/2$ and by substituting into the Riccati pseudopotential (9), (10) derive the BT for cylindrical KdV given in [20]. Note that if we choose $\Lambda = \lambda\tau$ then the Lax pair and DT for KdV are obtained in the limit $\tau \to \infty$.

Remarks

In the above derivation of the Lax pair and DT no mention was made of Painlevé analysis, of truncating a Painlevé expansion, or even of a "singular manifold equation." We simply set the coefficients of different powers of Y equal to zero, and solved for the coefficients U_i of the expansion (8) and a, b, c of the Lax pair in terms of a second solution u of the PDE and of course a spectral parameter (introduced as a constant of integration). The only residual elements of the Painlevé analysis are the leading order analysis, which determines the degree of the polynomial in Y.

The leading order analysis also determines the form of the DT (1), if we were to use this. Since in this example this assumption would be correct, this would have made the above calculations even easier. However the validity of this assumption depends on the coordinates in which we write our PDE, and the form chosen for the Lax pair; it is possible to obtain an expression for the DT as a polynomial in the pseudopotential which is of the form (1) only after a gauge transformation, which then changes the form of the Lax pair [17]. However, as we shall see, when (1) does hold it can prove very useful.

We now consider two extensions of this approach. First, we consider the question of deriving higher order Lax pairs, which can now be undertaken without any reference to Painlevé analysis. Second, we consider a generalisation of the form of Lax pair sought by including a gauge freedom to be fixed by the DT.

3.2 Higher order Lax pairs

For higher order Lax pairs, for example third order

$$\eta_{xxx} = a\eta_x + b\eta, \tag{23}$$
$$\eta_t = c\eta_{xx} + d\eta_x + e\eta, \tag{24}$$

or fourth order

$$\eta_{xxxx} = a\eta_{xx} + b\eta_x + c\eta, \tag{25}$$
$$\eta_t = d\eta_{xxx} + e\eta_{xx} + f\eta_x + g\eta, \tag{26}$$

we seek a solution of the PDE as a polynomial in a corresponding Riccati pseudopotential, respectively two (Y_1, Y_2) or three (Y_1, Y_2, Y_3) component. A Riccati system for a generalisation of (23), (24) is given later; the Riccati system for (25), (26) is too long to give here but is easy to derive. Full details are given in [17]. The cross-derivative conditions on (25), (26) are

$$(\eta_{xxxx})_t - (\eta_t)_{xxxx} \equiv X_3\eta_{xxx} + X_2\eta_{xx} + X_1\eta_x + X_0\eta, \tag{27}$$

where

$$X_3 \equiv -(2ae + 3a_xd + 5ad_x + 3bd + d_{xxx} + 4e_{xx} + 6f_x + 4g)_x = 0, \tag{28}$$

$$\begin{aligned} X_2 \equiv\ & a_t - (3aa_xd + 2a^2d_x + a_{xxx}d + 4a_{xx}d_x + 6a_xd_{xx} + 4ad_{xxx} \\ & +a_{xx}e + 4a_xe_x + 5ae_{xx} + a_xf + 2af_x + 3b_{xx}d + 8b_xd_x + 6bd_{xx} \\ & +2b_xe + 3be_x + 3c_xd + 4cd_x + e_{xxxx} + 4f_{xxx} + 6g_{xx}) = 0, \end{aligned} \tag{29}$$

$$\begin{aligned} X_1 \equiv\ & b_t - (3a_xbd + 2abd_x + b_{xxx}d + 4b_{xx}d_x + 6b_xd_{xx} + 4bd_{xxx} \\ & +b_{xx}e + 4b_xe_x + 6be_{xx} + b_xf + 3bf_x + 3c_{xx}d + 8c_xd_x + 6cd_{xx} \\ & +2c_xe + 4ce_x - af_{xx} - 2ag_x + f_{xxxx} + 4g_{xxx}) = 0, \end{aligned} \tag{30}$$

$$\begin{aligned} X_0 \equiv\ & c_t - (3a_xcd + 2acd_x + c_{xxx}d + 4c_{xx}d_x + 6c_xd_{xx} + 4cd_{xxx} + c_{xx}e \\ & +4c_xe_x + 6ce_{xx} + c_xf + 4cf_x - bg_x - ag_{xx} + g_{xxxx}) = 0, \end{aligned} \tag{31}$$

which are of course the same as those on the Riccati system in Y_1, Y_2, Y_3 [17].

As an example we consider the PDE

$$E[U] \equiv (U_{xxt} + U_x U_t)_x + c_{20}U_{xx} + c_{11}U_{xt} + c_{02}U_{tt} = 0, \qquad c_{02} \neq 0. \qquad (32)$$

This equation is a reduction to 1+1 dimensions of a special case of the Nizhnik-Novikov-Veselov (NNV) equation [21, 22, 23], and also of an equation due to Ito [24]; the Lax pair for the latter and so also for this reduction was derived using a Hirota bilinear BT in [25]. This fourth order Lax pair is also given in [26], derived using the WTC approach by Estévez; this derivation can be deduced from that given in [27], which involves adding and subtracting terms at different levels of φ, for the above-mentioned special case of NNV.

In the space available here we simply report that this fourth order Lax pair can be derived by seeking a solution of (32) in the form

$$U = U_0 Y_1 + U_1, \qquad (33)$$

where $Y_1 = \eta_x / \eta$ is the first component of the Riccati pseudopotential in Y_1, Y_2 and Y_3 corresponding to (25), (26), and setting coefficients of the resulting polynomial in Y_1, Y_2 and Y_3 equal to zero independently. Doing this we obtain the solution [17]

$$U_0 = 6, \qquad U_1 = u, \qquad (34)$$

$$a = -(c_{11} + u_x), \qquad b = \lambda - u_{xx}, \qquad c = c_{02}\left(\frac{c_{20}}{3} + \frac{u_t}{3}\right), \qquad (35)$$

$$d = -\frac{1}{c_{02}}, \qquad e = 0, \qquad f = -\frac{c_{11} + u_x}{c_{02}}, \qquad g = 0, \qquad (36)$$

where λ is a constant of integration and u a second solution of (32); $E[u] = 0$. Thus we obtain the Lax pair and the DT

$$U = 6(\log\eta)_x + u. \qquad (37)$$

The cross-derivative conditions (28)—(31) then become

$$X_3 \equiv 0, \qquad X_2 \equiv 0, \qquad X_1 \equiv 0, \qquad X_0 \equiv \frac{1}{3}F[u] = 0. \qquad (38)$$

Full details of the above calculation are given in [17].

3.3   A modified Boussinesq equation

We now consider a modified Boussinesq (Mikhailov-Shabat) equation,

$$E[P, Q] \equiv -P_t + \left(Q - \frac{3}{2}\kappa^2 P^2\right)_x = 0, \qquad (39)$$

$$F[P, Q] \equiv -Q_t - 3\kappa^2\left(P_{xx} - PQ + \kappa^2 P^3\right)_x = 0. \qquad (40)$$

We seek a third order Lax pair, but of a form more general than (23), (24);

$$\eta_{xxx} = a\eta_{xx} + b\eta_x + c\eta, \tag{41}$$

$$\eta_t = d\eta_{xx} + e\eta_x + f\eta. \tag{42}$$

The inclusion of the extra term in (41) corresponds to a gauge freedom; we will see that this can be fixed by assuming that the the DT is given by (1),

$$P = \frac{2}{\kappa}Y_1 + p, \tag{43}$$

$$Q = -6Y_{1,x} + q, \tag{44}$$

where $Y_1 = \eta_x/\eta$. Corresponding to (41), (42) we take the Riccati system

$$Y_{1,x} = -Y_1^2 + Y_2 + b, \tag{45}$$

$$Y_{2,x} = -Y_1 Y_2 + aY_2 + ab + c - b_x, \tag{46}$$

$$Y_{1,t} = -eY_1^2 - dY_1 Y_2 + e_x Y_1 + (ad + d_x + e)Y_2$$
$$+ (abd + be + cd + bd_x + f_x), \tag{47}$$

$$Y_{2,t} = -dY_2^2 - eY_1 Y_2 + (abd + cd + b_x d + 2bd_x + 2f_x + e_{xx})Y_1$$
$$+ (a^2 d + a_x d + 2ad_x + ae + 2e_x - bd + d_{xx})Y_2$$
$$+ (a^2 bd + acd + abe - a_x bd + abd_x + ae_{xx} + 2af_x + ce - b_x e$$
$$- c_x d - cd_x - 3b_x d_x - b_{xx} d - 2bd_{xx} - e_{xxx} - 2f_{xx}). \tag{48}$$

The cross-derivative conditions on the above Lax pair are

$$(\eta_{xxx})_t - (\eta_t)_{xxx} \equiv X_2 \eta_{xx} + X_1 \eta_x + X_0 \eta = 0, \tag{49}$$

with

$$X_2 \equiv a_t - (a^2 d + 2bd + a_x d + 2ad_x + ae + d_{xx} + 3e_x + 3f)_x = 0, \tag{50}$$

$$X_1 \equiv b_t - (2a_x bd + abd_x - 2af_x - ae_{xx} + b_x e + 2be_x + b_{xx} d + 3b_x d_x$$
$$+ 3bd_{xx} + 2c_x d + 3cd_x + e_{xxx} + 3f_{xx}) = 0, \tag{51}$$

$$X_0 \equiv c_t - (2a_x cd + acd_x - af_{xx} + 3ce_x + c_x e + 3cd_{xx} + 3c_x d_x + c_{xx} d$$
$$- bf_x + f_{xxx}) = 0, \tag{52}$$

these being the same of course as those on (45)—(48).

Substitution of (43), (44) into (39), (40) gives

$$E \equiv \sum_{i=0}^{2} \sum_{j=0}^{1} E_{i,j} Y_1^i Y_2^j = 0, \tag{53}$$

$$F \equiv \sum_{i=0}^{3} \sum_{j=0}^{2} F_{i,j} Y_1^i Y_2^j = 0, \tag{54}$$

and insisting that p, q are a second solution of (39), (40),

$$E[p, q] = 0, \qquad F[p, q] = 0, \tag{55}$$

leads to the solution of $E_{i,j} = 0$, $F_{i,j} = 0$, $X_0 = 0$, $X_1 = 0$, $X_2 = 0$ as

$$a = -\frac{3}{2}\kappa p, \qquad b = \frac{1}{4}\left(q - 3\kappa^2 p^2 - 3\kappa p_x\right), \qquad c = \lambda, \tag{56}$$

$$d = -3\kappa, \qquad e = -3\kappa^2 p, \qquad f = 0. \tag{57}$$

with λ a constant of integration [17]. The cross-derivative conditions become

$$X_2 \equiv \frac{3}{2}\kappa E[p, q] = 0, \tag{58}$$

$$X_1 \equiv -\frac{1}{4}F[p, q] + \frac{3}{2}\kappa^2 p E[p, q] + \frac{3}{4}\kappa(E[p, q])_x = 0, \tag{59}$$

$$X_0 \equiv 0. \tag{60}$$

Thus we see that the gauge freedom allowed in the Lax pair is fixed by the DT (43), (44). The above Lax pair differs from that given in [28], which is in fact equivalent to a second order Lax pair; this is discussed further in [17]. Note that again coefficients $E_{i,j}$, $F_{i,j}$ have been set equal to zero independently.

In (43) κ may take either sign; we can take a double iteration to obtain

$$P = \frac{2}{\kappa}(\log(\eta_1/\eta_2))_x + \tilde{p}, \qquad Q = -6(\log(\eta_1\eta_2))_{xx} + \tilde{q}, \tag{61}$$

which may be compared with the "two singular manifold" expansion used in [13] to try to obtain the Lax pair of (39), (40). We also note that we can exploit this choice of sign, together with the Miura map, to obtain from (43), (44) the DT of the Boussinesq equation (again details are in [17]):

$$\frac{1}{3\kappa^2}V_{tt} + (V_{xx} + 2V^2)_{xx}, \qquad V = 3(\log(\psi))_{xx} + v. \tag{62}$$

4　Conclusions

The natural extension of constant-level truncation in Painlevé analysis is to seek an expansion as a polynomial in a Riccati pseudopotential, the number of components of which depends on the order of the Lax pair being sought. The connection with Painlevé analysis then becomes a secondary consideration. The assumption of the form of the DT can ease calculations for those equations for which it holds; it can also be useful in fixing a gauge freedom allowed in the Lax pair. It is interesting that moving away from Painlevé analysis leads to a technique related to prolongation, but with the algebra being chosen in advance: this will be explored more fully in [17] (see also [4]). Also in [17] we give full details of all the above calculations. Of course further extensions of this approach may still be necessary.

Appendix A: the invariant Painlevé analysis

The invariant analysis [5] uses as expansion variable a function χ given in terms of the WTC singular manifold $\varphi - \varphi_0$ by

$$\chi = \left(\frac{\varphi_x}{\varphi - \varphi_0} - \frac{\varphi_{xx}}{2\varphi_x} \right)^{-1}. \tag{63}$$

The gradient of χ is given by

$$\chi_x = 1 + \frac{1}{2} S \chi^2, \tag{64}$$

$$\chi_t = -C + C_x \chi - \frac{1}{2}(C_{xx} + CS)\chi^2, \tag{65}$$

where the homographic invariants S and C are given by

$$S = \left(\frac{\varphi_{xx}}{\varphi_x} \right)_x - \frac{1}{2}\left(\frac{\varphi_{xx}}{\varphi_x} \right)^2, \qquad C = -\frac{\varphi_t}{\varphi_x}. \tag{66}$$

The cross-derivative condition on (64), (65) is

$$S_t + C_{xxx} + 2C_x S + C S_x = 0, \tag{67}$$

and is identically satisfied in terms of $\varphi - \varphi_0$. This invariant analysis builds in a resummation of the original WTC Painlevé expansion, and has the effect of greatly shortening the expressions for the coefficients of the expansion.

References

[1] J. Weiss, M. Tabor and G. Carnevale, J. Math. Phys. **24**, 522-526 (1983).

[2] *Soliton Theory: a Survey of Results*, ed. A. P. Fordy (Manchester University Press, 1990).

[3] J. Weiss, J. Math. Phys. **24**, 1405-1413 (1983).

[4] M. Musette and R. Conte, J. Math. Phys. **32**, 1450-1457 (1991).

[5] R. Conte, Phys. Lett. A **140**, 383-390 (1989).

[6] A. Pickering, J. Phys. A **26**, 4395-4405 (1993).

[7] M. Musette and R. Conte, J. Phys. A **27**, 3895-3913 (1994).

[8] R. Conte, M. Musette and A. Pickering, J. Phys. A **28**, 179-187 (1995).

[9] A. Pickering, J. Math. Phys. **37** 1894-1927 (1996).

[10] L. J. F. Broer, Appl. Sci. Res. **31**, 377-395 (1975).

[11] D. J. Kaup, Prog. Theor. Phys. **54**, 72-78 (1975).

[12] D. J. Kaup, Prog. Theor. Phys. **54**, 396-408 (1975).

[13] P. G. Estévez, P. R. Gordoa, L. Martínez Alonso and E. M. Reus, J. Phys. A **26**, 1915-1925 (1993).

[14] M. Musette and R. Conte, in *Nonlinear Evolution Equations and Dynamical Systems*, eds. M. Boiti, M. Martina and F. Pempinelli (World Scientific, Singapore, 1992), 161-170.

[15] P. R. Gordoa and P. G. Estévez, Teoret. i Matem. Fiz. **99**, 370-376 (1994).

[16] A. C. Newell, M. Tabor and Y. B. Zeng, Physica D **29**, 1-68 (1987).

[17] A. P. Fordy and A. Pickering, "On a direct approach to finding Lax pairs," in preparation (1996).

[18] F. Calogero and A. Degasperis, Lett. Nuovo Cimento **23**, 150-154 (1978).

[19] R. Hirota, Phys. Lett. A **71**, 393-394 (1979).

[20] J. J. C. Nimmo and D. G. Crighton, Phys. Lett. A **82**, 211-214 (1981).

[21] L. P. Nizhnik, Sov. Phys. Dokl. **25**, 706-708 (1980).

[22] A. P. Veselov and S. P. Novikov, Sov. Math. Dokl. **30**, 588-591 (1984).

[23] A. P. Veselov and S. P. Novikov, Sov. Math. Dokl. **30**, 705-708 (1984).

[24] M. Ito, J. Phys. Soc. Japan **49**, 771-778 (1980).

[25] M. Musette, in *Painlevé Transcendents. Their Asymptotics and Physical Applications*, eds. D. Levi and P. Winternitz (Plenum, New York, 1992), 197-209.

[26] E. M. Mansfield and P. A. Clarkson, "Symmetries and exact solutions for a 2+1 dimensional shallow water wave equation," preprint UKC/IMS/95/6.

[27] P. G. Estévez, and S. B. Leble, Acta Appl. Math. **39**, 277-294 (1995).

[28] H. Flaschka, A. C. Newell and M. Tabor, in *What is Integrability?*, ed. V. E. Zakharov (Springer-Verlag, Berlin, Heidelberg, 1991), 73-114.

Symplectic Runge-Kutta Schemes

W. Oevel

Department of Mathematics,
University of Paderborn,
33095 Paderborn, Germany
Email: walter@uni-paderborn.de

Abstract

An introduction to the theory of symplectic numerical integrators of Runge-Kutta type is presented.

1 Introduction

Recent years have seen a shift in paradigm away from classical considerations which motivated the construction of numerical methods for ordinary differential equations. Traditionally the focus has been on the stability of difference schemes for dissipative systems on compact time intervals. Modern research is instead shifting in emphasis towards the preservation of invariants and the reproduction of correct qualitative features [9, 11, 21]. Some examples of these new challenges include the solution of differential equations on manifolds and the preservation of geometric properties such as isospectral flows [4, 5, 10].

Perhaps the most prevalent, and therefore important, class of physical problems which challenge the new breeds of numerical methods are Hamiltonian systems. The equations of motion for a conservative classical mechanical system with n degrees of freedom may be written as a set of $2n$ first order differential equations in Hamiltonian form:

$$\frac{d}{dt} \begin{pmatrix} \mathbf{q} \\ \mathbf{p} \end{pmatrix} = \begin{pmatrix} \nabla_{\mathbf{p}} H(t, \mathbf{q}, \mathbf{p}) \\ -\nabla_{\mathbf{q}} H(t, \mathbf{q}, \mathbf{p}) \end{pmatrix}. \tag{1.1}$$

The Hamiltonian function $H(t, \mathbf{q}, \mathbf{p}) \in C^1(\mathbb{R} \times \mathbb{R}^{2n})$ represents the total energy. Many Hamiltonian problems of practical interest are of separable form:

$$H(t, \mathbf{q}, \mathbf{p}) = T(t, \mathbf{p}) + V(t, \mathbf{q}), \tag{1.2}$$

where the term $V(t, \mathbf{q})$ is referred to as the potential energy. Most, but not all, of the problems of interest in this class involve an autonomous function of momenta. In particular, Newton's laws of motion with a conservative force field

$$\frac{d^2\mathbf{q}}{dt^2} = -\nabla_{\mathbf{q}} V(t, \mathbf{q})$$

can be modeled using a Hamiltonian with the kinetic energy $T(t, \mathbf{p}) = \mathbf{p}^T\mathbf{p}/2$.

The solution of a dynamical system

$$\frac{d\mathbf{y}}{dt} = \mathbf{f}(\mathbf{y}) , \quad \mathbf{y}(t_0) = \mathbf{y}_0 \tag{1.3}$$

on $\mathbb{R}^N$ with some (smooth) vector field $\mathbf{f} : \mathbb{R}^N \to \mathbb{R}^N$ may be regarded as a 1-parameter family of maps $F_h : \mathbb{R}^N \to \mathbb{R}^N$, the flow of the system. It formally maps the initial condition to the phase vector after the time step $t_0 \to t_0 + h$: $\mathbf{y}(t_0 + h) = F_h(\mathbf{y}_0)$ is the exact solution of the initial value problem. Note, that there is no loss in generality assuming the dynamics to be autonomous. Any system of the form $d\mathbf{y}/dt = \mathbf{f}(t, \mathbf{y})$ may be rewritten as the autonomous system

$$\frac{d}{dt}\begin{pmatrix} t \\ \mathbf{y} \end{pmatrix} = \begin{pmatrix} 1 \\ \mathbf{f}(t, \mathbf{y}) \end{pmatrix} \tag{1.4}$$

on the extended phase space $\mathbb{R} \times \mathbb{R}^N$. The class of Hamiltonian systems (1.1) is distinguished among the set of all dynamical systems on $\mathbb{R}^{2n}$ by the fact that the flow map $F_h : (\mathbf{q}, \mathbf{p}) \to (\mathbf{Q}, \mathbf{P})$ is canonical:

$$d\mathbf{Q} \wedge d\mathbf{P} = d\mathbf{q} \wedge d\mathbf{p} .$$

Equivalently,

$$F_h'(\mathbf{y}) \begin{pmatrix} 0 & \mathbf{I} \\ -\mathbf{I} & 0 \end{pmatrix} F_h'^T(\mathbf{y}) = \begin{pmatrix} 0 & \mathbf{I} \\ -\mathbf{I} & 0 \end{pmatrix}, \tag{1.5}$$

where $F_h'(\mathbf{y})$ is the Jacobian matrix and $\mathbf{I}$ is the identity map on $\mathbb{R}^n$. Consequently, Hamiltonian systems are equipped with two invariants: the energy (a scalar conservation law) and the tensor valued symplectic form $d\mathbf{q} \wedge d\mathbf{p}$. Certain qualitative features are implied by the canonicity of the flow. The most obvious one is the preservation of phase volume (Liouville's theorem), which results from the determinant of (1.5): $\det(F_h'(\mathbf{y})) = 1$.

Numerically, the exact flow map F_h is approximated by a 1-parameter family of maps I_h. The integrator I_h is called **symplectic**, if it also satisfies (1.5). This way all qualitative features of the exact flow induced by (1.5) are inherited by the numerical solution: the symplectic class of numerical methods recreate phase trajectories for Hamiltonian systems more faithfully than standard techniques. However, symplectic integrators are not designed to preserve scalar invariants such as the energy.

The idea of symplectic integration seems to go back to DeVogelaere in 1956. Systematic research in the construction and application of symplectic methods started in the early '80s. Ruth [16] first published symplectic methods for separable problems (1.2). These methods were based on classical generating function techniques [7] and thus only applicable to Hamiltonian problems. The methods of Ruth have since been shown to be in direct correspondence with more established numerical techniques, variants of Runge-Kutta (RK) methods [1, 15, 18, 24]. Reviews on symplectic integration can be found in [18] or [19]. An excellent introduction to this field is given by the recent book of Sanz-Serna and Calvo [22].

Here we review results on symplectic RK methods. In section 2 some elements of Butcher's theory of RK methods are presented. Such schemes are symplectic if the parameters of the methods satisfy a remarkably simple algebraic relation. Unfortunately, it turns out that all symplectic methods are necessarily implicit which makes them quite costly in practical applications. In section 3 the more general family of partitioned RK methods is discussed. It includes schemes which are explicit when applied to Hamiltonian problems of separable form (1.2). Second order equations such as Newtonian ODEs may be solved by RK methods of Nyström type. These will be discussed in section 4.

2   Runge-Kutta (RK) methods

The basic observation behind all numerical approximations of the exact solution of (1.3) is the fact that its Taylor expansion

$$\mathbf{y}(t+h) = \mathbf{y}(t) + h\frac{d}{dt}\mathbf{y}(t) + \frac{h^2}{2!}\frac{d^2}{dt^2}\mathbf{y}(t) + \cdots$$

can be computed algorithmically up to any order: $d\mathbf{y}/dt = \mathbf{f}(\mathbf{y})$ and

$$\frac{d^n\mathbf{y}}{dt^n} = \frac{d^{n-1}}{dt^{n-1}}\,\mathbf{f}(\mathbf{y}) = (\mathbf{f}(\mathbf{y})\cdot\nabla_\mathbf{y})^{n-1}\,\mathbf{f}(\mathbf{y})\,, \quad n = 1, 2, \dots\,,$$

i.e.,

$$
\begin{aligned}
F_h(\mathbf{y}) &= \mathbf{y} + h\mathbf{f} + \frac{h^2}{2!}\,\mathbf{f}'[\mathbf{f}] + \frac{h^3}{3!}\left(\mathbf{f}''[\mathbf{f},\mathbf{f}] + \mathbf{f}'[\mathbf{f}'[\mathbf{f}]]\right) \\
&+ \frac{h^4}{4!}\left(\mathbf{f}'''[\mathbf{f},\mathbf{f},\mathbf{f}] + 3\,\mathbf{f}''[\mathbf{f}'[\mathbf{f}],\mathbf{f}] + \mathbf{f}'[\mathbf{f}''[\mathbf{f},\mathbf{f}]] + \mathbf{f}'[\mathbf{f}'[\mathbf{f}'[\mathbf{f}]]]\right) + \cdots\,,
\end{aligned}
$$

(2.6)

where $\mathbf{f}'[\mathbf{v}] = (\mathbf{v} \cdot \nabla_{\mathbf{y}})\mathbf{f}(\mathbf{y})$, $\mathbf{f}''[\mathbf{v}, \mathbf{w}] = (\mathbf{v} \cdot \nabla_{\mathbf{y}})(\mathbf{w} \cdot \nabla_{\mathbf{y}})\mathbf{f}(\mathbf{y})$ etc. A numerical approximation I_h of the flow F_h is said to be of **order** p, if its expansion w.r.t. the stepsize h matches the series (2.6) "term by term" through order h^p. Clearly, integrators of arbitrary order are available by cutting off (2.6) after h^p, leaving a map which can be computed in finite time (the "Taylor scheme of order p"). However, in any time step all partial derivatives of $\mathbf{f}$ w.r.t. $\mathbf{y}$ through order $p - 1$ need to be evaluated. This makes the Taylor schemes much too expensive for systems with many degrees of freedom.

A traditional family of numerical integrators are methods of RK type [2, 3], which try to approximate (2.6) by evaluating the vector field $\mathbf{f}$ at several neighbouring points without using partial derivatives:

Definition: *An s-stage RK scheme for solving $dy/dt = \mathbf{f}(\mathbf{y})$ is a map of the form*

$$I_h(\mathbf{y}) = \mathbf{y} + h \sum_{j=1}^{s} b_j \, \mathbf{f}(\mathbf{y}_j) \,,$$

where the "internal stages" $\mathbf{y}_j$ are determined as the solutions of a system of equations of the form

$$\mathbf{y}_i = \mathbf{y} + h \sum_{j=1}^{s} a_{ij} \, \mathbf{f}(\mathbf{y}_j) \,, \quad i = 1, \ldots, s \,.$$

For non-autonomous systems the time step $t \to t + h$ is given by applying the RK map to (1.4). One finds

$$I_h(t, \mathbf{y}) = \mathbf{y} + h \sum_{j=1}^{s} b_j \, \mathbf{f}(t + c_j h, \mathbf{y}_j)$$

with $\mathbf{y}_i = \mathbf{y} + h \sum_{j=1}^{s} a_{ij} \, \mathbf{f}(t + c_j h, \mathbf{y}_j)$ and $c_i = \sum_{j=1}^{s} a_{ij}$. The parameters of the method are presented in a "Butcher table"

$$
\begin{array}{c|ccc}
c_1 & a_{11} & \cdots & a_{1s} \\
\vdots & \vdots & \ddots & \vdots \\
c_s & a_{s1} & \cdots & a_{ss} \\
\hline
 & b_1 & \cdots & b_s
\end{array}
\qquad \text{with} \qquad c_i = \sum_{j=1}^{s} a_{ij} \,.
$$

The scheme is **explicit**, if (a_{ij}) is a strictly lower diagonal matrix. In this case the internal stages are evaluated recursively without the need to solve an equation. For implicit methods the system defining the internal stages needs to be solved numerically, either by a Newton scheme or by a fixed point iteration.

For the investigation of the order of the RK map one needs the Taylor expansion of I_h in order to match the terms in (2.6). Following Butcher's classical theory [3] the terms of the expansion may be encoded by graphical objects ("rooted trees" $\rho\tau$). One finds a representation of I_h similar to (2.6), where each term is multiplied with a factor which is a polynomial of the Butcher parameters (the "weight" $\Phi(\rho\tau)$). Matching the series of I_h and F_h one finds that the RK method is of order p for arbitrary systems (1.3), if and only if the order equations $\Phi(\rho\tau) = 1/\gamma(\rho\tau)$ hold for all rooted trees $\rho\tau$ through p vertices. The "density" γ, stemming from the coefficients of the terms in (2.6), is some combinatorical quantity (a rational number) associated with the trees. A modified formulation of the order theory, more adequate for symplectic and self-adjoint methods, is developed in [23].

The first 8 of these order equations, necessary and sufficient for order 4, are given in table 1. For higher orders the complexity of the order conditions becomes awesome: the number of equations that have to be satisfied by the parameters grows exponentially with the desired order:

order p	1	2	3	4	5	6	7	8	9	10	...	20
# of order equations	1	2	4	8	17	37	85	200	486	1205	...	20 247 374

For a given stage number s, the number of free parameters (a_{ij}), (b_j) is $s(s+1)$. The order equations represent a highly overdetermined system of polynomial equations, many of which are dependent. Surprisingly, with an appropriate choice of the parameters one can always attain order $2s$:

Theorem [3]: *Let $c_1, \ldots, c_s$ be the roots of the Legendre polynomial $P_s(c) = \frac{d^s}{dc^s} c^s (1-c)^s$. The s-stage RK scheme defined by*

$$\sum_{j=1}^{s} b_j c_j^{k-1} = \frac{1}{k}, \quad \sum_{j=1}^{s} a_{ij} c_j^{k-1} = \frac{c_i^k}{k}, \quad i,k = 1,\ldots,s$$

is of order $2s$.

Solving these linear equations for the parameters (b_j) and (a_{ij}) one finds the "Gauß-Legendre schemes". The 1-stage scheme of order 2

$$\begin{array}{c|c} \frac{1}{2} & \frac{1}{2} \\ \hline & 1 \end{array}$$

order	term	$\rho\tau$	$\Phi(\rho\tau) = \dfrac{1}{\gamma(\rho\tau)}$
1	$\mathbf{f}$	(tree)	$\sum\limits_{j=1}^{s} b_j = 1$
2	$\mathbf{f'[f]}$	(tree)	$\sum\limits_{j=1}^{s} b_j c_j = \dfrac{1}{2}$
3	$\mathbf{f''[f,f]}$	(tree)	$\sum\limits_{j=1}^{s} b_j c_j^2 = \dfrac{1}{3}$
3	$\mathbf{f'[f'[f]]}$	(tree)	$\sum\limits_{j=1}^{s}\sum\limits_{k=1}^{s} b_j\, a_{jk}\, c_k = \dfrac{1}{6}$
4	$\mathbf{f'''[f,f,f]}$	(tree)	$\sum\limits_{j=1}^{s} b_j c_j^3 = \dfrac{1}{4}$
4	$\mathbf{f''[f'[f],f]}$	(tree)	$\sum\limits_{j=1}^{s}\sum\limits_{k=1}^{s} b_j\, c_j\, a_{jk}\, c_k = \dfrac{1}{8}$
4	$\mathbf{f'[f''[f,f]]}$	(tree)	$\sum\limits_{j=1}^{s}\sum\limits_{k=1}^{s} b_j\, a_{jk}\, c_k^2 = \dfrac{1}{12}$
4	$\mathbf{f'[f'[f'[f]]]}$	(tree)	$\sum\limits_{j=1}^{s}\sum\limits_{k=1}^{s}\sum\limits_{l=1}^{s} b_j\, a_{jk}\, a_{kl}\, c_l = \dfrac{1}{24}$

Table 1: The order equations through order 4.

is the classical implicit mid-point rule, for which the time step $\mathbf{y} \to \mathbf{Y} = I_h(\mathbf{y})$ is determined by solving the equation

$$\mathbf{Y} = \mathbf{y} + h\,\mathbf{f}\!\left(\frac{1}{2}(\mathbf{y}+\mathbf{Y})\right).$$

The 2-stage scheme of order 4 is

$$
\begin{array}{c|cc}
\frac{1}{2}-\frac{\sqrt{3}}{6} & \frac{1}{4} & \frac{1}{4}-\frac{\sqrt{3}}{6} \\
\frac{1}{2}+\frac{\sqrt{3}}{6} & \frac{1}{4}+\frac{\sqrt{3}}{6} & \frac{1}{4} \\
\hline
 & \frac{1}{2} & \frac{1}{2}
\end{array}
$$

The 3-stage scheme of order 6 is

$$
\begin{array}{c|ccc}
\frac{1}{2} - \frac{\sqrt{15}}{10} & \frac{5}{36} & \frac{2}{9} - \frac{\sqrt{15}}{15} & \frac{5}{36} - \frac{\sqrt{15}}{30} \\[2mm]
\frac{1}{2} & \frac{5}{36} + \frac{\sqrt{15}}{24} & \frac{2}{9} & \frac{5}{36} - \frac{\sqrt{15}}{24} \\[2mm]
\frac{1}{2} + \frac{\sqrt{15}}{10} & \frac{5}{36} + \frac{\sqrt{15}}{30} & \frac{2}{9} + \frac{\sqrt{15}}{15} & \frac{5}{36} \\[2mm]
\hline
& \frac{5}{18} & \frac{4}{9} & \frac{5}{18}
\end{array}
$$

Around 1988, Lasagni [12], Sanz-Serna [17] and Suris [24] independently discovered that Runge-Kutta methods can be symplectic and the condition has since been proven to also be necessary. The map I_h corresponding to a Butcher-tableau $(A, \mathbf{b})$ is symplectic for all Hamiltonian systems if and only if

$$
b_i\, a_{ij} + b_j\, a_{ji} - b_i\, b_j = 0, \quad i, j = 1, \ldots, s . \tag{2.7}
$$

A proof, using differential forms, can be found in Sanz-Serna's recent survey article [18]. We note that consistent symplectic RK schemes are necessarily implicit: not all diagonal elements $a_{ii} = b_i/2$ can vanish.

The constraints (2.7) among the Butcher parameters have to be satisfied in addition to the order conditions. It turns out that the symplecticity condition does not impose any order barriers on the methods. On the contrary, on the submanifold of parameters satisfying (2.7) many of the order equations become dependent, whence symplecticity actually aids in the construction of high order methods. In fact, the Gauß-Legendre schemes given in the theorem above are symplectic. Another class of interesting symplectic methods are the diagonally implicit schemes with Butcher tables of the form

$$
\begin{array}{c|cccccc}
c_1 & \frac{1}{2} b_1 & & & & & 0 \\[1mm]
c_2 & b_1 & \frac{1}{2} b_2 & & & & \\[1mm]
c_3 & b_1 & b_2 & \ddots & & & \\[1mm]
\vdots & \vdots & & \ddots & \ddots & \frac{1}{2} b_{s-1} & \\[1mm]
c_s & b_1 & b_2 & \cdots & b_{s-1} & \frac{1}{2} b_s & \\[1mm]
\hline
& b_1 & b_2 & \cdots & b_{s-1} & b_s &
\end{array}
$$

The simplest example is the implicit mid-point rule. A 3-stage scheme of order 4 with

$$
b_1 = b_3 = \frac{1}{3} \left(2 + 2^{1/3} + 2^{-1/3} \right), \quad b_2 = 1 - 2\, b_1
$$

was studied in [6].

A complete classification all symplectic RK schemes enjoying the additional property $I_{-h} = (I_h)^{-1}$ ("self-adjointness" or "symmetry" of the method) through $s = 6$ stages and order $p = 12$ is given in [13]. As an example we represent the most general 5-stage symplectic symmetric RK method of order 8. It is given by

$$
\begin{array}{c|ccccc}
\frac{1}{2}+\tilde{c}_1 & \frac{b_1}{2} & \left(\frac{1}{2}-s_{21}\right)b_2 & \left(\frac{1}{2}-s_{31}\right)b_3 & \left(\frac{1}{2}-s_{41}\right)b_2 & \left(\frac{1}{2}-s_{51}\right)b_1 \\
\frac{1}{2}+\tilde{c}_2 & \left(\frac{1}{2}+s_{21}\right)b_1 & \frac{b_2}{2} & \left(\frac{1}{2}-s_{32}\right)b_3 & \left(\frac{1}{2}-s_{42}\right)b_2 & \left(\frac{1}{2}-s_{41}\right)b_1 \\
\frac{1}{2} & \left(\frac{1}{2}+s_{31}\right)b_1 & \left(\frac{1}{2}+s_{32}\right)b_2 & \frac{b_3}{2} & \left(\frac{1}{2}-s_{32}\right)b_2 & \left(\frac{1}{2}-s_{31}\right)b_1 \\
\frac{1}{2}-\tilde{c}_2 & \left(\frac{1}{2}+s_{41}\right)b_1 & \left(\frac{1}{2}+s_{42}\right)b_2 & \left(\frac{1}{2}+s_{32}\right)b_3 & \frac{b_2}{2} & \left(\frac{1}{2}-s_{21}\right)b_1 \\
\frac{1}{2}-\tilde{c}_1 & \left(\frac{1}{2}+s_{51}\right)b_1 & \left(\frac{1}{2}+s_{41}\right)b_2 & \left(\frac{1}{2}+s_{31}\right)b_3 & \left(\frac{1}{2}+s_{21}\right)b_2 & \frac{b_1}{2} \\
\hline
& b_1 & b_2 & b_3 & b_2 & b_1
\end{array}
$$

with

$$
b_1 = \frac{1}{480}\,\frac{3-20\,\tilde{c}_2^2}{\tilde{c}_1^2\,(\tilde{c}_1^2-\tilde{c}_2^2)}\,, \quad b_2 = \frac{1}{480}\,\frac{3-20\,\tilde{c}_1^2}{\tilde{c}_2^2\,(\tilde{c}_2^2-\tilde{c}_1^2)}\,, \quad b_3 = \frac{4}{9} - \frac{1}{420\,\tilde{c}_1^2\,\tilde{c}_2^2}\,,
$$

$$
s_{21} = (\tilde{c}_2 - \tilde{c}_1)\left(10\,\tilde{c}_1\,\tilde{c}_2 + \frac{3}{2} + \sigma\,b_3\right)\,, \quad s_{31} = \tilde{c}_1\left(-\frac{3}{2} + 2\,b_2\,\frac{\tilde{c}_1^2-\tilde{c}_2^2}{\tilde{c}_1^2}\,\sigma\right)\,,
$$

$$
s_{41} = (\tilde{c}_1 + \tilde{c}_2)\left(10\,\tilde{c}_1\,\tilde{c}_2 - \frac{3}{2} - \sigma\,b_3\right)\,, \quad s_{32} = \tilde{c}_2\left(-\frac{3}{2} + 2\,b_1\,\frac{\tilde{c}_2^2-\tilde{c}_1^2}{\tilde{c}_2^2}\,\sigma\right)\,,
$$

$$
s_{42} = \frac{\tilde{c}_2}{b_2}\left(-\frac{1}{3} + \frac{1}{280\,\tilde{c}_2^2\,(\tilde{c}_2^2-\tilde{c}_1^2)} + 2\,b_1\,b_3\,\frac{\tilde{c}_1^2}{\tilde{c}_2^2}\,\sigma\right)\,,
$$

$$
s_{51} = \frac{\tilde{c}_1}{b_1}\left(-\frac{1}{3} + \frac{1}{280\,\tilde{c}_1^2\,(\tilde{c}_1^2-\tilde{c}_2^2)} + 2\,b_2\,b_3\,\frac{\tilde{c}_2^2}{\tilde{c}_1^2}\,\sigma\right)
$$

involving 3 parameters σ und $\tilde{c}_1, \tilde{c}_2$. In this parametrization the 200 order equations for order 8 collapse to one single equation:

$$
(20\,\tilde{c}_1^2 - 3)\,(20\,\tilde{c}_2^2 - 3) = -\frac{12}{7}\,,
$$

whence one obtains a two-parameter family of methods.

3   Partitioned Runge-Kutta (PRK) Methods

Consider dynamical systems on $\mathbb{R}^{2n}$ of the special form

$$
\frac{d}{dt}\begin{pmatrix} \mathbf{q} \\ \mathbf{p} \end{pmatrix} = \begin{pmatrix} \mathbf{f}(t,\mathbf{p}) \\ \mathbf{F}(t,\mathbf{q}) \end{pmatrix} \tag{3.8}
$$

with $\mathbf{q}, \mathbf{p} \in \mathbb{R}^n$ and (smooth) vector fields $\mathbf{f}, \mathbf{F} : \mathbb{R} \times \mathbb{R}^n \to \mathbb{R}^n$. Separable Hamiltonians (1.2) lead to Hamiltonian systems

$$\frac{d}{dt} \begin{pmatrix} \mathbf{q} \\ \mathbf{p} \end{pmatrix} = \begin{pmatrix} \nabla_{\mathbf{p}} T(t, \mathbf{p}) \\ -\nabla_{\mathbf{q}} V(t, \mathbf{q}) \end{pmatrix} \qquad (3.9)$$

of this type.

A PRK scheme is an integrator of the form

$$I_h \left(\begin{pmatrix} \mathbf{q} \\ \mathbf{p} \end{pmatrix} \right) = \begin{pmatrix} \mathbf{q} + h \sum_{j=1}^{s} b_j\, \mathbf{f}(t + C_j h, \mathbf{p}_j) \\ \mathbf{p} + h \sum_{j=1}^{s} B_j\, \mathbf{F}(t + c_j h, \mathbf{q}_j) \end{pmatrix} \qquad (3.10)$$

with the internal stages $\mathbf{q}_i, \mathbf{p}_i$ defined by

$$\mathbf{q}_i = \mathbf{q} + h \sum_{j=1}^{s} a_{ij}\, \mathbf{f}(t + C_j h, \mathbf{p}_j)\,, \quad \mathbf{p}_i = \mathbf{p} + h \sum_{j=1}^{s} A_{ij}\, \mathbf{F}(t + c_j h, \mathbf{q}_j) \quad (3.11)$$

for $i = 1, \ldots, s$. It consists of two different RK methods applied to each set of variables $\mathbf{q}$ and $\mathbf{p}$, respectively. We note that explicit methods are obtained by Butcher tables of the form

$$
\begin{array}{c|cccc}
0 & 0 & & & 0 \\
c_2 & a_{21} & \ddots & & \\
\vdots & \vdots & \ddots & \ddots & \\
c_s & a_{s1} & \cdots & a_{s,s-1} & 0 \\
\hline
 & b_1 & \cdots & \cdots & b_s
\end{array}
\qquad
\begin{array}{c|cccc}
C_1 & A_{11} & & & 0 \\
\vdots & \vdots & \ddots & & \\
\vdots & \vdots & \ddots & \ddots & \\
C_s & A_{s1} & \cdots & \cdots & A_{ss} \\
\hline
 & B_1 & \cdots & \cdots & B_s
\end{array}
$$

with $c_i = \sum_{j=1}^{s} a_{ij}$, $C_i = \sum_{j=1}^{s} a_{ij}$.

Sanz-Serna [20] and Suris [25] independently showed (see also [1]) that the PRK scheme (3.10)/(3.11) applied to a separable Hamiltonian system (3.9) defines a symplectic map I_h if

$$b_i A_{ij} + B_j a_{ji} = b_i B_j\,, \quad i, j = 1, \ldots, s\,.$$

Remarkably, there exist explicit symplectic PRK schemes characterized by

$$
\begin{array}{c|cccccc}
0 & 0 & & & & & 0 \\
c_1 & b_1 & 0 & & & & \\
c_2 & b_1 & b_2 & 0 & & & \\
\vdots & \vdots & & \ddots & \ddots & \ddots & \\
c_s & b_1 & b_2 & \cdots & b_{s-1} & 0 & \\
\hline
 & b_1 & b_2 & \cdots & b_{s-1} & b_s
\end{array}
\qquad
\begin{array}{c|cccccc}
C_1 & B_1 & & & & & 0 \\
C_2 & B_1 & B_2 & & & & \\
C_3 & B_1 & B_2 & B_3 & & & \\
\vdots & \vdots & & \ddots & \ddots & \ddots & \\
C_s & B_1 & B_2 & \cdots & B_{s-1} & B_s & \\
\hline
 & B_1 & B_2 & \cdots & B_{s-1} & B_s
\end{array}
$$

The order theory for PRK methods may be formulated in terms of coloured rooted trees [8, 14]. A list of of explicit symplectic schemes may be found in [22]. In particular, a 4-stage scheme satisfying the order conditions through order 4 is given by

$$b_1 = b_3 = \frac{1}{3}\left(2 + 2^{1/3} + 2^{-1/3}\right), \quad b_2 = 1 - 2\,b_1, \quad b_4 = 0,$$

$$B_1 = B_4 = \frac{b_1}{2}, \quad B_2 = B_3 = \frac{b_1 + b_2}{2}.$$

4 Runge-Kutta-Nyström (RKN) Methods

Consider the 2nd order system

$$\frac{d^2\mathbf{q}}{dt^2} = \mathbf{F}(t,\mathbf{q}), \quad t \in \mathbb{R}, \quad \mathbf{q} \in \mathbb{R}^n$$

with some (smooth) vector field $\mathbf{F} : \mathbb{R} \times \mathbb{R}^n \to \mathbb{R}^n$. The PRK step (3.10) applied to (3.8) with $\mathbf{f}(\mathbf{p}) = \mathbf{p}$ leads to integrators of the form

$$I_h\left(\begin{pmatrix}\mathbf{q}\\\mathbf{p}\end{pmatrix}\right) = \begin{pmatrix}\mathbf{q} + h\sum_{j=1}^{s} b_j\,\mathbf{p}_j\\[2mm]\mathbf{p} + h\sum_{j=1}^{s} B_j\,\mathbf{F}(t+c_jh,\mathbf{q}_j)\end{pmatrix}$$

with the internal stages

$$\mathbf{q}_i = \mathbf{q} + h\sum_{j=1}^{s} a_{ij}\,\mathbf{p}_j, \quad \mathbf{p}_i = \mathbf{p} + h\sum_{j=1}^{s} A_{ij}\,\mathbf{F}(t+c_jh,\mathbf{q}_j), \quad i = 1,\dots,s.$$

With

$$c_i = \sum_{j=1}^{s} a_{ij}, \quad \alpha_{ij} = \sum_{k=1}^{s} a_{ik}\,A_{kj}, \quad \beta_j = \sum_{i=1}^{s} b_i\,A_{ij}$$

elimination of the internal stages $\mathbf{p}_i$ in conjunction with the consistency assumption $\sum_j b_j = 1$ yields the family of RKN integrators

$$I_h\left(\begin{pmatrix}\mathbf{q}\\\mathbf{p}\end{pmatrix}\right) = \begin{pmatrix}\mathbf{q} + h\,\mathbf{p} + h^2\sum_{j=1}^{s}\beta_j\,\mathbf{F}(t+c_jh,\mathbf{q}_j)\\[2mm]\mathbf{p} + h\sum_{j=1}^{s} B_j\,\mathbf{F}(t+c_jh,\mathbf{q}_j)\end{pmatrix} \qquad (4.12)$$

with the s internal stages $\mathbf{q}_i$ defined by the algebraic equations

$$\mathbf{q}_i = \mathbf{q} + h\,c_i\,\mathbf{p} + h^2 \sum_{j=1}^{s} \alpha_{ij}\,\mathbf{F}(t + c_j h, \mathbf{q}_j)\,, \quad i = 1, \ldots, s\,. \tag{4.13}$$

Its Butcher table is given by

c_1	α_{11}	$\cdots$	α_{1s}
$\vdots$	$\vdots$	$\ddots$	$\vdots$
c_s	α_{s1}	$\cdots$	α_{ss}
	β_1	$\cdots$	β_s
	B_1	$\cdots$	B_s

Explicit methods are obtained for strictly lower triangular matrices (α_{ij}).

Suris [24] showed that the RKN method (4.12)/(4.13) is symplectic when applied to a Hamiltonian system $d^2\,\mathbf{q}/dt^2 = -\nabla_{\mathbf{q}} V(t, \mathbf{q})$, if

$$\beta_j = B_j(1 - c_j)\,, \quad B_i\,\alpha_{ij} - B_j\,\alpha_{ji} = B_i\,B_j\,(c_i - c_j)$$

for $i, j = 1, \ldots, s$. Explicit symplectic methods are characterized by

$$\beta_j = B_j(1 - c_j)\,, \quad \alpha_{ij} = \begin{cases} B_j\,(c_i - c_j) & \text{for } 1 \le j < i \le s\,, \\ 0 & \text{for } 1 \le i \le j \le s\,. \end{cases}$$

Examples of explicit symplectic RKN schemes are given in [22].

References

[1] L. ABIA AND J.M. SANZ-SERNA, *Partitioned Runge-Kutta Methods for Separable Hamiltonian Problems*, Math. Comp. 60, No. 202 (1993), pp. 617–634.

[2] P. ALBRECHT, *The Runge-Kutta Theory in a Nutshell*, SIAM J. Numer. Anal., 33 (1996) (to appear).

[3] J.C. BUTCHER, *The Numerical Analysis of Ordinary Differential Equations*, John Wiley and Sons, Chichester, 1987.

[4] M.P. CALVO, A. ISERLES AND A. ZANNA, *Runge-Kutta methods for orthogonal and isospectral flows*, Appl. Num. Maths. (to appear).

[5] L. DIECI, B. RUSSELL AND E. VAN DER VLECK, *Unitary integrators and Applications to Continuous Orthonormalization Techniques*, SIAM J. Numer. Anal., 31 (1994), pp. 261–281.

[6] J. DE FRUTOS AND J. M. SANZ-SERNA, *An easily implementable fourth-order method for the time integration of wave problems*, J. Comput. Phys., 103 (1992), pp. 160–168.

[7] H. GOLDSTEIN, *Classical Mechanics*, Second edition, Addison-Wesley, London, 1980.

[8] E. HAIRER, S. P. NØRSETT AND G. WANNER, *Solving Ordinary Differential Equations I: Nonstiff Problems*, 2nd Edition, Springer, Berlin, 1993.

[9] A. ISERLES, *Beyond the classical theory of computational ordinary differential equations*, in State of the Art in Numerical Analysis (to appear).

[10] A. ISERLES, *Numerical Methods on (and off) Manifolds*, DAMPT Tech. Rep. 1996/NA12, University of Cambridge.

[11] A. ISERLES AND A. ZANNA, *Qualitative numerical analysis of ordinary differential equations*, To appear in *Lectures in Applied Mathematics*, J. Renegar, A. Shub and S. Smale eds., Amer. Math. Soc. Providence, RI.

[12] F.M. LASAGNI, *Canonical Runge–Kutta methods*, ZAMP, 39 (1988), pp. 952–953.

[13] W. OEVEL AND M. SOFRONIOU, *Symplectic Runge-Kutta Schemes II: Classification of Symmetric Methods*, preprint, 1996.

[14] W. OEVEL AND M. SOFRONIOU, *Modified Order Theory for partitioned Runge-Kutta and Runge-Kutta-Nyström methos*, preprint, 1996.

[15] D. OKUNBOR AND R.D. SKEEL, *Explicit Canonical Methods for Hamiltonian Systems*, Math. Comp. 59, No. 200 (1992), pp. 439–455.

[16] R.D. RUTH, *A canonical integration technique*, IEEE Trans. on Nuclear Science **30** (1983), pp. 2669–2671.

[17] J. M. SANZ-SERNA, *Runge-Kutta schemes for Hamiltonian systems*, BIT, 28 (1988), pp. 877–883.

[18] J. M. SANZ-SERNA, *Symplectic integrators for Hamiltonian problems: an overview*, Acta Numerica, 1 (1992), pp. 243–286.

[19] J.M. SANZ-SERNA, *Symplectic Runge-Kutta and related methods: recent results*, Physica D, **60** (1992), pp. 293–302.

[20] J.M. SANZ-SERNA, *The numerical integration of Hamiltonian systems*, In *Computational Ordinary Differential Equations*, J.R. Cash and I. Gladwell eds., Clarendon Press, Oxford (1992), pp. 437–449.

[21] J.M. SANZ-SERNA, *Geometric Integration*, in State of the Art in Numerical Analysis (to appear).

[22] J. M. SANZ-SERNA AND M. P. CALVO, *Numerical Hamiltonian Problems*, Chapman and Hall, London, 1994.

[23] M. SOFRONIOU AND W. OEVEL, *Symplectic Runge-Kutta Schemes I: Order Conditions*, SIAM J. Num. Anal., 34–5 (1997), to appear.

[24] Y.B. SURIS, *On the conservation of symplectic structure in the course of numerical integration of Hamiltonian systems*, in: *Numerical Solutions of Differential Equations*, S. Filippov (Ed.), Moscow, Keldysh Inst. of Appl. Math., 1988, pp. 148-160. (In Russian.); *Canonical transformations generated by methods of Runge–Kutta type for the numerical integration of the system* $x'' = -\partial U/\partial x$, Zh. Vychisl. Mat. i Mat. Fiz., 29 (1989), pp. 202–211. (In Russian.); translated to English in U.S.S.R Comp. Maths. Math. Phys., 29 (1989), pp. 138–144.

[25] Y.B. SURIS, *Hamiltonian methods of Runge-Kutta type and their variational interpretation*, Math. Model., **2** (1990), pp. 78–87 (In Russian).

Chapter 8

Cellular Automata

Soliton Cellular Automata

C.R. Gilson

Department of Mathematics, University of Glasgow

Glasgow G12 8QW, Scotland, UK.

Email: Claire@maths.gla.ac.uk

Abstract

We shall review some of the past work on cellular automata and its connection with solitons. We suggest possible extensions of this work to higher dimensions.

1 Introduction

Cellular automata have been studied for many years. Much of the work has been connected *not* to coherent structures but rather to pattern formation, chaos, biological systems, dynamical systems and computer oriented applications to name but a few areas [1]. There had been some discussion of coherent structures, for example in 'the game of life' [2] and Margolus' billiard ball models [3].

In 1986 Park, Steiglitz and Thurston [4] proposed a *filter automaton*, this is a type of automaton where newly calculated lattice sites are put to use immediately. Some of these automata support coherent structures and these structures have good interaction properties. This led to the idea of a *solitary wave or particle* and a *soliton* in such a discrete system. A *solitary wave* is taken to be a periodic pattern of non-zero cell values which propagates with a finite constan t velocity. A collision is described as solitonic if these particles retain there structure after a collision.

Fokas, Papadopoulou, Saridakis and Ablowitz [5, 6, 7] examined this kind of cellular automaton in detail and looked at the 'solitonic' interactions within this system, deriving several analytic results about the system. In 1990 Takahashi and Satsuma also proposed an automaton [8] with soliton solutions, this system is particularly interesting as its only solutions are solitons. In sections 2 and 3 we shall discuss these two systems (PS&T's and T&S's). In section 4 we shall loo k at possible ways of moving into higher dimensions. In general this turns out to be a difficult problem, the simplicity of the 1-dimensional

systems appears to be lost. There are some two dimensional examples which
have been proposed these include syst ems which are direct products of one
dimensional systems and systems where a given site is allowed to take matrix values. For several of these systems it is not difficult to find coherent
structures which behave well when moving on their own, but often things go
wrong under interactions. In section 5 we shall draw a few conclusions. The
approach taken through out this paper is phenomenological, with the main
emphasis on looking at how these automata work, rather than a rigorous
mathematical analysis.

2   The Cellular Automaton of Park, Steiglitz and Thurston

2.1   A Cellular Automaton

A simple cellular automaton (CA) [1] consists of a discrete time evolution
of an infinite row of sites each taking the value 0 or 1. The discrete time
evolution is usually governed by a simple rule. For example, If the value of a
site at position i at time t is given by a_i^t then the rule may take the form

$$a_i^{t+1} = \phi(a_{i-r}^t, a_{i-r+1}^t, \ldots\ldots a_{i-1}^t, a_i^t, \ldots\ldots, a_{i+r}^t). \tag{1}$$

More generally a site may have any finite number of values and the rule could
also depend upon other time steps.

2.2   The CA of Park, Steiglitz and Thurston

Park, Steiglitz and Thurston (PS&T) [4] look at a so called filter automaton.
In such an automaton, the next state rule is of the form

$$a_i^{t+1} = \phi(a_{i-r}^{t+1}, a_{i-r+1}^{t+1}, \ldots\ldots, a_{i-1}^{t+1}, a_i^t, \ldots\ldots, a_{i+r}^t). \tag{2}$$

Thus the next state is computed using the newly updated values $a_{i-r}^{t+1}, \ldots, a_{i-1}^{t+1}$,
rather than $a_{i-r}^t, \ldots, a_{i-1}^t$. To prevent ambiguity in computing the evolution
of the automaton, an assumption is made that far enough to the left, all the
lattice points are zero.

RULE: Let a_j^t be the value of the j-th lattice point at time t. Calculating
from left to right,

$$a_i^{t+1} = \begin{cases} 1 & \text{if } S \text{ is even,} \\ 0 & \text{if } S \text{ is odd or zero,} \end{cases} \tag{3}$$

where

$$S = \sum_{j=1}^{r} a_{i-j}^{t+1} + \sum_{j=0}^{r} a_{i+j}^{t}, \tag{4}$$

r is called the *radius* and takes the value of a positive integer.

Consider an example, represented in (5), where the radius is 3. To find the value in the box with #, you add together the values of all the boxes with +'s, if the result is even and non-zero then you let the value in the # box, a_i^{t+1}, be 1, otherwise take it as 0.

$$
\begin{array}{l}
t \;\longrightarrow \\
t+1 \longrightarrow
\end{array}
\quad
\begin{array}{|c|c|c|c|c|c|c|c|}
\hline
 & & & & + & + & + & + \\
\hline
\end{array}
\tag{5}
$$

For convenience of notation we shall shift the spatial coordinates by a distance r and look at the correspondingly altered S

$$S = \sum_{j=1}^{r} a_{i-j}^{t+1} + \sum_{j=0}^{r} a_{i-j}^{t}. \tag{6}$$

The diagram corresponding to (5) is now

$$
\begin{array}{|c|c|c|c|}
\hline
+ & + & + & + \\
\hline
+ & + & + & \# \\
\hline
\end{array}
\tag{7}
$$

2.3   The Fast Rule Theorem (FRT)

For this particular automaton there is an easy way of calculating the next lattice point value, this was introduced by Fokas et al. [5].

2.3.1   FRT

STEP 1: Working from left to right, put a box around the left most '1' and then every $(r+1)$th lattice site, (unless $\exists\,(r+1)$ 0's after a given box, if this happens put a box around the first available '1' after the 0's and continue as before).

STEP 2: Change all the boxed sites $1 \to 0$, $0 \to 1$, and leave the unboxed sites alone.

Below is an example of this process:

$t=0$	0	$\boxed{1}$	1	0	1	$\boxed{0}$	1	0	0	$\boxed{0}$	0	0	0
$t=1$	0	0	$\boxed{1}$	0	1	1	$\boxed{1}$	0	0	1	$\boxed{0}$	0	0
$t=2$	0	0	0	0	$\boxed{1}$	1	0	0	$\boxed{0}$	1	1	0	$\boxed{0}$
$t=3$	0	0	0	0	0	1	0	0	1	1	1	0	1

Starting with certain patterns of 0's and 1's at a time $t = 0$ patterns that repeat after a certain number of time steps can be produced. In the examples below a '1' has been represented by a '$\star$' or a '$\bullet$' and a zero by a '.'

The above pattern can be seen to repeat every three time steps. The collection of 1's can be regarded as a particle moving with a speed 4/3, ie; it shifts 4 sites in 3 units of time. Below we have a second example, where it takes eight time steps for the pattern to repeat. In this time the pattern has shifted 16 sites to the right. Thus we may think of this as a particle with speed $16/8 = 2$ sites per unit of time.

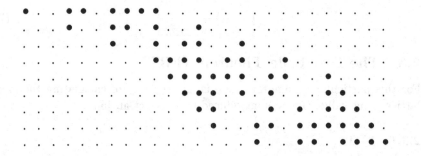

Not only do these particles propagate on their own as 'coherent structures' but, as they travel at different speeds, they can be made to collide with each other. Some collisions are solitonic in nature, ie; the particles will emerge from a collision u nscathed apart from a phase shift. This type of behaviour resembles that of solitons.

2.4 Formulation as a Difference Equation

This kind of cellular automaton derived from a rule can usually be formulated as a difference equation, although its form will not be unique. Here we may take

$$a_n^{t+1} = \left\{ \sum_{j=0}^{r} a_{n-j}^t - \sum_{j=1}^{r} a_{n-j}^{t+1} + \max[a_{n-r}^t \cdots a_n^t; a_{n-r}^{t+1} \cdots a_{n-1}^{t+1}] \right\} \mod 2. \tag{8}$$

The system, as we have seen above, supports soliton–like structures. However, not all configurations display this behaviour. In general we find structures breaking up into smaller parts analogously to the evolution of non-soliton initial conditions fo r soliton equations. In some circumstances there can be a loss of energy from the system, for instance if you take $r = 3$ and look at an initial set up where the 1's are spaced out by three 0's then we find at the second time step the 1's have disappeared.

$$t = 0 \rightarrow \quad . \ \boxed{\bullet} \ . \ . \ . \ \boxed{\bullet} \ . \ . \ .$$
$$t = 1 \rightarrow \quad . \quad . \ . \ . \ . \ . \ . \ . \ .$$

A small adaptation in the system introduced by Z.Jiang [9] serves to reduce energy losses. In place of (8) he introduced

$$a_n^{t+1} = \left\{ \sum_{j=0}^{r} a_{n-j}^t - \sum_{j=1}^{r} a_{n-j}^{t+1} + \max[a_{n-r-1}^t \cdots a_n^t; a_{n-r}^{t+1} \cdots a_{n-1}^{t+1}] \right\} \mod 2. \tag{9}$$

The difference equation now considers the a_{n-r-1}^t entry in the lattice, this has the effect of 'plugging up' the energy leakage. In the above example the $\bullet$ gets reintroduced into the system at each time step.

3 The Box and Ball System

Takahashi and Satsuma proposed a very simple and elegant cellular automaton [8] which contains only solitons, this takes the form of a so called 'box and ball system'. It is a $1 + 1$ dimensional system with the rule:

$$u_j^{t+1} = \begin{cases} 1 & \text{if } u_j^t = 0 \text{ and } \sum_{i=-\infty}^{j-1} u_i^{t+1} > \sum_{i=-\infty}^{j-1} u_i^{t+1} \\ 0 & \text{otherwise.} \end{cases} \tag{10}$$

This corresponds to taking a line of boxes, either containing a ball *or* empty, for instance:

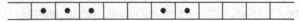

The rule corresponds to taking the leftmost ball out of its box and putting it in the first empty box to its right, then taking the new leftmost ball (as long as it has not already been moved at this time step) and moving it to the first empty box to its right. One continues with this process until each ball has been moved, this represents one time step. Applying this rule to the above example gives

It is easily seen from the way the process is defined that, if the number of balls represents the energy, then there is *no* energy loss from the system (as the balls can only be moved *along* the rows and cannot be removed from the system). In addition to this, a row of balls next to each other, without any empty boxes between them represents a particle, the larger the particle the faster it moves. For instance consider a four ball particle:

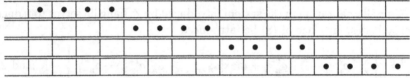

When we look at interactions of 'particles' we find that after an interaction the particles re-form but will be phase-shifted. This is very like what one would expect in a collision of solitons in a continuous system. Due to the conservation of energy and particle conservation these 'particles' can be thought of as solitons.

In the diagram on the following page there is an interaction of 4 particles, formed from 5,3,2 and 1 balls respectively. They have phase shifts due to the interaction and they retain their shapes after the interaction. The diagram has been printed on i ts side in order to fit on the page.

3.1   Associated Difference Equation

This automaton can be constructed as a difference equation, this is most easily done in terms of a new variable. Firstly we shall introduce a function,

$$F(n) = \begin{cases} n & \text{for } n \neq 0 \\ 1 & \text{for } n = 0. \end{cases} \tag{11}$$

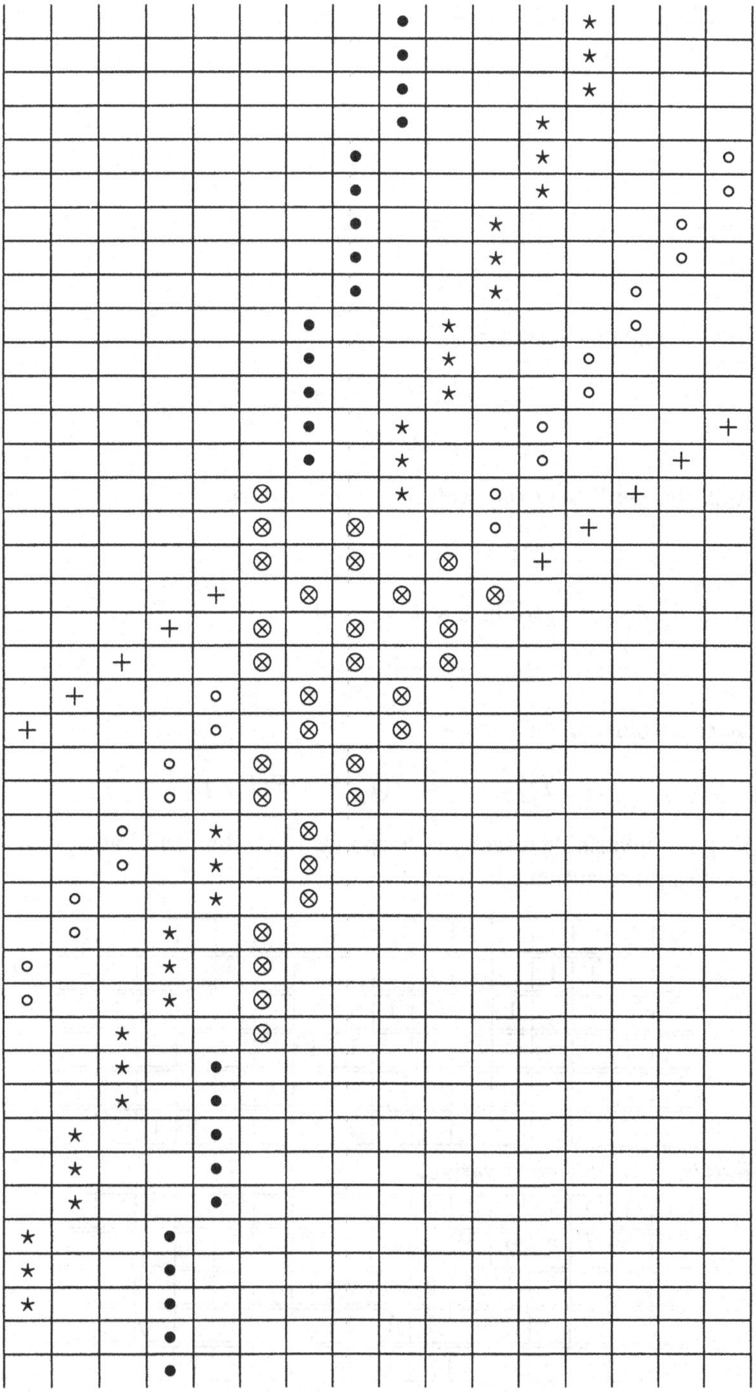

Pattern chart (16 columns wide). Symbols: ● (filled dot), ★ (star), ○ (open circle), ⊗ (crossed circle), + (plus).

1	2	3	4	5	6	7	8	9	10	11	12	13	14	15	16
								●				★			
								●				★			
								●				★			
								●			★				
							●				★				○
							●				★				○
							●			★				○	
							●			★				○	
							●			★			○		
						●			★				○		
						●			★			○			
						●			★			○			
						●		★			○				+
						●		★			○			+	
					⊗			★		○			+		
					⊗		⊗			○		+			
					⊗		⊗		⊗		+				
				+		⊗		⊗		⊗					
			+		⊗		⊗		⊗						
		+			⊗		⊗		⊗						
	+		○		⊗		⊗								
+			○		⊗		⊗								
		○		⊗		⊗									
		○		⊗		⊗									
	○		★		⊗										
	○		★		⊗										
○			★		⊗										
○			★		⊗										
	○	★		⊗											
		★		⊗											
		★		⊗											
	★		●												
	★		●												
	★		●												
	★		●												
★			●												
★			●												
★			●												
			●												
			●												

This F will be the non-linear part to the equation. The new variable we shall introduce is a partial sum S_j^t,

$$S_j^t = \sum_{i=-\infty}^{j} u_i^t \tag{12}$$

where u_i^t is the value of the ith entry at the time step t

$$u_i^t = 0 \quad \text{or} \quad 1. \tag{13}$$

Note that from the construction,

$$u_j^t = S_j^t - S_{j-1}^t \tag{14}$$

$$S_j^t \geq S_j^{t+1}. \tag{15}$$

The equation describing the evolution of this system is

$$S_{j+1}^{t+1} - S_j^t = 1 - F(S_{j+1}^t - S_j^{t+1}). \tag{16}$$

Alternatively we can introduce a new variable

$$T_j^t = S_{j+1}^t - S_j^{t+1} \tag{17}$$

so that the evolution equation is

$$T_{j+1}^{t+1} - T_j^t = F(T_j^{t+1}) + F(T_{j+1}^t). \tag{18}$$

Using the u variable directly corresponds to the box and ball system, thus a two soliton interaction;

1	1				1										
	1	1			1										
		1	1		1										
			1		1	1									
				1			1	1							
					1			1	1						

becomes in terms of the T variable

1	2	2	1		1	1									
	1	2	2	1	1	1									
		1	2	2	2	2	1								
				1	1	1	2	2	1						
					1	1		1	2	2	1				

4 Higher Dimensional Box and Ball Systems

The system we have been looking at is a $1+1$ dimensional system, the dimensions being discrete time and space. There are many possible ways in which one could try to construct a $2+1$ dimensional box and ball system. Mostly these constructions will not give soliton type solutions. We are going to look at a system which is perhaps *not* truly $2+1$ dimensional but extends the $1+1$ system of Takahashi and Satsuma. This particular system is interesting in that it has a wide variety of solutions.

Instead of taking a row of boxes we shall take a grid of boxes (infinite in extent) so that now the balls have two possible spatial directions to move in. We also need a rule to decide how to move the balls at a particular time step, obviously there are many possibilities. We have chosen one which produces coherent structures that propagate, some of which are preserved under collisions with other such structures. The construction of this system is as follows:

- Start with some initial configuration of balls in the boxes.

- Take the leftmost and uppermost ball, move it to the right until it reaches an empty box, then move it down until it reaches an empty box. This will be the new position of this ball.

- Work through all the balls in this fashion, taking the leftmost and then uppermost ball each time (as long as it has not already been moved at this time step).

- When all the balls have been moved once this gives the configuration after 1 time step.

The example below shows how one configuration of four balls is moved during a time step:

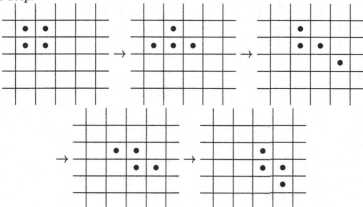

If we label the boxes by the positions of the balls at a particular time step we find that starting with certain initial configurations, the structure undergoes a cycle of shape changes, returning to its original form after a characteristic number of tim e steps. Four simple examples are given below:

0	0																
0	0	1															
		1	1														
			1	2													
			2	2	2												
				3	3												
					3	4											
					3	4	4										
0								5	4								
0	0							5	5								
		1	1					5		6							
			1	2				6	6	6	7						
				2	2						7	7					
					3	3					7	8	8				
					3	4							8	8			
						4	4								9	9	
							5	5							9	9	
	0							5	6								
0	0	0						6	6								
	0	1	1							7	7						
		1	1							7	8						
		1	2	2						8	8						
		4	2	2							9	9					
				3	3						9						
0	0						3	3	4								
	1						3	4	4								
	1	2						4	4								
			2				5	5	5								
				3	3		5	5	6								
					4				6	6							
					4	5			6	6	7						
							5				7	7	7				
								6	6			7	8				
										8			8	8	8		
										8	9		8	9	9		
											9					9	9
																9	

The initial configuration of these particular structures, has been carefully chosen so that the pattern repeats. These structures can be thought of as solitary waves. If you start with some less specific initial configurations then usually the balls will break up in to smaller coherent structures.

To consider these cyclic structures to be solitons they need to reappear after interactions. In general this seems quite difficult to achieve although examples do exist. In general small particles tend to be more stable than large particles. For instan ce if you collide a 2×2 particle with a 1×2 particle then after such a collision they both emerge unscathed. There are actually three types of phase shift occurring here. Firstly, after the collision the particles may have shifted where in their cycle they are. Secondly, how far they have traveled during the collision is different from how far they would individually have traveled if there had been no collision. Thirdly, in general when looking at one particle on its own, it is restr ained to move down a particular diagonal, but after a collision with another particle that diagonal may have changed.

There are, also non-solitonic interactions, in which there is an exchange of energy. For example a five ball particle and a three ball particle can be collided to produce a two ball particle and a six ball particle as shown below.

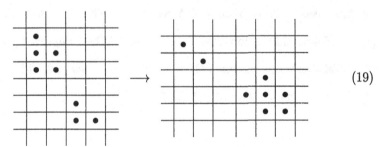

$$(19)$$

5   Conclusions and Discussion

We have seen that there are several cellular automata which exhibit soliton type structures. These automata are usually formulated in terms of a differ- ence equation or a simple rule for determining the value of the next lattice site. In the case of Tak ahashi & Satsuma's work the automaton is defined using a simple rule of moving balls along lines of boxes. Some of the known systems are related to known continuous (integrable) equations while others appear to have little connection to continuous systems.

ns.

In two spatial dimensions some cellular automata systems are known, but at present only very few. Again there are a variety of approaches for de- scribing such systems and in this paper we looked at a 'box and ball' type approach. In some of the other tw o dimensional approaches the soliton-type solutions found take the form of plane wave solitons, rather than localized lump solutions. This is reminiscent of the behaviour found in certain (2+1) dimensional partial differential equations.

References

[1] S.Wolfram (1986) *Theory and Applications of Cellular Automata*, World Scientific, Singapore.

[2] E.R.Berlekamp, J.H.Conway & R.K.Guy (1984) *Winning ways for your mathematical plays*, **vol.2**, Academic Press

[3] N.Margolus (1984) *Physica D*, **10**, 81.

[4] J.K.Park, K.Steiglitz & W.P.Thurston (1986) *Physica D*, **19**, 423.

[5] A.S.Fokas, E.Papadopoulou, Y.Saridakis & M.J.Ablowitz (1989) *Studies in Applied Math.*, **81**, 153.

[6] A.S.Fokas, E.Papadopoulou & Y.Saridakis (1990) *Phys. Lett. A*, **147**, 369.

[7] A.S.Fokas, E.Papadopoulou & Y.Saridakis (1990) *Physica D*, **41**, 297.

[8] D.Takahashi & J.Satsuma (1990) *J.Phys.Soc.Japan*, **59**, 3514.

[9] Z. Jiang (1992) *J.Phys.A*, **25**, 3369.

Painlevé Equations and Cellular Automata

B. Grammaticos[†] and A. Ramani [†]

[†] LPN, Université Paris VII
Tour 24-14, 5^{ème}étage
75251 Paris, France

[‡] CPT, Ecole Polytechnique
CNRS, UPR 14
91128 Palaiseau, France

Abstract

We carry the discretisation of the Painlevé equations one step further by introducing systems that are cellular automata in the sense that the dependent variables take only integer values. This ultra-discretisation procedure is based on recent progress that led to a systematic way for the construction of ultra-discrete systems starting from known discrete forms. The ultra-discrete Painlevé equations have properties characteristic of the continuous and discrete Painlevé's, like coalescence cascades, particular solutions and auto-Bäcklund relations.

1 Introduction

Cellular automata are usually thought of as hyper-simplified evolution equations where only the fundamental features of the dynamics are retained. Painlevé equations are the simplest nontrivial integrable equations which encapsulate all the characteristics of integrable systems. This work will attempt to blend these two interesting ingredients and produce a new kind of systems: cellular automata that behave like Painlevé equations.

Why are cellular automata (CA) interesting? The key words are simplicity and richness. Since CA take values in a finite field (often $\{0,1\}$) or in $\mathbb{Z}$, the numerical computations associated with their simulation are fast and reliable. Neither discretisation nor truncation errors can perturb the accuracy of the result. On the other hand, studies of even the simplest CA have revealed a very rich dynamical behaviour. Even in the 1+1 dimensional case

8. Cellular Automata

one finds behaviours ranging from integrability to chaos [1]. While the use
of CA in modelling generically chaotic equations has been extensive, the opposite is true as far as integrable equations are concerned. Relatively few
studies exist to date on integrable cellular automata [2]. Still, to be fair, one
must point out that the question has been addressed and some interesting
answers proposed. Unfortunately, most of these approaches depended on the
ingeniousness of their inventors and no systematic method in the domain of
integrable CA seemed to exist. The situation has changed radically recently
with the proposal of the ultra-discrete limit by the Tokyo-Kyoto group [3].
They have shown how one can systematically construct a cellular automaton
analog of a given discrete evolution equaion. Moreover, their method is of a
wide applicability and can be adapted to a large variety of systems. It was
thus natural to ask whether one could apply it to the Painlevé equations.

Why did we choose the Painlevé equations? The reason is that these
equations are the simplest integrable systems one can construct where the
integrability is not obtained through a trivial linearisation. This fact conveys
to the Painlevé equations a host of interesting properties that mirror the
properties of higher dimensional soliton equations. One other reason why the
Painlevé equations were chosen is that we have now reached a point where our
knowledge on discrete Painlevé equations, although not complete, is adequate
[4]. Thus we have at our disposal the fundamental discrete objects out of
which we can construct cellular automata.

In what follows we shall show how one can apply systematically the ultra-
discrete limit to discrete Painlevé equations and recover ultra-discrete Painlevé equations. Morover, once the latter are obtained, we shall show that they
have properties that are characteristic of Painlevé equations: they organize
themselves in coalescence cascades, possess whole families of special solutions
as well as special auto-Bäcklund relations.

2 Introducing the ultra-discrete limit

The ultra-discrete limit as a systematic way to derive cellular automata from
discrete evolution equations was proposed by Matsukidaira, Satsuma, Takahashi and Tokihiro in [3]. In order to illustrate the method we shall present
two simple examples (due to Hirota [5]). Let us start with the discrete Burgers
equation:

$$x_n^{m+1} - x_n^m = x_n^m(x_{n+1}^m - x_n^{m+1}) \tag{2.1}$$

This equation is indeed the discrete analog of Burgers (a fact that can be
assessed through the continuous limit). It satisfies the singularity confinement
integrability criterion and is integrable through linearisation. Putting $x_n^m = F_{n+1}^m/F_n^m$ reduces the equation to a linear one: $F_n^{m+1} = F_n^m + F_{n+1}^m$. In order

to proceed to the ultra-discretisation we start by rewriting (2.1) as:

$$\frac{x_n^{m+1}}{x_n^m} = \frac{1 + x_{n+1}^m}{1 + x_n^m} \tag{2.2}$$

Next, we introduce the dependent variable transformation $x = e^{X/\epsilon}$. We have:

$$X_n^{m+1} - X_n^m = \epsilon \log(1 + e^{X_{n+1}^m/\epsilon}) - \epsilon \log(1 + e^{X_n^m/\epsilon}) \tag{2.3}$$

The ultra-discrete limit corresponds to $\epsilon \to 0$ and is based on the well-known result:

$$\lim_{\epsilon \to 0} \epsilon \log(1 + e^{a/\epsilon}) = \max(0, a) = (a + |a|)/2 \equiv (a)_+ \tag{2.4}$$

We find thus the ultra-disrete Burgers equation:

$$X_n^{m+1} - X_n^m = (X_{n+1}^m)_+ - (X_n^m)_+ \tag{2.5}$$

The important remark is that the function $(a)_+$, called truncated power function, is equal to 0 for $a \leq 0$ and equal to a for $a \geq 0$. Thus, if a is integer, the value of $(a)_+$ is also integer. This explains why the ultra-discrete limit allows one to obtain a cellular automaton starting from a discrete equation. The ultra-discrete equation may well be defined in $\mathbb{R}$, but if the initial values and parameters are in $\mathbb{Z}$, then the iterated values remain in $\mathbb{Z}$.

In a way analogous to Burgers, one can find the ultra-discrete limit of the KdV equation. Starting from:

$$x_n^{m+1} - x_n^m = x_n^m x_{n-1}^m - x_n^{m+1} x_{n+1}^{m+1} \tag{2.6}$$

one finds the ultra-discrete limit:

$$X_n^{m+1} - X_n^m = (X_{n-1}^m)_+ - (X_{n+1}^{m+1})_+ \tag{2.7}$$

It is quite remarkable that both the ultra-discrete Burgers and KdV conserve the essential properties of their continuous (and discrete) analogs. The first equation has unidirectional travelling shockwaves, while the second exhibits multisoliton solutions.

More properties of the ultra-discrete limit can be easily derived. One can show, for example that $\lim_{\epsilon \to 0} \epsilon \log(e^{a/\epsilon} + e^{b/\epsilon}) = \max(a, b)$ and the generalisation to n terms is straightforward. Another useful property is $(a)_+ = a + (-a)_+$, which will be used in the following sections.

3   The ultra-discrete Painlevé equations

Before deriving the ultra-discrete forms of the Painlevé equations we must make an important remark. The necessary condition for the ultra-discretisation procedure to be applicable is that the discrete dependent variable be positive, since we are defining the ultra-discrete variable through a logarithm. This means that out of the solutions of a given discrete equation only the ones satisfying the positivity requirement will survive in the ultra-discretisation. This may turn out to be a serious restriction for most discrete systems. The exceptions are systems that have a multiplicative form where the positivity can be easily ensured.

In the case of discrete Painlevé equations (dP's), multiplicative forms are known for all of them [6] and can be easily be used for the ultra-discretisation [7]. Let us work out in detail the case of d-P_I. We shall treat together three different forms:

$$x_n^\sigma x_{n+1} x_{n-1} = 1 + \lambda^n x_n \tag{3.1}$$

where $\sigma = 0, 1, 2$. Putting $x = e^{X/\epsilon}$ and $\lambda = e^{1/\epsilon}$ we have:

$$X_{n+1} + X_{n-1} + \sigma X_n = \epsilon \log(1 + e^{X_n + n/\epsilon}) \tag{3.2}$$

and the limit $\epsilon \to 0$ leads to:

$$X_{n+1} + X_{n-1} + \sigma X_n = (X_n + n)_+ \tag{3.3}$$

This is the ultra-discrete form of the P_I equation or u-P_I. Similarly starting from the discrete canonical [8] forms of P_{II} and P_{III}:

d-P_{II-1}

$$x_{n+1} x_{n-1} = \frac{\alpha(x_n + \lambda^n)}{x_n(1 + \lambda^n x_n)} \tag{3.4}$$

d-P_{II-2}

$$x_{n+1} x_{n-1} = \frac{\alpha(x_n + \lambda^n)}{1 + \lambda^n x_n} \tag{3.5}$$

d-P_{III}

$$x_{n+1} x_{n-1} = \frac{(x_n + \alpha\lambda^n)(x_n + \lambda^n/\alpha)}{(1 + \beta x_n \lambda^n)(1 + x_n \lambda^n/\beta)} \tag{3.6}$$

we find their ultra-discrete equivalent:

u-P_{II-1}

$$X_{n+1} + X_{n-1} = a + (n - X_n)_+ - (n + X_n)_+ \tag{3.7}$$

u-P_{II-2}

$$X_{n+1} + X_{n-1} - X_n = a + (n - X_n)_+ - (n + X_n)_+ \tag{3.8}$$

u-P_{III}

$$X_{n+1} + X_{n-1} - 2X_n = (n + a - X_n)_+ + (n - a - X_n)_+$$
$$- (X_n + b + n)_+ - (X_n - b + n)_+ \tag{3.9}$$

(In the ultra-discrete limits above we have used the transformations $\alpha = e^{a/\epsilon}$ and $\beta = e^{b/\epsilon}$). For the higher Painlevé equations the appropriate ultra-discrete form is a system [8]. Let us show this in the case of P_V. We start with the expression:

$$(x_{n+1}x_n - 1)(x_{n-1}x_n - 1) = \frac{(1 + \alpha x_n)(1 + x_n/\alpha)(1 + \beta x_n)(1 + x_n/\beta)}{(1 + \gamma\lambda^n x_n)(1 + \lambda^n x_n/\gamma)} \quad (3.10)$$

Next we introduce a variable $y_n = x_{n+1}x_n - 1$ and write d-P_V as a purely multiplicative system:

$$x_{n+1}x_n = 1 + y_n \quad (3.11a)$$

$$y_n y_{n-1} = \frac{(1 + \alpha x_n)(1 + x_n/\alpha)(1 + \beta x_n)(1 + x_n/\beta)}{(1 + \gamma\lambda^n x_n)(1 + \lambda^n x_n/\gamma)} \quad (3.11b)$$

The ultra-discretisation is now straightforward. We obtain:

$$X_{n+1} + X_n = (Y_n)_+ \quad (3.12a)$$

$$Y_n + Y_{n-1} = (X_n - a)_+ + (X_n + a)_+ + (X_n - b)_+ + (X_n + b)_+ \\ -(X_n + c + n)_+ - (X_n - c + n)_+ \quad (3.12b)$$

In a similar way one can obtain ultra-discrete forms for d-P_{IV} and d-P_{VI}.

4 Properties of the ultra-discrete Painlevé equations

The first property one checks for the Painlevé equations (and this was indeed what we did in the discrete case) is whether they are organized in coalescence cascades. It turns out that this is a property of the u-P's as well.

Let us start with u-P_{III} and introduce a large parameter Ω. The coalescence limits are obtained through $\Omega \to +\infty$. We put $a = \Omega + \alpha$, $b = \Omega - \alpha$, $n = m + \Omega$, $X_n = Z_m - \alpha$. At the limit $\Omega \to +\infty$ we find exactly u-P_{II-1} for the variable Z and parameter 4α. Starting form from u-P_{II-1} we obtain u-P_I ($\sigma = 2$) by putting $a = 4\Omega$, $n = \Omega - m$ and $X_n = Z_m + \Omega$. It turns out that starting from u-P_{II-1} we can also obtain u-P_I ($\sigma = 1$). In this case we must take $a = -2\Omega$, $n = -m - \Omega$ and $X_n = (2Z_m - m)/3 - \Omega$. On the other hand u-$P_{II-2}$ belongs to a different coalescence cascade and is not related to u-P_{III}. Still, u-P_{II-2} is related to u-P_I ($\sigma = 1$) which can be obtained through $a = 3\Omega$, $n = \Omega - m$ and $X_n = \Omega + Z_m$. Just as in the case of u-P_{II-1}, here also we can find a second coalescence. Putting $a = -\Omega$, $n = -\Omega - m$ and $X_n = (Z_m - m)/2 - \Omega$ we obtain at the limit $\Omega \to +\infty$ equation u-P_I ($\sigma = 0$).

Another interesting property of the u-P's is the fact that they possess special solutions [8,9]. The latter play the same role as the multisoliton solutions for the integrable evolution equations. Solutions of both continuous and discrete Painlevé equations have been obtained (for particular values of their parameters). They are either rational or expressed in terms of special functions. Special solutions exist also for the ultra-discrete Painlevé equations (except for u-P_I which is parameter-free). Let us start with u-P_{II-1}. We find readily that for $a = 0$, a solution $X = 0$ exists. The higher solutions for (3.4) are multi-step ones. For $a = 4p$, with integer p, when n is large positive, a constant solution exists where X is equal to p while a constant solution with $X = 2p$ exists when n is large negative. The remarkable fact is that that these constant "half solutions" do really join to form a solution of (3.4) with p jumps from the value $X = 2p$ to $X = p$. The first jump of $-p/|p|$ occurs at $n_0 = 1 - 2|p|$ and we have successive jumps at $n = n_0 + 3k$, $k = 0, 1, 2, \ldots, |p| - 1$. For u-$P_{II-2}$ multistep solutions exist for $a = 3p$ with integer p. When $n << 0$, a possible asymptotic behaviour of X is $X = a = 3p$, while for $n >> 0$ we find $X = p$. The multistep solution is the following: $X = 3p$ up to $n = 1 - 3|p|$ then we have $|p|$ times the elementary pattern of two jumps of $-p/|p|$ followed by two steps with constant value. However the last two steps of the last pattern blend into the asymptotic value $X = p$ which is therefore obtained already at point $|p| - 1$. More complicated solutions exist for the higher uP's.

Another property of the continuous and discrete Painlevé equations is the fact that they are characterised by a host of interrelations. These relations are of two kinds: either they connect the solution of one d-P to that of some other one (Miura transformation) or they connect the solutions of a given d-P corresponding to different values of the parameters (auto-Bäcklund or Schlesinger transformations). Analogous relations exist for the ultra-discrete Painlevé equations. We shall illustrate this in the case of u-P_{II-1}. We start with the discrete P_{II} (obtained from (3.4) after a straightforward gauge transformation):

$$x_{n-1}x_{n+1} = \alpha\frac{1 + z/x_n}{1 + zx_n} \tag{4.1}$$

where $z = \lambda^n$. We have given the auto-Bäcklund transformation of (4.1) in [8] (although in a slightly different gauge). It transforms the solution x of d-P_{II} with parameter α, to the solution $\tilde{x}$ corresponding to a parameter $\tilde{\alpha} = \alpha/\lambda^4$. We have:

$$\tilde{x}_n = \frac{x_n x_{n-1} + \alpha(1 + x_n/z)/\lambda}{x_n(x_n x_{n-1} + \lambda x_n/z + \lambda)} = \frac{\alpha(1 + x_{n+1}(x_n + 1/z)/\lambda)}{x_n(\alpha + \lambda x_{n+1}(x_n + 1/z))} \tag{4.2}$$

where the expressions of $\tilde{x}$ are equivalent, as a consequence of (4.1).

The ultra-discrete limit of (4.1) is just u-P_{II-1}. By applying the same procedure one can give the ultra-discrete form of the auto-Bäcklund:

$$\tilde{X}_n = -X_n + \max(X_n + X_{n-1}, a - 1, X_n + a - n - 1)$$
$$- \max(X_n + X_{n-1}, X_n - n + 1, 1)$$
$$= a - X_n + \max(X_n + X_{n+1} - 1, X_{n+1} - n - 1, 0)$$
$$- \max(X_n + X_{n+1} + 1, X_{n+1} - n + 1, a) \quad (4.3)$$

Using (4.3) we can construct the multistep solutions of u-P_{II-1} using $X = 0$ (for $a = 0$) as seed solution.

Auto-Bäcklund transformations can be established for the higher ultra-discrete Painlevé equations. The starting point in every case is the auto-Bäcklund for the corresponding discrete system. In fact this is the only limitation for finding auto-Bäcklund's for the uP's: our knowledge of the corresponding transformations for the dP's is still fragmentary.

5 Integrability of the ultra-discrete systems

In the previous sections we have presented the ultra-discrete Painlevé equations that extend the Painlevé transcendental equations to the domain of cellular automata. It is clear from the results we presented that these new objects, the uP's, have special properties that draw a close parallel with their discrete and continuous analogs. Still, one very important question remains: are these equations indeed integrable? And if yes, in which sense? Is there something analogous to the Lax-pair linearisation for ultra-discrete systems? We do not know the answer to these questions (which makes the study of ultra-discrete systems all the more interesting).

One other question concerning the ultra-discrete Painlevé equations is the one related to some underlying integrability criterion. Continuous Painlevé equations satisfy the Painlevé property, and discrete Painlevé equations satisfy the singularity confinement criterion. How about ultra-discrete Painlevé equations? Clearly any criterion based on singularities will be inoperable here. The situation is analogous to polynomial mappings where no singularity can appear. There, the criterion for integrability is based on arguments of growth of the degree of the iterate. Veselov [10] has shown that among mappings of the form $x_{n+1} - 2x_n + x_{n-1} = P(x_n)$ with polynomial $P(x)$, only the linear one has non-exponential growth of the degree of the polynomial that results from the iteration of the initial conditions. Let us apply such a low-growth criterion to a family of ultra-discrete P_I equations of the form:

$$X_{n+1} + \sigma X_n + X_{n-1} = (X_n + \phi(n))_+ \quad (5.1)$$

The three u-P_I obtained in section 3 correspond to $\sigma = 0, 1, 2$. What is the condition for X not to grow exponentially towards $\pm\infty$? We ask simply that the polynomials $r^2 + \sigma r + 1$ and $r^2 + (\sigma - 1)r + 1$ have complex roots (otherwise exponential growth ensues). The only *integer* values of σ satisfying this criterion are $\sigma = -1, 0, 1, 2$. We remark that the three values mentioned above are exactly retrieved with, in addition, the value $\sigma = -1$. A close inspection of this mapping shows that it is also integrable: it is just a form of an ultra-discrete P_{III}, obtained from the discrete system $x_{n+1}x_{n-1} = x_n(1 + \lambda^n x_n)$. We have applied the low-growth criterion to other cases like u-P_{II} and u-P_{III} and in every case the results of the growth analysis correspond to the already obtained integrable cases [8].

However low-growth is not as powerful an integrability criterion as we would have liked. In particular the inhomogeneous terms (ϕ in equation (5.1)) cannot be fixed by this argument. Any $\phi(n)$ would satisfy this requirement. Of course, when we ask for more properties (like the existence of special solutions) to be satisfied, then $\phi(n)$ is fixed in an unambiguous way. The linear stability argument can only be a first requirement. In [8] we have seen, in our analysis of the Riccati equation that it is also important that the "backward" evolution be defined. Still, the slow-growth criterion, coupled to the existence (and uniqueness?) of a preimage are not sufficiently strong in order to fix in an unambiguous way the form of an integrable ultra-discrete equation. In fact the only reliable method at our disposal to produce integrable ultra-discrete systems is to start from an integrable *discrete* system (the singularity confinement [11] integrability criterion is fully operative in this case) and apply systematically the ultra-discretization procedure.

6   Acknowledgements

The research that led to this work (and to several related publications) was made possible because J. Satsuma managed to convince one of the authors (BG) that cellular automata are indeed interesting. The enlightened author can only express his gratitude to Satsuma-sensei for this advice.

The work presented here started with exploratory calculations during the stay of the same author in India (were enlightenment took place). K.M. Tamizhmani participated actively during this first phase of the work. Upon return to France the full team (of two!) complemented by Y. Ohta and D. Takahashi (the "Japanese storm" according to Satsuma-sensei) sank their teeth into this juicy subject and managed to produce all the results presented here (and many more!) in record time.

To all our valiant collaborators we wish to express our sincerest thanks.

References

[1] S. Wolfram, *Theory and applications of cellular automata*, World Scientific (Singapore) 1986.

[2] Y. Pomeau, J. Phys. A 17 (1984) L415.
K. Park, K. Steiglitz and W.P. Thurston, Physica 19D (1986) 423.
T.S. Papatheodorou, M.J. Ablowitz and Y.G. Saridakis, Stud. Appl. Math. 79 (1988) 173.
T.S Papatheodorou and A.S. Fokas, Stud. Appl. Math. 80 (1989) 165.
M. Bruschi, P.M. Santini and O. Ragnisco, Phys. Lett. A 169 (1992) 151.
A. Bobenko, M. Bordemann, C. Gunn and U. Pinkall, Commun. Math. Phys. 158 (1993) 127.
D. Takahashi and J. Satsuma, J. Phys. Soc. Jpn. 59 (1990) 3514.

[3] T. Tokihiro, D. Takahashi, J. Matsukidaira and J. Satsuma, *From soliton equations to integrable Cellular Automata through a limiting procedure*, preprint (1996).
D. Takahashi and J. Matsukidaira, Phys. Lett. A 209 (1995) 184.
J. Matsukidaira, J. Satsuma, D. Takahashi T. Tokihiro and M. Torii, *Toda-type Cellular Automaton and its N-soliton solution*, in preparation.

[4] B. Grammaticos, F. Nijhoff and A. Ramani, Discrete Painlevé equations, course at the Cargèse 96 summer school on Painlevé equations.

[5] R. Hirota, private communication.

[6] A. Ramani and B. Grammaticos, Physica A 228 (1996) 160.

[7] B. Grammaticos, Y. Ohta, A. Ramani, D. Takahashi, K.M. Tamizhmani, *Cellular automata and ultra-discrete Painlevé equations*, Preprint 96.

[8] A. Ramani, D. Takahashi, B. Grammaticos, Y. Ohta, *The ultimate discretisation of the Painlevé equations*, preprint 96.

[9] D. Takahashi, Y. Ohta, A. Ramani, B. Grammaticos, in preparation.

[10] A.P. Veselov, Comm. Math. Phys. 145 (1992) 181.

[11] B. Grammaticos, A. Ramani and V. Papageorgiou, Phys. Rev. Lett. 67 (1991) 1825.

2+1 dimensional soliton cellular automaton

S. Moriwaki∗, A. Nagai†, J. Satsuma†, T. Tokihiro†,
M. Torii†, D. Takahashi‡ and J. Matsukidaira‡

∗Nippon MOTOROLA Ltd. Second Design Section,
Pager Subscriber Unit Product Design Department,
Paging Products Devision
Minami-Azabu 3-20-1, Minato-ku, Tokyo 106, Japan
†Graduate School of Mathematical Sciences,
University of Tokyo,
Komaba 3-8-1, Meguro-ku, Tokyo 153, Japan
‡Department of Applied Mathematics and Informatics,
Ryukoku University,
Yokotani 1-5, Seta, Ooe-cho, Ohtsu 520-21, Japan

Abstract

A 2+1 dimensional soliton cellular automaton is derived from discrete analogue of a generalized Toda equation through the procedure
of the so-called ultra-discretization. Its soliton solution and the time
evolution are also discussed.

1 Introduction

Recently, discrete soliton systems have attracted much attention. Among
them, "ultra-discrete" soliton equations, in which dependent variables as well
as independent variables take discrete values, have been actively studied. One
of the most important ultra-discrete soliton systems is the so-called "soliton
cellular automaton", or SCA for short [1]. This is 1(space) + 1(time) dimensional and two-valued (0 and 1). The time evolution of the value of the j−th

cell at time t, u_j^t, is given by

$$u_j^{t+1} = \begin{cases} 1 & \text{if } u_j^t = 0 \ \& \ \sum_{i=-\infty}^{j-1} u_i^t > \sum_{i=-\infty}^{j-1} u_i^{t+1} \\ 0 & \text{otherwise} \end{cases} \qquad (1)$$

A remarkable feature of equation (1) is that any state consists only of solitons, interacting in the same manner as KdV solitons. Moreover it possesses an abundant combinatoric structure and an infinite number of conserved quantities [2]. Quite recently, a direct connection between the SCA and the Lotka-Volterra equation, which is considered as one integrable discretization of the KdV equation, has been clarified [3, 4]. A key to the discretization, which we call the "ultra-discretization" in this context, is the following formula:

$$\lim_{\varepsilon \to +0} \varepsilon \log(1 + e^{X/\varepsilon}) = F(X) = \max[0, X]. \qquad (2)$$

The purpose of this paper is to present a 2+1 dimensional SCA derived from ultra-discretization of the 2+1 dimensional Toda equation. In section 2, we derive the 2+1 dimensional SCA through ultra-discretization of a special case of DAGTE, i.e. discrete analogue of a generalized Toda equation. In section 3, we discuss soliton solution for the 2+1 dimensional SCA and its time evolution. Concluding remarks are given in section 4.

2 Discrete analogue of a generalized Toda equation and the 2+1 dimensional SCA

We start with the following difference equation proposed by Hirota [5],

$$[Z_1 \exp(D_1) + Z_2 \exp(D_2) + Z_3 \exp(D_3)] f \cdot f = 0, \qquad (3)$$

where $Z_i (i = 1, 2, 3)$ are arbitrary parameters and $D_i (i = 1, 2, 3)$ stand for Hirota's derivatives [6] with respect to variables of the unknown function f. Equation (3) is called discrete analogue of a generalized Toda equation, or Hirota-Miwa equation. Many soliton equations are obtained by taking proper limit of equation (3) [5].

In this paper, we consider a particular case of equation (3),

$$\left\{ \exp(D_t) - \delta^2 \exp(D_x) - (1 - \delta^2) \exp(D_y) \right\} f \cdot f = 0, \qquad (4)$$

or equivalently,

$$\begin{aligned} f(t-1, x, y) f(t+1, x, y) - \delta^2 f(t, x-1, y) f(t, x+1, y) \\ - (1 - \delta^2) f(t, x, y+1) f(t, x, y-1) = 0. \end{aligned} \qquad (5)$$

Equation (5) reduces to a discrete analogue of the 2+1 dimensional Toda equation [7, 8],

$$\begin{cases} V(l+1,m,n) - V(l,m,n) &= I(l,m+1,n)V(l+1,m,n) \\ &\quad -I(l,m,n+1)V(l,m,n), \qquad (6) \\ I(l,m+1,n) - I(l,m+1,n) &= V(l+1,m,n-1) - V(l,m,n). \end{cases}$$

through the following independent and dependent variable transformations,

$$l = \frac{-x-y-t+1}{2}, \; m = \frac{x-y+t+1}{2}, \; n = x \qquad (7)$$

$$V(l,m,n) = \frac{\tau(l+1,m,n+1)\tau(l,m+1,n-1)}{\tau(l,m+1,n)\tau(l,m,n)} \qquad (8)$$

$$I(l,m,n) = \frac{1}{\delta}\left\{(1-\delta^2)\frac{\tau(l+1,m,n)\tau(l,m,n-1)}{\tau(l,m,n)\tau(l+1,m,n-1)} - 1\right\} \qquad (9)$$

$$\tau(l+\frac{1}{2}, m+\frac{1}{2}, n) = f(x,y,z). \qquad (10)$$

Let us derive an ultra-discrete version of equation (5). The dependent variable transformation,

$$f(t,x,y) = \exp[S(t,x,y)] \qquad (11)$$

yields

$$\exp[\Delta_t^2 S(t,x,y)] - \delta^2 \exp[\Delta_x^2 S(t,x,y)] - (1-\delta^2)\exp[\Delta_y^2 S(t,x,y)] = 0, \quad (12)$$

or equivalently,

$$\exp[(\Delta_t^2 - \Delta_y^2)S(t,x,y)] = (1-\delta^2)\left(1 + \frac{\delta^2}{1-\delta^2}\exp[(\Delta_x^2 - \Delta_y^2)S(t,x,y)]\right). \qquad (13)$$

Each operator Δ_t, Δ_x and Δ_y represents central difference operator defined, for example, by

$$\Delta_t^2 S(t,x,y) = S(t+1,x,y) - 2S(t,x,y) + S(t-1,x,y). \qquad (14)$$

Taking a logarithm of equation (13) and operating $(\Delta_x^2 - \Delta_y^2)$, we have

$$(\Delta_t^2 - \Delta_y^2)u(t,x,y) = (\Delta_x^2 - \Delta_y^2)\log\left(1 + \frac{\delta^2}{1-\delta^2}\exp[u(t,x,y)]\right), \qquad (15)$$

where

$$u(t,x,y) = (\Delta_x^2 - \Delta_y^2)S(t,x,y). \qquad (16)$$

We finally take an ultra-discrete limit of equation (15). Putting

$$u(t, x, y) = \frac{v_\varepsilon(t, x, y)}{\varepsilon}, \quad \frac{\delta^2}{1 - \delta^2} = e^{-\frac{\theta_0}{\varepsilon}}, \tag{17}$$

and taking the small limit of ε, we obtain the following equation,

$$(\Delta_t^2 - \Delta_y^2)v(t, x, y) = (\Delta_x^2 - \Delta_y^2)F(v(t, x, y) - \theta_0), \tag{18}$$
$$F(X) = \max[0, X]. \tag{19}$$

We have rewritten $\lim_{\varepsilon \to +0} v_\varepsilon(t, x, y)$ as $v(t, x, y)$ in equation (18). We call the ultra-discrete system satisfying the above equation (18) the *2+1 dimensional SCA*.

3   Soliton solution for the 2+1 dimensional SCA

In this section, we discuss soliton solution for the 2+1 dimensional SCA governed by equation (18). Since we have derived equation (18) by taking an ultra-discrete limit of bilinear equation (5), we may well consider that the soliton solution for equation (18) is also obtained by ultra-discretization of that for equation (5).

We first consider one-soliton solution. The bilinear equation (5) admits one-soliton solution given by

$$f(t, x, y) = 1 + e^\eta, \quad \eta = px + qy + \omega t, \tag{20}$$

where the set of parameters (p, q, ω) satisfies a dispersion relation,

$$(e^{-\omega} + e^\omega) - \delta^2(e^{-p} + e^p) - (1 - \delta^2)(e^{-q} + e^{-q}) = 0. \tag{21}$$

Following the procedures given by eqs. (11) and (16), we have

$$u(t, x, y) = \log(1 + e^{\eta+p}) + \log(1 + e^{\eta-p}) - \log(1 + e^{\eta+q}) - \log(1 + e^{\eta-q}). \tag{22}$$

In order to take an ultra-discrete limit, we introduce new variables as

$$\varepsilon p = P, \quad \varepsilon q = Q, \quad \varepsilon\omega = \Omega, \quad K = Px + Qy + \Omega t, \tag{23}$$
$$v_\varepsilon(t, x, y) = \varepsilon u(t, x, y). \tag{24}$$

Taking a limit $\varepsilon \to +0$, we obtain

$$v(t, x, y) = F(K + P) + F(K - P) - F(K + Q) - F(K - Q). \tag{25}$$

The dispersion relation (21) reduces, through the same limiting procedure, to

$$|\Omega| = \max[|P|, |Q| + \theta_0] - \max[0, \theta_0]. \qquad (26)$$

Next we construct two-soliton solution. Equation (5) possesses two-soliton solution written as

$$f(t, x, y) = 1 + e^{\eta_1} + e^{\eta_2} + e^{\eta_1 + \eta_2 + \theta_{12}}, \; \eta_i = p_i x + q_i y + \omega_i t \; (i = 1, 2), \quad (27)$$

$$(e^{-\omega_i} + e^{\omega_i}) - \delta^2(e^{-p_i} + e^{p_i}) - (1 - \delta^2)(e^{-q_i} + e^{-q_i}) = 0 \; (i = 1, 2). \quad (28)$$

The variable θ_{12} stands for a phase shift and is determined by the following relation:

$$e^{\theta_{12}} = -\frac{(e^{-\omega_1 + \omega_2} + e^{\omega_1 - \omega_2}) - \delta^2(e^{-p_1 + p_2} + e^{p_1 - p_2}) - (1 - \delta^2)(e^{-q_1 + q_2} + e^{q_1 - q_2})}{(e^{\omega_1 + \omega_2} + e^{-\omega_1 - \omega_2}) - \delta^2(e^{p_1 + p_2} + e^{-p_1 - p_2}) - (1 - \delta^2)(e^{q_1 + q_2} + e^{-q_1 - q_2})}.$$
$$(29)$$

Introducing new variables as

$$\varepsilon p_i = P_i, \; \varepsilon q_i = Q_i, \; \varepsilon \omega_i = \Omega_i, \; K_i = P_i x + Q_i y + \Omega_i t, \; (i = 1, 2) \qquad (30)$$

$$v_\varepsilon(t, x, y) = \varepsilon u(t, x, y), \; \varepsilon \theta_{12} = \Theta_{12}, \qquad (31)$$

and taking the same limit $\varepsilon \to +0$, we have

$$\begin{aligned}
v(t, x, y) &= \max[0, K_1 + P_1, K_2 + P_2, K_1 + K_2 + P_1 + P_2 + \Theta_{12}] \\
&\quad + \max[0, K_1 - P_1, K_2 - P_2, K_1 + K_2 - P_1 - P_2 + \Theta_{12}] \\
&\quad - \max[0, K_1 + Q_1, K_2 + Q_2, K_1 + K_2 + Q_1 + Q_2 + \Theta_{12}] \\
&\quad - \max[0, K_1 - Q_1, K_2 - Q_2, K_1 + K_2 - Q_1 - Q_2 + \Theta_{12}] (32) \\
|\Omega_i| &= \max[|P_i|, |Q_i| + \theta_0] - \max[0, \theta_0] \; (i = 1, 2). \qquad (33)
\end{aligned}$$

Phase shift term Θ_{12} is determined by

$$\begin{aligned}
\max\,[&\Theta_{12} + \max[0, \theta_0] + |\Omega_1 + \Omega_2|, \max[0, \theta_0] + |\Omega_1 - \Omega_2|] \\
= \max\,[&\Theta_{12} + |P_1 + P_2|, \Theta_{12} + \theta_0 + |Q_1 + Q_2|, \\
&|P_1 - P_2|, \theta_0 + |Q_1 - Q_2|]\,, \qquad (34)
\end{aligned}$$

which is obtained through the same limit $\varepsilon \to +0$ in equation (29).

It should be noted that N-soliton solution can also be found through the same limiting procedure. This is given by

$$\begin{aligned}
v(t, x, y) &= \max_{\mu=0,1} \left[\max_{i=1,2,\cdots,N} [\mu_i(K_i + P_i)], \max_{i<j}[\mu_i \mu_j \Theta_{ij}] \right] \\
&\quad + \max_{\mu=0,1} \left[\max_{i=1,2,\cdots,N} [\mu_i(K_i - P_i)], \max_{i<j}[\mu_i \mu_j \Theta_{ij}] \right]
\end{aligned}$$

$$- \max_{\mu=0,1} \left[\max_{i=1,2,\cdots,N} [\mu_i(K_i + Q_i)], \max_{i<j} [\mu_i \mu_j \Theta_{ij}] \right]$$

$$- \max_{\mu=0,1} \left[\max_{i=1,2,\cdots,N} [\mu_i(K_i - Q_i)], \max_{i<j} [\mu_i \mu_j \Theta_{ij}] \right], \qquad (35)$$

$$K_i = P_i x + Q_i y + \Omega_i t,$$

$$|\Omega_i| = \max[|P_i|, |Q_i| + \theta_0] - \max[0, \theta_0]. \qquad (36)$$

Each phase shift term $\Theta_{ij} (1 \leq i < j \leq N)$ satisfies the relation,

$$\max [\Theta_{ij} + \max[0, \theta_0] + |\Omega_i + \Omega_j|, \max[0, \theta_0] + |\Omega_i - \Omega_j|]$$
$$= \max [\Theta_{ij} + |P_i + P_j|, \Theta_{ij} + \theta_0 + |Q_i + Q_j|,$$
$$|P_i - P_j|, \theta_0 + |Q_i - Q_j|]. \qquad (37)$$

4 Concluding Remarks

We have shown that a 2+1 dimensional SCA is obtained by taking an ultra-discrete limit of the 2+1 dimensional Toda equation. We have also found its soliton solutions. It is a future problem to construct an ultra-discrete version of other kind of solutions, for example, rational, molecular and quasi-periodic solutions.

References

[1] D. Takahashi and J. Satsuma, J. Phys. Soc. Jpn. **59** (1990) 3514.

[2] M. Torii, D. Takahashi and J. Satsuma, Physica D **92** (1996) 209.

[3] D. Takahashi and J. Matsukidaira, Phys. Lett. A **209** (1995) 184.

[4] T. Tokihiro, D. Takahashi, J. Matsukidaira and J. Satsuma, Phys. Rev. Lett. **76** (1996) 3247.

[5] R. Hirota, J. Phys. Soc. Jpn. **45** (1981) 3785.

[6] R. Hirota, *Direct Method in Soliton Theory* (Iwanami, Tokyo, 1992) [in Japanese].

[7] R. Hirota, M. Ito and F. Kako, Prog. Theor. Phys. Suppl. **94** (1988) 42.

[8] R. Hirota, S. Tsujimoto and T. Imai, RIMS Kokyuroku **822** (1992) 144.

A. Time evolutions of one- and two-solitons

We here show one- and two-soliton solutions and their time evolutions in the following figures. Figure 1 displays one soliton solution with parameters $P = 5, Q = 1$ and $\theta_0 = 2$. Figures 2 and 3 demonstrate two-soliton solution at $t = -4, -3$ and its time evolution, respectively, with parameters $P_1 = 6, Q_1 = 1, P_2 = 6, Q_2 = 5$ and $\theta_0 = 2$.

Figures 2 and 3 demonstrate two-soliton solution at $t = -4, -3$ and its time evolution, respectively, with parameters $P_1 = 6$, $Q_1 = 1$, $P_2 = 6$, $Q_2 = 5$ and $\theta_0 = 2$.

8	0000032000000000
7	0000023000000000
6	0000014000000000
5	0000004000000000
4	0000004100000000
3	0000003200000000
2	0000002300000000
1	0000001400000000
0	0000000400000000
-1	0000000410000000
-2	0000000320000000
-3	0000000230000000
-4	0000000140000000
-5	0000000040000000
-6	0000000041000000
-7	0000000032000000
y x	012345678

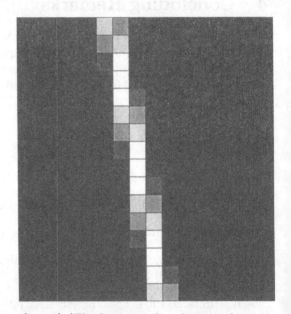

Figure 1: One-solition solution ($t = 0$) (The horizontal and vertical axes represent x and y coordinates, respectively. At the bottom of the left-side figure, negative values of x coordinate are expressed as "." for convenience' sake.)

```
              t=-4                                    t=-3

15  0000000000000042000000000000000    15  0000000000000024000000000000000
14  0000000000000033000000000000000    14  1000000000000015000000000000000
13  1000000000000024000000000000000    13  1100000000000050000000000000000
12  1100000000000015000000000000000    12  0110000000000051000000000000000
11  0110000000000005000000000000000    11  0011000000000042000000000000000
10  0011000000000051000000000000000    10  0001100000000033000000000000000
 9  0001100000000042000000000000000     9  0000100000000024000000000000000
 8  0000100000000033000000000000000     8  0000110000000015000000000000000
 7  0000110000000024000000000000000     7  0000011000000050000000000000000
 6  0000011000000015000000000000000     6  0000001100000051000000000000000
 5  0000001100000005000000000000000     5  0000000110000042000000000000000
 4  0000000110000051000000000000000     4  0000000011000033000000000000000
 3  0000000011000042000000000000000     3  0000000001000024000000000000000
 2  0000000001000033000000000000000     2  0000000001100015000000000000000
 1  0000000001100024000000000000000     1  0000000000110005000000000000000
 0  0000000000110015000000000000000     0  0000000000011000510000000000000
-1  0000000000011000050000000000000    -1  0000000000001100420000000000000
-2  0000000000001100051000000000000    -2  0000000000000110330000000000000
-3  0000000000000110042000000000000    -3  0000000000000010240000000000000
-4  0000000000000010033000000000000    -4  0000000000000011150000000000000
-5  0000000000000011024000000000000    -5  0000000000000001150000000000000
-6  0000000000000001115000000000000    -6  0000000000000000161000000000000
-7  0000000000000000115000000000000    -7  0000000000000000052000000000000
-8  0000000000000000161000000000000    -8  0000000000000000043100000000000
-9  0000000000000000052000000000000    -9  0000000000000000033110000000000
-10 0000000000000000043100000000000   -10  0000000000000000024010000000000
-11 0000000000000000033100000000000   -11  0000000000000000015011000000000
-12 0000000000000000024110000000000   -12  0000000000000000005001100000000
-13 0000000000000000015011000000000   -13  0000000000000000051001100000000
-14 0000000000000000005001100000000   -14  0000000000000000042000110000000
 yx.............0123456789******      yx.............0123456789******
```

Figure 2: Two-solition solution ($t = -4$, -3) (At the bottom, negative values of x coordinate are expressed as "." and values greater than 10 are also done as "*" for convenience sake.)

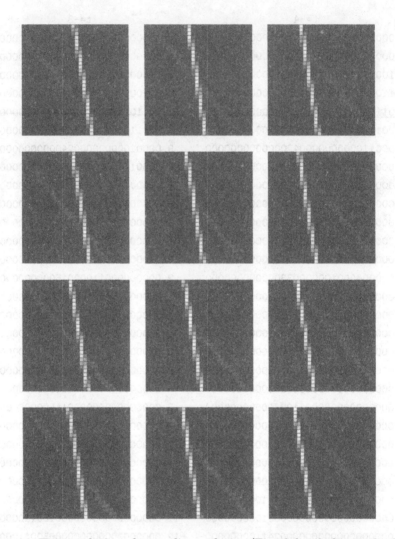

Figure 3: Time evolution of two-soliton solution (Four left-side figures display, from top to bottom, values of $v(t, x, y)$ at $t = -4, -3, -2. -1$. four central figures at $t = 0, 1, 2, 3$ and four right-hand figures at $t = 4, 5, 6, 7$.)

Chapter 9

q-Special Functions and
q-Difference Equations

Fourier–Gauss Transforms of
q-Exponential and q-Bessel Functions

N.M. Atakishiyev *

Instituto de Matematicas, UNAM

Apartado Postal 139-B

62191 Cuernavaca, Morelos, Mexico

Email: natig@matcuer.unam.mx

Abstract

The properties of a family of q-exponential functions, which depend on an extra parameter α, are examined. These functions have a well-defined meaning for both $0 < |q| < 1$ and $|q| > 1$ cases if only $\alpha \in [0,1]$. It is shown that Fourier–Gauss transformations relate any two members of this family with different values of the parameter α. As a consequence, this yields the corresponding relations between Jackson's q-Bessel functions $J_\nu^{(i)}(z;q)$, $i = 1,2$, and $J_\nu^{(3)}(z;q)$. The Fourier–Gauss transformation properties of the continuous q-Hermite polynomials and associated q-coherent states are also briefly discussed.

Let us consider the one-parameter family of q-exponential functions

$$E_q^{(\alpha)}(z) = \sum_{n=0}^{\infty} \frac{q^{\alpha n^2/2}}{(q;q)_n} z^n \tag{1}$$

with $\alpha \in \Re$ [1]–[3]. The q-shifted factorial $(q;q)_n$ in (1) is defined as $(z;q)_0 = 1$ and $(z;q)_n = \prod_{j=0}^{n-1}(1-zq^j)$, $n = 1,2,3,\ldots$. Consequently, in the limit when $q \to 1$ we have

$$\lim_{q \to 1} E_q^{(\alpha)}((1-q)z) = e^z. \tag{2}$$

Two particular cases of this family with $\alpha = 0$ and $\alpha = 1$ are well known: they are the q-exponential function

$$e_q(z) = E_q^{(0)}(z) = \sum_{n=0}^{\infty} \frac{z^n}{(q,q)_n} \tag{3}$$

and its reciprocal

$$E_q(z) = e_q^{-1}(z) = E_q^{(1)}\left(-q^{-1/2}\,z\right) = \sum_{n=0}^{\infty} \frac{q^{n(n-1)/2}}{(q;q)_n}\,(-z)^n\,, \tag{4}$$

respectively [4]. Another particular example of (1) corresponds to the value $\alpha = 1/2$ and is

$$E_q^{(1/2)}(z) = \mathcal{E}_q(-;0,z) = \varepsilon_q(z)\,, \tag{5}$$

where $\mathcal{E}_q(z;a,b)$ is a two-parameter q-exponential function, introduced in [5]. The q-exponential function (5) is closely related to Jackson's q-Bessel function $J_\nu^{(3)}(z;q)$ [6, 7].

We start with the observation that by the ratio test the infinite series in (1) is convergent for $0 < |q| < 1$ and arbitrary complex z only if the parameter α is positive: $0 < \alpha < \infty$. The case $\alpha = 0$ is a little bit more involved, but $E_q^{(0)}(z) = e_q(z)$ and the properties of the q-exponential function (3) are well studied [1, 4, 8].

The series (1) converges also for $1 < |q| < \infty$ and arbitrary complex z, provided that $-\infty < \alpha < 1$. Thus, the q-exponential functions (1) are well defined for both $0 < |q| < 1$ and $1 < |q| < \infty$, if the parameter α belongs to the line segment $[0,1]$. Observe that the inversion formula

$$(q^{-1};q^{-1})_n = (-1)^n\, q^{-n(n+1)/2}\,(q;q)_n \tag{6}$$

leads to the relation

$$E_{q^{-1}}^{(\alpha)}(z) = E_q^{(1-\alpha)}\left(-q^{1/2}z\right). \tag{7}$$

When $\alpha = 0$, (11) reproduces the known relation [1, 8]

$$e_{q^{-1}}(z) = E_q(qz) \tag{8}$$

between the q-exponential functions (3) and (4) for $0 < |q| < 1$ and $1 < |q| < \infty$ (or *vice versa*), respectively.

It is not hard to show that Fourier–Gauss transforms for the q-exponential functions (1) have the form [3]

$$E_q^{(\alpha+\nu^2/2)}\left(te^{-\nu\kappa x}\right)e^{-x^2/2} = \frac{1}{\sqrt{2\pi}}\int_{-\infty}^{\infty} e^{ixy-y^2/2}\,E_q^{(\alpha)}\left(te^{i\nu\kappa y}\right)dy\,, \tag{9}$$

provided that $-\alpha \le \nu^2/2 \le 1-\alpha$ and $q = \exp(-2\kappa^2)$. Indeed, to evaluate the right-hand side of (9) one only needs to use the definition (1) with $z = te^{i\kappa\nu y}$ and to integrate this sum termwise by the Fourier transform

$$\frac{1}{\sqrt{2\pi}}\int_{-\infty}^{\infty} e^{ixy-y^2/2}\,dy = e^{-x^2/2} \tag{10}$$

for the Gauss exponential function $\exp(-x^2/2)$.

For $\alpha = 0$ and $\nu = \sqrt{2}$ the Fourier–Gauss transform (9) gives the following relation between the q-exponential functions $e_q(z)$ and $E_q(z)$

$$E_q(-q^{1/2}\, te^{-\sqrt{2}\kappa x})\, e^{-x^2/2} = \frac{1}{\sqrt{2\pi}} \int\limits_{-\infty}^{\infty} e^{ixy-y^2/2}\, e_q(te^{i\sqrt{2}\kappa y})\, dy\,, \qquad (11)$$

which is a particular case of Ramanujan's integral with a complex parameter [9]-[11].

An interesting point is the possibility of representing $E_q^{(\alpha)}(z)$ as an infinite product. We know that in the particular cases of $\alpha = 0$ and $\alpha = 1$ the affirmative answer to this question is given by Euler's formulas [4]

$$e_q(z) = (z;q)_\infty^{-1}\,, \qquad\qquad E_q(z) = (z;q)_\infty\,, \qquad (12)$$

for the q-exponential functions (3) and (4). Observe that Euler's formulas (12) follow from the functional relations

$$e_q(qz) = (1-z)\, e_q(z)\,, \qquad\qquad E_q(z) = (1-z)\, E_q(qz)\,, \qquad (13)$$

and the side conditions $e_q(0) = E_q(0) = 1$. One may try to combine one of the relations (13) with (9) in order to get the corresponding representation at least for the parameter $\alpha = 1/2$. But in both cases this results in the functional relation (*cf.* [12])

$$\varepsilon_q(qz) = \varepsilon_q(z) - q^{1/4}\, z\, \varepsilon_q(q^{1/2}z) \qquad (14)$$

with $z = te^{-\kappa x}$ and $z = te^{i\kappa x}$, respectively. Actually, this type of functional relation holds for arbitrary z and $\alpha \in [0,1]$ (not for $\alpha = 1/2$ only !) and has the form

$$E_q^{(\alpha)}(qz) = E_q^{(\alpha)}(z) - q^{\alpha/2}z\, E_q^{(\alpha)}(q^\alpha z)\,. \qquad (15)$$

The validity of (15) can be readily verified by using the definition (1). When $\alpha = 0$ and $\alpha = 1$, from (15) follow the functional relations (13) for the q-exponential functions $e_q(z)$ and $E_q(z)$, respectively.

The relation (15) is equivalent to a q-difference equation. Indeed, in terms of the Jackson q-difference operator [13]

$$\Delta f(z) = \frac{f(z) - f(qz)}{z(1-q)} \qquad (16)$$

it may be represented as

$$\Delta E_q^{(\alpha)}(z) = \frac{q^{\alpha/2}}{1-q}\, E_q^{(\alpha)}(q^\alpha z)\,. \qquad (15')$$

Note also that applying the functional relations (13) n times in succession leads to

$$e_q(q^n z) = (z; q)_n \, e_q(z) \,, \qquad\qquad E_q(z) = (z; q)_n \, E_q(q^n z) \,, \qquad (17)$$

respectively. These simple formulae turn out to be very useful in proving the orthogonality of the classical q-polynomials with respect to measures, containing q-exponential functions $e_q(z)$ and $E_q(z)$ [14]–[17]. In an analogous manner, from (15) it follows at once that

$$E_q^{(\alpha)}(q^n z) = \sum_{k=0}^{n} \begin{bmatrix} n \\ k \end{bmatrix}_q (-z)^k \, q^{[(\alpha+1)k-1]k/2} \, E_q^{(\alpha)}(q^{\alpha k} \, z) \,, \qquad (18)$$

where $\begin{bmatrix} n \\ k \end{bmatrix}_q = \dfrac{(q; q)_n}{(q; q)_k \, (q; q)_{n-k}}$ are the q-binomial coefficients. This relation is easily verified by induction, upon using the following property [4] of the q-binomial coefficients

$$\begin{bmatrix} n \\ k \end{bmatrix}_q + q^{n-k+1} \begin{bmatrix} n \\ k-1 \end{bmatrix}_q = \begin{bmatrix} n+1 \\ k \end{bmatrix}_q \,.$$

As we have mentioned above, the q-exponential function $E_q^{(1/2)}(z) = \varepsilon_q(z)$ is related to the third q-Bessel function [6, 7]

$$J_\nu^{(3)}(z; q) := \frac{1}{(q; q)_\nu} \sum_{n=0}^{\infty} \frac{(-1)^n \, q^{n(n+1)/2}}{(q^{\nu+1}, q; q)_n} \, z^{2n+\nu} \,, \qquad (19)$$

where $(a_1, \ldots a_n; q)_k = \prod_{j=1}^{n}(a_j; q)_k$ is the product of q-shifted factorials. Therefore the representation (9) suggests that $J_\nu^{(3)}(z; q)$ can be related by a Fourier–Gauss transform to two other q-Bessel functions $J_\nu^{(1)}(z; q)$ and $J_\nu^{(2)}(z; q)$, introduced also by F.H. Jackson in [18]. We remind the reader that these q-Bessel functions, interrelated by a replacement of the base $q \to q^{-1}$, are defined [19] as

$$J_\nu^{(1)}(z; q) := \frac{1}{(q; q)_\nu} \sum_{n=0}^{\infty} \frac{(-1)^n \, (z/2)^{2n+\nu}}{(q^{\nu+1}, q; q)_n} \,, \qquad (20a)$$

$$J_\nu^{(2)}(z; q) := \frac{1}{(q; q)_\nu} \sum_{n=0}^{\infty} \frac{(-1)^n \, q^{n(n+\nu)}}{(q^{\nu+1}, q; q)_n} \, (z/2)^{2n+\nu} \,, \qquad (20b)$$

$$J_\nu^{(2)}(z; q) = E_q(-z^2/4) \, J_\nu^{(1)}(z; q), \qquad |z| < 2 \,. \qquad (20c)$$

Using the Fourier transform (10) for the Gauss exponential function $\exp(-x^2/2)$ and the definitions (19) and (20a), it is easy to verify that

$$J_\nu^{(3)}(q^{(\nu-1)/4} t e^{-\kappa x}; q) \, e^{-x^2} = q^{\nu(\nu-2)/8} \int_{-\infty}^{\infty} \frac{dy}{\sqrt{\pi}} \, J_\nu^{(1)}(2t e^{i\kappa y}; q) \, e^{2ixy-y^2} \,. \qquad (21)$$

From (20c) it also follows that

$$J_\nu^{(3)}(q^{(\nu-1)/4}te^{-\kappa x};q)e^{-x^2} = q^{\nu(\nu-2)/8} \int\limits_{-\infty}^{\infty} \frac{dy}{\sqrt{\pi}} J_\nu^{(2)}(2te^{i\kappa y};q)e_q(-t^2e^{2i\kappa y})e^{2ixy-y^2},$$

(22)

where $|t| < 1$.

Now we can turn our attention to the Fourier–Gauss transformation properties of the continuous q-Hermite polynomials and associated q-coherent states [20].

One realization of the q-harmonic oscillator [21, 22] is built on the finite interval $x \in [-1, 1]$ in terms of the continuous q-Hermite polynomials $H_n(x|q)$ [23]. The factorization of the difference equation for $H_n(x|q)$ yields explicit lowering $b(x|q)$ and raising $b^+(x|q)$ operators, which satisfy the q-Heisenberg commutation relation [24]. As in the case of the well-known non-relativistic quantum-mechanical oscillator, one can also construct coherent states for this q-deformed system. They are defined as eigenfunctions of the lowering operator $b(x|q)$ and involve the infinite series of the form [24]

$$I(x,t;q) = \sum_{n=0}^{\infty} \frac{q^{n^2/4}t^n}{(q;q)_n} H_n(x|q).$$

(23)

In the limit when the parameter $q \equiv \exp(-2\kappa^2)$ tends to 1 (and, consequently, the parameter $\kappa \to 0$), we have

$$\lim_{q\to 1^-} \kappa^{-n} H_n(\sin \kappa s|q) = H_n(s),$$

(24)

where $H_n(s)$ are the classical Hermite polynomials. Therefore

$$\lim_{q\to 1^-} I(\sin \kappa s, 2\kappa\tau;q) = \sum_{n=0}^{\infty} \frac{\tau^n}{n!} H_n(s) = e^{2s\tau-\tau^2}$$

(25)

and we recover the coherent states for the non-relativistic linear oscillator, according to the generating function in (25). Observe also that a summand in the Rogers generating function for the continuous q-Hermite polynomials [25]

$$\sum_{n=0}^{\infty} \frac{t^n}{(q;q)_n} H_n(\cos \theta|q) = e_q(te^{i\theta})\, e_q(te^{-i\theta}), \qquad |t| < 1,$$

(26)

falls short by the factor $q^{n^2/4}$ in order to express $I(x,t;q)$ in terms of the q-exponential function $e_q(z)$. Actually, as suggested in [5] (see also [26, 27, 20]), the series (23) can be used to define yet another q-exponential function

$$\varepsilon_q(\sin \gamma;\tau) = \sum_{n=0}^{\infty} \frac{\tau^n}{(q;q)_n} q^{n^2/4} (q^{(1-n)/2}e^{-i\gamma}, -q^{(1-n)/2}e^{i\gamma};q)_n,$$

(27)

different from $e_q(z)$ and its reciprocal $E_q(z)$.

From (27) and the Fourier expansion

$$H_n(\sin \kappa x | q) = i^n \sum_{k=0}^{\infty} (-1)^k \begin{bmatrix} n \\ k \end{bmatrix}_q e^{i(2k-n)\kappa x} \tag{28}$$

for the q-Hermite polynomials [23] it follows that

$$I(x,t;q) = \sum_{n=0}^{\infty} \frac{q^{n^2/4} t^n}{(q;q)_n} H_n(x|q) = E_{q^2}(t^2)\, \varepsilon_q(x;-it) . \tag{29}$$

It turns out that the function $\varepsilon_q(x;\tau)$ can be expressed as a Fourier-Gauss transform of the product of two q-exponential functions $E_q(z)$. Indeed, the continuous q-Hermite $H_n(x|q)$ and q^{-1}-Hermite $h_n(x|q)$ polynomials are known to be related to each other by the Fourier-Gauss transform [28]

$$H_n(\sin \kappa x | q)\, e^{-x^2/2} = \frac{i^n}{\sqrt{2\pi}} q^{n^2/4} \int_{-\infty}^{\infty} h_n(\sinh \kappa y)\, e^{-ixy-y^2/2}\, dy . \tag{30}$$

Substituting (30) into the left-hand side of (29) and summing over n with the aid of the generating function [29, 17]

$$\sum_{n=0}^{\infty} \frac{t^n q^{n(n-1)/2}}{(q;q)_n} h_n(\sinh \kappa x | q) = E_q(te^{-\kappa x})\, E_q(-te^{\kappa x}) \tag{31}$$

for the q^{-1}-Hermite polynomials $h_n(x|q)$, one obtains

$$\varepsilon_q(\sin \kappa x; \tau)\, E_{q^2}(-\tau^2)\, e^{-x^2/2}$$

$$= \frac{1}{\sqrt{2\pi}} \int_{-\infty}^{\infty} e^{-ixy-y^2/2}\, E_q(\tau q^{1/2} e^{\kappa y})\, E_q(-\tau q^{1/2} e^{-\kappa y})\, dy. \tag{32}$$

It is interesting to compare (32) with Ramanujan's integral [9]-[11]

$$\frac{1}{\sqrt{2\pi}} \int_{-\infty}^{\infty} E_q(-aq^{1/2} e^{\sqrt{2}\kappa y})\, E_q(-bq^{1/2} e^{-\sqrt{2}\kappa y})\, e^{-ixy-y^2/2}\, dy$$

$$= E_q(ab)\, e_q(ae^{-i\sqrt{2}\kappa x})\, e_q(be^{i\sqrt{2}\kappa x})\, e^{-x^2/2} . \tag{33}$$

As we have seen, the classical Fourier transform with q-independent kernel turns out to be very useful in revealing close relations between various q-extensions of the exponential function e^z, as well as among different types of q-Bessel functions.

Acknowledgements

Discussions with A.B. Balantekin, P. Feinsilver, F. Leyvraz, and K.B. Wolf are gratefully acknowledged. This work is partially supported by the UNAM-DGAPA Project IN106595.

References

[1] H. Exton. *q-Hypergeometric Functions and Applications*, Ellis Horwood, Chichester, 1983 .

[2] R. Floreanini, J. LeTourneux, and L. Vinet. *J. Phys. A: Math. Gen.*, **28**, No.10, pp. L287–L293, 1995.

[3] N.M. Atakishiyev. *J. Phys. A: Math. Gen.*, **29**, No.10, pp. L223–L227, 1996.

[4] G. Gasper and M. Rahman. *Basic Hypergeometric Functions*, Cambridge University Press, Cambridge, 1990.

[5] M.E.H. Ismail and R. Zhang. *Adv. Math.*, **109**, pp.1–33, 1994.

[6] F.H. Jackson. *Proc. London Math. Soc.*, **2**, pp.192–220, 1905.

[7] M.E.H. Ismail. Some properties of Jackson's third *q*-Bessel function, to appear.

[8] N.M. Atakishiyev. *Proceedings of the Workshop on Symmetries and Integrability of Difference Equations* (Estérel, Canada, May 22–29, 1994), Eds. D. Levi, L. Vinet and P. Winternitz, CRM Proceedings and Lecture Notes, **9**, pp.21–27, American Mathematical Society, Rhode Island, Providence, 1996.

[9] S. Ramanujan. *Collected Papers*, Eds. G.H. Hardy, P.V. Seshu Aiyar and B.M. Wilson , Cambridge University Press, Cambridge, 1927; reprinted, Chelsea, New York, 1959.

[10] R. Askey. *Proc. Amer. Math. Soc.*, **85**, No.2, pp.192–194, 1982.

[11] N.M. Atakishiyev and P. Feinsilver. *Proceedings of the IV Wigner Symposium* (Guadalajara, Mexico, August 7–11, 1995), Eds. N.M.i Atakishiyev, T.H. Seligman and K.B. Wolf, pp.406–412, World Scientific, Singapore, 1996.

[12] C.A. Nelson and M.G. Gartley. *J. Phys. A: Math. Gen.*, **27** , pp.3857–3881, 1994.

[13] F.H. Jackson. *Trans. Roy. Soc. Edin.*, **46**, pp.253–281, 1908.

[14] G.E. Andrews and R.A. Askey. Lecture Notes in Mathematics, **1171**, pp.36–63, Springer-Verlag, Berlin, 1984.

[15] R.A. Askey and J.A. Wilson. *Mem. Amer. Math. Soc.i*, **54**, pp.1–55, 1985.

[16] C. Berg and M.E.H. Ismail. *Can. J. Math.*, **48**, pp.43–63, 1996.

[17] N.M. Atakishiyev. *Teor. Math. Phys.*, **105**, pp.1500–1508, 1995.

[18] F.H. Jackson. *Trans. Roy. Soc. Edin.*, **41**, pp.1–28, 1904.

[19] M.E.H. Ismail. *J. Math. Anal. Appl.*, **86**, No.1, pp.1–19, 1982.

[20] N.M. Atakishiyev and P.i Feinsilver. *J. Phys. A: Math. Gen.*, **29**, pp.1659–1664, 1996.

[21] A.J. Macfarlane. *J. Phys. A: Math. Gen.*, **22**, pp.4581–4588, 1989.

[22] L.C. Biedenharn. *J. Phys. A: Math. Gen.*, **22**, pp.L873–L878, 1989.

[23] R.A. Askey and M.E.H. Ismail. *Studies in Pure Mathematics*, Ed. P. Erdös, pp. 55–78, Birkhäuser, Boston, Massachusetts, 1983.

[24] N.M. Atakishiyev and S.K. Suslov. *Theor. Math. Phys.*, **85**, pp.1055–1062, 1991.

[25] L.J. Rogers. *Proc. London Math. Soc.*, **25**, pp.318–343, 1894.

[26] R. Floreanini and L. Vinet. *J. Math. Phys.*, **36**, pp.3800–3813, 1995.

[27] R. Floreanini, J. LeTourneux and L. Vinet. *J. Math. Phys.*, **36**, pp.5091–5097, 1995.

[28] N.M. Atakishiyev and Sh.M. Nagiyev. *Theor. Math. Phys.*, **98**, pp.162–166, 1994.

[29] M.E.H. Ismail and D.R. Masson. *Trans. Amer. Math. Soc.*, **346**, pp.63–116, 1994.

The Wilson Bispectral Involution: Some Elementary Examples

F. Alberto Grünbaum[†*]and Luc Haine[‡†]
[†] Department of Mathematics
University of California
Berkeley, CA 94720, U.S.A.
[‡] Department of Mathematics
Université Catholique de Louvain
1348 Louvain-la-Neuve, Belgium

1   Introduction

The notion of a second order bispectral differential operator was introduced in [DG].

One says that such an operator $L = -d^2/dx^2 + V(x)$ is bispectral if there exists a family of its eigenfunctions

$$L(x, d/dx)f(x, z) = z^2 f(x, z)$$

that are also eigenfunctions of a differential operator B in the spectral parameter z, i.e.

$$B(z, d/dz)f(x, z) = \Theta(x)f(x, z).$$

In [DG] one finds a complete solution to the classification problem in the case when L is a second order differential operator; namely, L is either the Bessel operator, the Airy operator, or it is obtained by a finite number of applications of the (rational) Darboux process starting from either $V = 0$ or $V = -1/(4x^2)$. The corresponding result when x and z are of different type, i.e. $x \in \mathbb{Z}$, $z \in \mathbb{R}$ and L, B are of order 2 has been recently given in [GH1].

---

[*]The first author was supported in part by NSF Grant # DMS94-00097 and by AFOSR under Contract FDF49620-96-1-0127.

[†]The second author is a Research Associate for FNRS.

In a fundamental paper G. Wilson [W] considered the natural extension of this problem to that of classifying bispectral commutative algebras of differential operators. In this paper Wilson obtains a complete description of all (maximal) rank one commutative algebras of bispectral differential operators.

One of the crucial novel ingredients in [W] is the notion of a "bispectral involution". The methods in [DG] and in [W] appear to be very different: the first one does not mention Sato's Grassmanians while the second does not mention the Darboux process. Later papers have established connections between these two methods. The usefulness of this approach has been amply demonstrated in a series of papers by A. Kasman and M. Rothstein, [KS], and by B. Bakalov, E. Horozov and M. Yakimov, [BHY1, BHY2, BHY3, BHY4]. It is indeed one of the main observations of [BHY4] that these two methods can be seen as parts of a general theory.

An "algorithmic version" of some of the ideas in [W] (as further elaborated in [KS], [BHY]) can be surmised as follows:

Consider a situation when one has a family of functions $\varphi(x, z)$ and $\psi(x, z)$ as well as differential operators $P(x, d/dx)$, $Q(x, d/dx)$, $P_b(z, d/dz)$ and $Q_b(z, d/dz)$ and functions $g(z)$, $f(z)$, $\theta(x)$ and $\nu(x)$ such that

$$\frac{1}{g(z)} P(x, d/dx)\varphi(x, z) = \frac{1}{\theta(x)} P_b(z, d/dz)\varphi(x, z) = \psi(x, z)$$

and

$$\frac{1}{f(z)} Q(x, d/dx)\psi(x, z) = \frac{1}{\nu(x)} Q_b(z, d/z)\psi(x, z) = \varphi(x, z).$$

In this case, it is clear that $\varphi(x, z)$ satisfies

$$QP\varphi(x, z) = f(z)g(z)\varphi(x, z)$$

as well as

$$Q_b P_b\varphi(x, z) = \theta(x)\nu(x)\varphi(x, z)$$

i.e., we have a "bispectral situation".

Now, this gives rise to a NEW bispectral situation in a very straightforward fashion: we have that

$$PQ\psi(x, z) = f(z)g(z)\psi(x, z)$$

as well as

$$P_b Q_b\psi(x, z) = \theta(x)\nu(x)\psi(x, z).$$

In this paper we will see instances when this construction gives rise to new bispectral situations out of "trivial" ones.

We will take $\varphi(x, z)$ to be an eigenfunction of the "trivial cases" [DG], i.e. Bessel or Airy.

We will (most of the time) consider the case when P and Q are FAMILIES of second order differential operators—depending on some free parameters—that are factors of either the square of the Airy operator or of the Bessel operator, i.e. we have

$$QP = (-D_x^2 + x)^2 \qquad \text{or} \qquad QP = (D_x^2 + c/x^2)^2$$

respectively.

The fact that QP and $Q_b P_b$ are a bispectral pair is a trivial observation: it will turn out that this second operator is a polynomial in the original Bessel or Airy operator in the spectral variable.

We will consider the case when the operators Q_b and P_b are also second order, in the Airy case we consider also a fourth order P_b, and in the last section a case of order three.

As we see, the main difference with the method in [DG] is that we no longer factorize the Bessel operator itself, but rather a polynomial in it. Related ideas are considered in [F] and [AvM].

The construction is entirely elementary and it is based on the method used in the rank two case discussed in section 5 of [DG]. Once the appropriate equations are obtained all that remains to do is to solve a sequence of linear ordinary differential equations. This is done in detail below to illustrate the simplicity of the method.

Other related constructions are given in these rank two cases in the series of papers [KS,BHY] mentioned above. In the language of [BHY4] we are able to obtain extensions of some of the examples in [DG] within the class of *monomial* Darboux transformations.

Included in this—as a very special case—will be the situation $P = Q =$ Airy or Bessel. This will correspond to the case $c_1 = 0$ in the Airy case and to the case $k_4 = 0$ in the Bessel case (see the expressions for P and Q given in sections 2.4 and 3.4).

We have managed to push these ideas in some "mixed cases' where the variables x and z are of different types. These results are reported in [GH2].

2 The Airy case

2.1 Introduction

Consider a function $\varphi(x, z)$ satisfying

$$(-D_x^2 + x)\varphi(x, z) = z\varphi(x, z).$$

This amounts to taking a function $f = f(\xi)$ that satisfies

$$D_\xi^2 f = \xi f$$

and putting $\varphi(x, z) = f(\xi)$ with $\xi = x - z$.

Note, for later use, that the same $\varphi(x, z)$ satisfies the equation

$$(-D_z^2 - z)\varphi(x, z) = x\varphi(x, z).$$

It follows that all higher order derivatives of f are expressed in terms of f and $f_1 = D_\xi f$. This is the basis of the method used in section 5 of [DG] and will be used throughout.

2.2   The operator P

We consider finding differential operators $P(x, d/dx)$ and $P_b(z, d/dz)$ such that

$$\frac{1}{z}P(x, d/dx)\varphi(x, z) = \frac{1}{\theta(x)}P_b(z, d/dz)\varphi(x, z)$$

for the function φ considered above.

If we try to get both P and P_b to be of order TWO in the form

$$P = D_x^2 + aD_x + b, \qquad P_b = eD_z^2 + gD_z + h$$

then the equations that have to be satisfied are

$$\frac{f(x - z) + af_1 + bf}{z} - \frac{ef(x - z) + fh - f_1g}{\theta} = 0.$$

The coefficient of f_1 gives

$$gz + a\theta = 0$$

forcing the result

$$g = \frac{c_1}{z}, \qquad \theta = -\frac{c_1}{a}$$

with c_1 an arbitrary constant.

When this is plugged into the coefficient of f we get

$$ez^2 - exz - hz + \frac{c_1z}{a} - \frac{c_1x}{a} - \frac{bc_1}{a} = 0.$$

A differentiation with respect to x followed by one with respect to z gives

$$-\frac{de}{dz}z - e - \frac{c_1}{a^2}\frac{da}{dx} = 0.$$

This forces

$$e = \frac{c_3}{z} + c_2$$

with c_3 and c_2 arbitrarily constants, and then we get a differential equation for a that gives

$$a = \frac{c_1}{c_2x - c_4c_1}$$

with yet another constant c_4.

When this is plugged back into the coefficient of f, we get

$$b = \frac{c_2 z^2 + (-h - c_4 c_1 + c_3)z - c_2 x^2 + (c_4 c_1 - c_3)x}{c_2 x - c_4 c_1}$$

and upon differentiation by z we get a differential equation for h that is solved by

$$h = c_2 z + \frac{c_5}{z} - c_4 c_1 + c_3$$

for an arbitrary constant c_5.

This in turn pins down b above as

$$b = -\frac{c_2 x^2 + (c_3 - c_4 c_1)x + c_5}{c_2 x - c_4 c_1}.$$

In summary, we have that P is given by

$$-\frac{(c_2 x^2 + (c_3 - c_4 c_1)x + c_5)}{c_2 x - c_4 c_1} + \frac{c_1 D_x}{c_2 x - c_4 c_1} + D_x^2$$

and P_b is given by

$$\left(c_2 z + \frac{c_5}{z} - c_4 c_1 + c_3\right) + \frac{c_1}{z} D_z + \left(\frac{c_3}{z} + c_2\right) D_z^2.$$

Define the function $\psi(x, z)$ by the relation

$$\psi(x, z) = \frac{1}{z} P(x, d/dx)\varphi(x, z)$$

with P as determined above.

2.3 Going for Q and Q_b

We now try to find a new pair of differential operators $Q(x, d/dx)$ and $Q_b(z, d/dz)$ such that on one hand

$$\varphi(x, z) = \frac{1}{z} Q\psi(x, z)$$

and on the other

$$\frac{1}{z} Q\psi(x, z) = \frac{1}{\nu(x)} Q_b \psi(x, z).$$

From the first requirement it follows that

$$QP\varphi(x, z) = (-D_x^2 + x)^2 \varphi(x, z)$$

and since P is already available one can thus determine Q by the condition

$$QP = (-D_x^2 + x)^2.$$

2.4   Finding Q

P was given earlier and we look for Q in the form

$$D_x^2 + c(x)D_x + d(x).$$

If Airy is the operator

$$-D_x^2 + x$$

the condition $QP = $ Airy^2 gives

$$c(x) = -\frac{c_1}{c_2 x - c_4 c_1}$$

and

$$d(x) = -\frac{c_2^2 x^3 + (-2c_4 c_2 c_1 - c_3 c_2)x^2}{c_2^2 x^2 - 2c_4 c_2 c_1 x + c_4^2 c_1^2}$$
$$- \frac{(-c_2 c_5 + c_4^2 c_1^2 + c_4 c_3 c_1)x + c_4 c_1 c_5 - c_1^2 - 2c_2 c_1}{c_2^2 x^2 - 2c_4 c_2 c_1 x + c_4^2 c_1^2}.$$

These conditions are not sufficient to insure the factorization above. For that end extra restrictions on the constants should hold. One can see for instance that at least one of the three constants c_2, c_2, c_3 vanishes.

If we take either of the first two equal to zero we are, at best, led to the trivial case

$$P = Q = \text{Airy}.$$

Making the choice

$$c_3 = 0$$

we are eventually led into the further restrictions

$$c_2 = -c_1, \qquad c_3 = 0, \qquad c_4 = -\frac{c_5^2}{c_1^2}$$

which result in

$$d(x) = -\frac{c_1^4 x^3 - 2c_1^2 c_5^2 x^2 + (c_5^4 + c_1^3 c_5)x - c_1 c_5^3 + c_1^4}{c_1^4 x^2 - 2c_1^2 c_5^2 x + c_5^4},$$

$$c(x) = \frac{c_1^2}{c_1^2 x - c_5^2}.$$

With these restrictions we get for P the expression

$$-\frac{(c_1^2 x^2 - c_5^2 x - c_2 c_5)}{c_1^2 x - c_5^2} - \frac{c_1^2 D_x}{c_1^2 x - c_5^2} + D_x^2$$

while Q is given by

$$-\frac{(c_1^4 x^3 - 2c_1^2 c_5^2 x^2 + (c_5^4 + c_1^3 c_5)x - c_1 c_5^3 + c_1^4)}{c_1^4 x^2 - 2c_1^2 c_5^2 x + c_5^4} + \frac{c_1^2 D_x}{c_1^2 x - c_5^2} + D_x^2.$$

Notice that we have a family depending on TWO FREE parameters.

If $c_1 = 0$ we recover the case $P = Q =$ Airy. When this is not the case one can use

$$y = c_1^2 x - c_5^2$$

as a new variable to rewrite P and Q.

2.5   The operator Q_b

Now try to solve for Q_b, starting from the Q given above, in such a way that

$$\frac{1}{z}Q(x, d/dx)\psi = \frac{1}{\nu(x)}Q_b(z, d/dz)\psi.$$

We will look for Q_b of order two.

Before we start we need an expression for ψ obtained by applying $z^{-1}P\varphi$ with the P above.

In terms of f and f_1 as above, one gets that ψ is given by

$$-\frac{(c_1^2 fx - fc_5^2)z - c_1 fc_5 + c_1^2 f_1}{(c_1^2 x - c_5^2)z}.$$

We check easily that $z^{-1}Q\psi = \varphi$.

Then we get expressions for ψ_z and ψ_{zz} and look for Q_b in the form

$$Q_b = uD_z^2 + vD_z + w.$$

The coefficient of f_1 gives

$$u = \frac{c_1^2 wz^2}{c_1^2 z^3 - c_5^2 z^2 + c_1 c_5 z - c_1^2}, \qquad v = \frac{c_1^2 wz}{c_1^2 z^3 - c_5^2 z^2 + c_1 c_5 z - c_1^2}$$

and going with this into the coefficient of f one gets as a necessary condition

$$\nu = -\frac{(c_1^2 wx - c_5^2 w)z^2}{c_1^2 z^3 - c_5^2 z^2 + c_1 c_5 z - c_1^2}.$$

We still need to determine $\nu(x)$ and $w(z)$.

Differentiating with respect to x above we can solve for w and then plug this expression back to get a differential equation in ν solved by

$$\nu = c_8(c_1^2 x - c_5^2).$$

When this is plugged into the expression above one gets the final ingredient

$$w = -\frac{c_1^2 c_8 z^3 - c_8 c_5^2 z^2 + c_1 c_8 c_5 z - c_1^2 c_8}{z^2}.$$

When these expression are substituted in $u(z)$ and $v(z)$ we get

$$u = -c_1^2 c_8, \qquad v = -\frac{c_1^2 c_8}{z}.$$

So, we have found Q_b as given below

$$-\frac{(c_1^2 c_8 z^3 - c_8 c_5^2 z^2 + c_1 c_8 c_5 z - c_1^2 c_8)}{z^2} - \frac{c_1^2 c_8 D_z}{z} - c_1^2 c_8 D_z^2.$$

With the restrictions on the constants $c_1, c_2, \ldots$ found earlier the original P_b takes the form

$$\left(-c_1 z + \frac{c_5}{z} + \frac{c_5^2}{c_1}\right) + \frac{c_1 D_z}{z} - c_1 D_z^2.$$

2.6   The product $Q_b P_b$

We consider now the issue of computing the product $Q_b P_b$.

Take the Airy operator (in the z variable)

$$fz + D_z^2 \tag{1}$$

and compute $w_2 \, \text{Airy}^2 + w_1 \, \text{Airy} + w_0 I$.

It turns out that $Q_b P_b$ is this constant coefficient polynomial in the Airy operator if we choose

$$w_2 = c_1^3, \qquad w_1 = -2c_1 c_5^2, \qquad w_0 = \frac{c_5^4}{c_1}.$$

2.7 P_b of order four

Here we allow for P_b of order four and repeat the computations given above for the same P.

We get

$$\theta = -\frac{(x - c_9)(c_1^2 x - c_5^2)}{c_1^2}$$

and for P_b we get

$$\frac{c_1^2 z^3 - c_5^2 z^2 - c_1^2 c_9 z^2 + c_9 c_5^2 z - c_1 c_5 z + c_1 c_9 c_5 - c_1^2}{c_1^2 z}$$

$$+\frac{(2c_1^2 z^2 - c_5^2 z - c_1^2 c_9 z - c_1 c_5) D_z^2}{c_1^2 z} + \frac{(z + c_9) D_z}{z} - \frac{D_z^3}{z} + D_z^4.$$

When we compute the operator Q_b we discover that it is of order TWO and it is given by

$$\frac{(c_1^2 z^3 - c_5^2 z^2 + c_1 c_5 z - c_1^2)}{c_1^2 z^2} + \frac{D_z}{z} + D_z^2$$

while the function $\nu(x)$ is

$$\nu = -\frac{t^2 x - 1}{t^2}.$$

We consider again the product $Q_b P_b$ and conclude that it is given by

$$w_3 \text{ Airy}^3 + w_2 \text{ Airy}^2 + w_1 \text{ Airy} + w_0 I$$

as long as you choose

$$w_3 = 1, \qquad w_2 = -\frac{2c_5^2 + c_1^2 c_9}{c_1^2}, \qquad w_1 = \frac{c_5^2(c_5^2 + 2c_1^2 c_9)}{c_1^4}, \qquad w_0 = -\frac{c_9 c_5^4}{c_1^4}.$$

One may suspect that the fourth order P_b given above and the earlier one given by

$$\left(-c_1 z + \frac{c_5}{z} + \frac{c_5^2}{c_1}\right) + \frac{c_1 D_z}{z} - c_1 D_z^2$$

might commute. However this is not the case.

On the other hand the products $P_b Q_b$ computed with these two different P_b and Q_b are given by

$$\frac{(c_1^4 z^6 - 2c_1^2 c_5^2 z^5 + c_5^4 z^4 - 2c_1^4 z^3 + 4c_1^3 c_5 z - 8c_1^4)}{c_1 z^4}$$

$$+\frac{2c_1^2(z(c_1 z^2 - 2c_5) + 4c_1)D_z}{z^3} + \frac{2c_1(z^2(c_1^2 z - c_5^2) - 2c_1^2)D_z^2}{z^2} + c_1^3 D_z^4$$

and

$$-\quad (c_1^4 z^9 - 2c_1^2 c_5^2 z^8 - c_1^4 c_9 z^8 + c_5^4 z^7 + 2c_1^2 c_9 c_5^2 z^7$$
$$-c_9 c_5^4 z^6 - c_1^4 z^6 + c_1^2 c_5^2 z^5 + 2c_1^4 c_9 z^5 + 3c_1^3 c_5 z^4$$
$$-2c_1 c_5^3 z^3 - 4c_1^3 c_9 c_5 z^3 - 9c_1^4 z^3 + 4c_1^2 c_5^2 z^2$$
$$+8c_1^4 c_9 z^2 + 36c_1^3 c_5 z - 144c_1^4)/(c_1^2 z^6)$$

$$-\quad (3c_1^4 z^6 - 4c_1^2 c_5^2 z^5 - 2c_1^4 c_9 z^5 + c_5^4 z^4$$
$$+2c_1^2 c_9 c_5^2 z^4 - 9c_1^4 z^3 + 2c_1^2 c_5^2 z^2 + 4c_1^4 c_9 z^2$$
$$+18c_1^3 c_5 z - 72c_1^4)D_z^2/(c_1^2 z^4)$$

$$-\quad (6c_1^3 z^6 - 4c_1 c_5^2 z^5 - 2c_1^3 c_9 z^5 - 6c_1^2 c_5 z^4 + 2c_1^3 z^3 + 4c_1^3 c_9 c_5 z^3$$
$$+9c_1^3 z^3 - 4c_1 c_5^2 z^2 - 8c_1^3 c_9 z^2 - 36c_1^2 c_5 z + 144c_1^3)D_z/(c_1 z^5)$$

$$-\quad \frac{(3c_1^2 z^3 - 2c_5^2 z^2 - c_1^2 c_9 z^2 - 6c_1^2)D_z^4}{z^2}$$

$$-\quad \frac{6c_1(z(c_1 z^2 - c_5) + 4c_1)D_z^3}{z^3} - c_1^2 D_z^6$$

and these two operators (sharing the common family of eigenfunctions $\psi(x, z)$) commute.

3   The Bessel case

3.1   Introduction

Take a function $f = f(\xi)$ that satisfies

$$D_\xi^2 f = (1 - c/\xi^2)f(\xi).$$

If we put $\xi = xz$ and define

$$\varphi(x, z) = f(\xi)$$

we get an eigenfunction of the Bessel operator, i.e.

$$(D_x^2 + c/x^2)\varphi(x, z) = z^2\varphi(x, z).$$

3.2   The operator P

We consider finding differential operators $P(x, d/dx)$ and $P_b(z, d/dz)$ such that

$$\frac{1}{z^2}P(x, d/dx)\varphi(x, z) = \frac{1}{\theta(x)}P_b(z, d/dz)\varphi(x, z)$$

for the function φ considered above.

If we try to get both P and P_b to be of order TWO in the form

$$P = D_x^2 + aD_x + b, \qquad P_b = eD_z^2 + gD_z + h$$

then the equations that have to be satisfied are

$$\frac{f_2 z^2 + af_1 z + bf}{z^2} - \frac{ef_2 x^2 + gf_1 x + fh}{\theta} = 0$$

where f_1, f_2 denote the derivatives of f of orders one and two respectively.

The coefficient of f_1 gives

$$a\theta - gxz = 0$$

which forces the solution

$$g = \frac{c_{10}}{z}, \qquad \theta = \frac{c_{10}x}{a}$$

with c_{10} an arbitrary constant.

Plugging into the coefficient of f we get

$$aex^3z^2 - c_{10}x^2z^2 + ahxz^2 - bc_{10}x^2 - acex + c_{10} = 0.$$

By differentiating with respect to z and solving for h, we get

$$\frac{de}{dz} = -\frac{(2a^2e + \frac{da}{dx}c_{10})x - ac_{10}}{a^2xz}.$$

By differentiating this with respect to z and to x and using the corresponding equations to eliminate $a(x)$ we get

$$\frac{d^2e}{dz^2} = -\frac{3}{z}\frac{de}{dz}$$

which gives

$$e = \frac{k_1}{z^2} + k_2.$$

Going with this into the differential equation for $e(z)$ above we get the Riccati type equation for $a(x)$

$$\frac{da}{dx} = -\frac{2a^2k_2x - ac_{10}}{c_{10}x}.$$

This equation is solved with the usual log-derivative trick to give

$$a = \frac{c_{10}k_4x}{k_2(k_4x^2 + k_5)}.$$

Going with these expressions for $a(x)$ and $e(z)$ into the coefficient for f above we get an expression for h, namely

$$h = \frac{k_2k_5x^2z^4 + ((bk_2 - k_1)k_4x^4 + bk_2k_5x^2 - ck_2k_5)z^2 + ck_1k_4x^2}{k_4x^4z^4}.$$

This expression above will not be a function of z alone unless b is chosen properly. This amounts to the requirement that

$$\frac{db}{dx} = -\frac{(2bk_2 - 2k_1)k_4x^4 + 2ck_2k_5}{k_2k_4x^5 + k_2k_5x^3}$$

which gives

$$b = \frac{1}{k_4x^2 + k_5}\left(k_1k_4x^2k_2 + \frac{ck_5}{x^2} + k_8\right).$$

Once this is taken into account we get

$$h = \frac{k_2k_5z^4 + k_2k_8z^2 + ck_1k_4}{k_4z^4}.$$

Define the function $\psi(x, z)$ by the relation

$$\psi(x, z) = \frac{1}{z^2}P(x, d/dx)\varphi(x, z)$$

with P as given above.

3.3   Going for Q and Q_b

We now try to find a new pair of differential operators $Q(x, d/dx)$ and $Q_b(z, d/dz)$ such that on one hand

$$\varphi(x, z) = \frac{1}{z} Q \psi(x, z)$$

and on the other

$$\frac{1}{z^2} Q \psi(x, z) = \frac{1}{\nu(x)} Q_b \psi(x, z).$$

From the first requirement it follows that

$$QP\varphi(x, z) = (D_x^2 + c/x^2)^2 \varphi(x, z)$$

and since P is already available one can thus determine Q by the condition

$$QP = (D_x^2 + c/x^2)^2.$$

3.4   Finding Q

Since P is given by

$$\frac{(k_1 k_4 x^4 + k_2 k_8 x^2 + c k_2 k_5)}{k_2 x^2 (k_4 x^2 + k_5)} + \frac{c_{10} k_4 x D_x}{k_2 (k_4 x^2 + k_5)} + D_x^2$$

if we look for Q in the form

$$D_x^2 + c(x) D_x + d(x)$$

we get that $c(x)$ is given by

$$c(x) = -\frac{c_{10} k_4 x}{k_2 k_4 x^2 + k_2 k_5}$$

and $d(x)$ by

$$\begin{aligned}
d(x) &= -(k_1 k_2 k_4^2 x^6 + k_2^2 k_4 k_8 x^4 + k_1 k_2 k_4 k_5 x^4 - 2c k_2^2 k_4^2 x^4 \\
&\quad - 2c_{10} k_2 k_4^2 x^4 - c_{10}^2 k_4^2 x^4 + k_2^2 k_5 k_8 x^2 - 3c k_2^2 k_4 k_5 x^2 \\
&\quad + 2c_{10} k_2 k_4 k_5 x^2 - c k_2^2 k_5^2)/(k_2^2 x^2 (k_4 x^2 + k_5)^2).
\end{aligned}$$

These conditions do not yet guarantee that $QP = $ Bessel^2.

They have to be supplemented by certain extra requirements on the—so far—free parameters c_{10}, k_1, k_2, etc.

In particular one condition that has to be imposed is

$$k_1 k_2 k_4 = 0.$$

Instead of analyzing all possible cases we settle for the choice

$$k_1 = 0, \qquad c_{10} = -2k_2$$

accompanied by the requirement that

$$k_8^2 - 2ck_4k_8 - 2k_4k_8 + c^2k_4^2 + 6ck_4^2 = 0.$$

With these conditions we end up getting that Q is given by

$$-\frac{(k_4k_8x^4 - 2ck_4^2x^4 + k_5k_8x^2 - 3ck_4k_5x^2 - 4k_4k_5x^2 - ck_5^2)}{x^2(k_4x^2 + k_5)^2} + \frac{2k_4xD_x}{k_4x^2 + k_5} + D_x^2$$

and P is given by

$$\frac{(k_8x^2 + ck_5)}{k_4x^4 + k_5x^2} - \frac{2k_4xD_x}{k_4x^2 + k_5} + D_x^2.$$

Again we have a TWO parameter family of operators P, Q.

If $k_4 = 0$ we get the case where $P = Q = $ Bessel since in that case we need to put $k_8 = 0$ too.

3.5   The operator Q_b

Now try to solve for Q_b starting from the Q given above in such a way that

$$\frac{1}{z^2}Q(x, d/dx)\psi = \frac{1}{\nu(x)}Q_b(z, d/dz)\psi.$$

We will look for Q_b of order two.

Before we start we need an expression for ψ obtained by applying $z^{-2}P\varphi$ with the P above.

The expression for ψ is obtained again by expanding everything in terms of $f = \varphi$ and its first derivative f_1, and one checks that

$$\frac{1}{z^2}Q\psi = \varphi.$$

Now one computes carefully ψ_z and ψ_{zz} in terms of φ and its first derivative and then sets out to compute Q_b in the form

$$Q_b = e(z)D_z^2 + d(z)D_z + g(z).$$

One gets from the coefficient of f_1

$$e = \frac{gk_4z^2}{k_5z^2 - k_8 + 2ck_4}, \qquad d = \frac{2gk_4z}{k_5z^2 - k_8 + 2ck_4}.$$

Plugging this into the coefficient for f we get the relation

$$
\begin{aligned}
\nu &= ((gk_4^2 x^4 + 2gk_4 k_5 x^2 + gk_5^2)z^4 - gk_8^2 + (2c+2)gk_4 k_8 \\
&+ (-c^2 - 6c)gk_4^2)/((k_4 k_5 x^2 + k_5^2)z^4 \\
&+ ((2ck_4^2 - k_4 k_8)x^2 - k_5 k_8 + 2ck_4 k_5)z^2).
\end{aligned}
$$

We need to find both ν and g. This is done by differentiating this relation with respect to x, solving for g and plugging back. One then gets an equation for ν if one remembers the very important relation

$$
k_8^2 = (2c+2)k_4 k_8 + (-c^2 - 6c)k_4^2.
$$

The equation for $\nu(x)$ is solved by

$$
\nu = k_9(k_4 x^2 + k_5)
$$

and then one obtains that

$$
g = \frac{k_5 k_9 z^2 + (2ck_4 - k_8)k_9}{z^2}.
$$

Replacing these expressions into $e(z)$ and $d(z)$ we get

$$
e = k_4 k_9, \qquad d = \frac{2k_4 k_9}{z}.
$$

3.6 The product $Q_b P_b$

Recall that Q_b is given by

$$
\frac{(k_5 k_9 z^2 + (2ck_4 - k_8)k_9)}{z^2} + \frac{2k_4 k_9 D_z}{z} + k_4 k_9 D_z^2
$$

while P_b is given by

$$
\frac{(k_2 k_5 z^4 + k_2 k_8 z^2)}{k_4 z^4} - \frac{2k_2 D_z}{z} + D_z^2.
$$

If we introduce the Bessel operator

$$
D_z^2 + \frac{c}{z^2}
$$

we can see that $Q_b P_b$ is given by

$$
w_2 \, \mathrm{Bessel}^2 + w_1 \, \mathrm{Bessel} + w_0 I
$$

if we make the choices

$$
w_2 = k_2 k_4 k_9, \qquad w_1 = 2k_2 k_5 k_9, \qquad w_0 = \frac{k_2 k_5^2 k_9}{k_4}.
$$

3.7   A special case: the first "even family case"

It is interesting to notice that we recover—as a very special case of the results above—the simplest examples in [DG]. This is done in these two last sections.

If we impose the conditions

$$k_8 = \frac{5k_4}{4}, \qquad c = \frac{1}{4}$$

which are compatible with the requirement

$$k_8^2 - 2ck_4k_8 - 2k_4k_8 + c^2k_4^2 + 6ck_4^2 = 0,$$

as well as $k_5/k_4 = t$, we get that the product PQ is nothing but the square of the Schroedinger operator

$$D_x^2 + \tfrac{1}{4}x^2 + 2\frac{d^2x}{dx^2}\log(x^2 + t).$$

On the other hand, with these choices, the product P_bQ_b turns out to be

$$k_2k_4k_9((D_z^2 - 15/4z^2)^2 + 2t(D_z^2 + \tfrac{1}{4}z^2) + t^2).$$

Now, these operators correspond exactly to the first example of the "even family" in [DG].

To get the first example in the Korteweg–deVries family we need to revisit the construction given above allowing—in the spirit of section 2.7—for a higher order P_b. This is the purpose of the next section. In the next section, however, we cannot rely on the method used so far, since we are dealing with a rank one example.

3.8 P_b of order three and the first KdV example

If we take for P and Q a pair of second order differential operators and require, in the spirit of the last example, that

$$QP = D_x^4 = (-D_x^2)^2$$

and

$$PQ = (-D_x^2 + 2d^2/dx^2\log(x^3 + t))^2$$

we conclude after some elementary integrations of the type performed earlier that

$$P = D_x^2 - \frac{3x^2}{x^3 + t}D_x + \frac{3x}{x^3 + t}$$

and

$$Q = D_x^2 + \frac{3x^2}{x^3 + t}D_x + \frac{9xt}{(x^3 + t)^2}.$$

We will look for P_b and Q_b in the form

$$eD_z^3 + fD_z^2 + gD_z + h.$$

Since we want QP to be the power of D_x we can take

$$\varphi(x, z) = e^{xz}$$

and then ψ is determined by

$$\psi(x, z) = \frac{1}{z^2} P\varphi(x, z)$$

which gives

$$\psi(x, z) = \frac{((x^3 + t)z^2 - 3x^2 z + 3x)e^{xz}}{(x^3 + t)z^2}.$$

Now the condition

$$\frac{1}{\theta(x)} P_b(z, d/dz)\varphi(x, z) = \psi(x, z)$$

gives for P_b the expression

$$P_b = D_z^3 - \frac{3}{z}D_z^2 + \frac{3}{z^2}D_z + t$$

along with $\theta(x) = x^3 + t$, while the requirement that

$$\frac{1}{\nu(x)} Q_b \psi(x, z) = \varphi(x, z)$$

results in $\nu(x) = \theta(x)$ and

$$Q_b = D_z^3 + \frac{3}{z}D_z^2 - \frac{3}{z^2}D_z + t.$$

One checks easily that $\theta(x)P$ and $z^2 P_b$ are in the Weyl algebra and are related by the BISPECTRAL INVOLUTION introduced in section 8 of [W].

For the product $Q_b P_b$ we get $(D_z^3 + t)^2$ while the product $P_b Q_b$ gives the six order differential operator

$$B_6 \equiv D_z^6 - \frac{18}{z^2}D_z^4 + \frac{2(tz^3 + 36)}{z^3}D_z^3 - \frac{144}{z^4}D_z^2 + \frac{144}{z^5}D_z + t^2$$

and we notice that this operator commutes with the "bispectral operator" given in [DG], namely

$$B_4 \equiv (D_z^2 - 6/z^2)^2 + 4tD_z.$$

One has $B_4^3 - B_6^2 = 8tD_z^9 + \dots$ making the algebra of all differential operators that commute with B_4 a "rank one" algebra.

References

[AvM]   Adler, M. and van Moerbeke, P., *Birkhoff strata, Bäcklund trans-formations, and regularization of isospectral operators*, Adv. Math. **108**, 140–240 (1994).

[BHY1]   Bakalov, B., Horozov, E., and Yakimov, M., *Highest weight modules of $W_{1+\infty}$, Darboux transformations and the bispectral problem*, to appear in Proc. Conf. Geom. and Math. Phys., Zlatograd 95, Bulgaria, q-alg/9601017.

[BHY2]   Bakalov, B., Horozov, E., and Yakimov, M., *Tau-functions as highest weight vectors for $W_{1+\infty}$ algebra*, Sofia preprint (1995), hep-th/9510211.

[BHY3]   Bakalov, B., Horozov, E., and Yakimov, M., *Bäcklund–Darboux transformations in Sato's Grassmannian*, Sofia preprint (1996), q-alg/9602010.

[BHY4]   Bakalov, B., Horozov, E., and Yakimov, M., *Bispectral algebras of commuting ordinary differential operators*, Sofia preprint (1996), q-alg/9602011.

[DG]   Duistermaat, J. J., and Grünbaum, F. A., *Differential equations in the spectral parameter*, Commun. Math. Phys. **103**, 177–240 (1986).

[F]   Fastré, J., *Bäcklund–Darboux transformations and W-algebras*, Doctoral Dissertation, University of Louvain, 1993.

[GH1]   Grünbaum, F. A. and Haine, L., *A theorem of Bochner*, revisited in Algebraic Aspects of Integrable Systems: In memory of Irene Dorfman, editors A. Fokas and I. M. Gelfand, Birkhäuser, 143–172 (1996).

[GH2]   Grünbaum, F. A. and Haine, L., *Wilson bispectral involution: an extension of the Krall polynomials.*

[KS]   Kasman, A., and Rothstein, M., *Bispectral Darboux transformations: the generalized Airy case*, Physica D (to appear), q-alg/9606018.

[W]   Wilson, G., *Bispectral commutative ordinary differential operators*, J. reine angew. Math. **442**, 177–204 (1993).

Factorisation of Macdonald polynomials

V.B. Kuznetsov [†*] and E.K. Sklyanin [‡†]
† Department of Applied Mathematical Studies,
University of Leeds,
Leeds LS2 9JT, UK
Email: vadim@amsta.leeds.ac.uk
‡ Research Institute for Mathematical Sciences,
Kyoto University,
Kyoto 606, Japan
Email: sklyanin@pdmi.ras.ru

1. Macdonald polynomials

Macdonald polynomials $P_\lambda(x; q, t)$ are orthogonal symmetric polynomials which are the natural multivariable generalisation of the *continuous q-ultraspherical polynomials* $C_n(x; \beta| q)$ [2] which, in their turn, constitute an important class of hypergeometric orthogonal polynomials in one variable. Polynomials $C_n(x; \beta|q)$ can be obtained from the general *Askey-Wilson polynomials* [3] through a specification of their four parameters (see, for instance, [9]), so that $C_n(x; \beta|q)$ depend only on one parameter β, apart from the degree n and the basic parameter q. In an analogous way, the Macdonald polynomials $P_\lambda(x; q, t)$ with one parameter t could be obtained as a limiting case of the 5-parameter Koornwinder's multivariable generalisation of the Askey-Wilson polynomials [10].

The main reference for the Macdonald polynomials is the book [17], Ch. VI, where they are called *symmetric functions with two parameters*. Let $\mathbb{K} = \mathbb{Q}(q, t)$ be the field of rational functions in two indeterminants q, t; $\mathbb{K}[x] = \mathbb{K}[x_1, \ldots, x_n]$ be the ring of polynomials in n variables $x = (x_1, \ldots, x_n)$ with coefficients in $\mathbb{K}$; and $\mathbb{K}[x]^W$ be the subring of all symmetric polynomials.

The Macdonald polynomials $P_\lambda(x) = P_\lambda(x; q, t)$ are symmetric polynomials labelled by the sequences $\lambda = \{0 \le \lambda_1 \le \lambda_2 \le \ldots \le \lambda_n\}$ of integers (dominant weights). They form a $\mathbb{K}$ basis of $\mathbb{K}[x]^W$ and are uniquely characterised as joint eigenvectors of the commuting q-difference operators H_k

$$H_k\, P_\lambda = h_{k;\lambda}\, P_\lambda\,, \qquad k = 1, \ldots, n\,, \tag{1.1}$$

normalised by the condition

$$P_\lambda = \sum_{\lambda' \preceq \lambda} \kappa_{\lambda\lambda'}\, m_{\lambda'}\,, \qquad \kappa_{\lambda\lambda} = 1\,, \qquad (\kappa_{\lambda\lambda'} \in \mathbb{K}) \tag{1.2}$$

where, for each μ, $m_\mu(x)$ stands for the monomial symmetric function: $m_\mu(x) = \sum x_1^{\nu_1} \ldots x_n^{\nu_n}$ with the sum taken over all distinct permutations ν of μ, and $\preceq$ is the dominance order on the dominant weights

$$\lambda' \preceq \lambda \quad \Longleftrightarrow \quad \left\{ |\lambda'| = |\lambda|\,;\quad \sum_{j=k}^{n} \lambda'_j \le \sum_{j=k}^{n} \lambda_j\,,\quad k = 2, \ldots, n \right\}.$$

The commuting operators H_i have the form:

$$H_i = \sum_{\substack{J \subset \{1,\ldots,n\} \\ |J|=i}} \left(\prod_{\substack{j \in J \\ k \in \{1,\ldots,n\}\backslash J}} v_{jk} \right) \left(\prod_{j \in J} T_{q,x_j} \right)\,, \qquad i = 1, \ldots, n\,, \tag{1.3}$$

where

$$v_{jk} = \frac{t^{1/2} x_j - t^{-1/2} x_k}{x_j - x_k}\,, \qquad |q| < 1,\ |t| < 1\,. \tag{1.4}$$

The T_{q,x_j} stands for the q-shift operator in the variable x_j: $(T_{q,x_j} f)(x_1, \ldots, x_n) = f(x_1, \ldots, qx_j, \ldots, x_n)$. The operators (1.3) were introduced for the first time in [21] and are the integrals of motion for the quantum Ruijsenaars model, which is a relativistic (or q-) analog of the trigonometric Calogero-Moser-Sutherland model. They are called sometimes Macdonald operators in the mathematical literature.

The corresponding eigenvalues $h_{k;\lambda}$ in (1.1) are

$$h_{k;\lambda} = \sum_{j_1 < \ldots < j_k} \mu_{j_1} \ldots \mu_{j_k}\,, \qquad \mu_j = q^{\lambda_j} t^{j - \frac{n+1}{2}}\,. \tag{1.5}$$

The polynomials P_λ are orthogonal

$$\frac{1}{(2\pi i)^n} \oint_{|x_1|=1} \frac{dx_1}{x_1} \ldots \oint_{|x_n|=1} \frac{dx_n}{x_n}\, \overline{P_\lambda(x; q, t)}\, P_{\lambda'}(x; q, t)\, \Delta(x) = 0\,, \qquad \lambda \ne \lambda'$$

$$\tag{1.6}$$

with respect to the weight

$$\Delta(x_1,\ldots,x_n) = \prod_{j \neq k} \frac{(x_j x_k^{-1}; q)_\infty}{(t x_j x_k^{-1}; q)_\infty}. \tag{1.7}$$

For instance, for $n = 3$,

$$m_{000} = 1,$$
$$m_{001} = x_1 + x_2 + x_3,$$
$$m_{011} = x_1 x_2 + x_1 x_3 + x_2 x_3,$$
$$m_{002} = x_1^2 + x_2^2 + x_3^2,$$
$$m_{111} = x_1 x_2 x_3,$$
$$m_{012} = x_1 x_2^2 + x_1^2 x_2 + x_1 x_3^2 + x_1^2 x_3 + x_2 x_3^2 + x_2^2 x_3,$$
$$m_{112} = x_1^2 x_2 x_3 + x_1 x_2^2 x_3 + x_1 x_2 x_3^2,$$
$$m_{022} = x_1^2 x_2^2 + x_1^2 x_3^2 + x_2^2 x_3^2,$$
$$m_{003} = x_1^3 + x_2^3 + x_3^3.$$

$$P_{000} = m_{000},$$
$$P_{001} = m_{001},$$
$$P_{011} = m_{011},$$
$$P_{002} = m_{002} + \frac{(1-t)(1+q)}{1-qt} m_{011},$$
$$P_{111} = m_{111},$$
$$P_{012} = m_{012} + \frac{(1-t)(q(2t+1)+t+2)}{1-qt^2} m_{111},$$
$$P_{112} = m_{112},$$
$$P_{022} = m_{022} + \frac{(1-t)(1+q)}{1-qt} m_{112},$$
$$P_{003} = m_{003} + \frac{(1-t)(1+q+q^2)}{1-q^2 t} m_{012}$$
$$+ \frac{(1-t)^2(1+q)(1+q+q^2)}{(1-qt)(1-q^2 t)} m_{111}.$$

Rodrigues type of formula for Macdonald polynomials was recently found in [11, 12, 16]. In the limit $q \uparrow 1$ Macdonald polynomials turn into *Jack polynomials* [17].

A multivariable function/polynomial can be called *special function* if it is some recognised (classical) special function in the case of one variable and if it is common eigenfunction of a complete set of commuting linear partial differential/difference operators defining a *quantum integrable system*. Macdonald polynomials *are* special functions in the above-mentioned sense. They diagonalise the integrals of motion H_j of the quantum Ruijsenaars system.

For any special function in many variables one can set up a general problem of its factorisation in terms of one-variable (special) functions. For the Macdonald polynomials P_λ this would mean finding a factorising integral operator M such that

$$M : P_\lambda(x; q, t) \mapsto \prod_{j=1}^{n} \Phi_{\lambda,j}(y_j).$$

For some particular special functions (or, in other words, for some particular quantum integrable systems) such an operator might simplify to a local transform which is a simple change of variables, from x to y. This happens for example in the case of ellipsoidal harmonics in $\mathbb{R}^n$ (see [18] and [23] for $n = 3$), correspondingly, in the case of quantum Neumann system, where the transform $x \mapsto y$ is the change of variables, from Cartesian to ellipsoidal.

As it is shown further on, in Section 3, the factorising operator for the (symmetric) Macdonald polynomials has to be non-local, i.e. to be some linear integral operator. Moreover, it is explicitly described in the first two non-trivial cases (when $n = 2$ and $n = 3$) in terms of the Askey-Wilson operator.

The formulated factorisation problem also makes sense in the limit to the Liouville integrable systems in classical mechanics, becoming there the well-known problem of separation of variables (SoV) in the Hamilton-Jacobi equation.

The kernel of the factorising integral operator M turns in this limit into the generating function of the separating canonical transformation.

2. Hypergeometric polynomials f_λ and φ_λ in one variable

In this Section we introduce the hypergeometric polynomials $f_\lambda/\varphi_\lambda$ of one variable each constituting a basis which is conjugated to Jack/Macdonald polynomials with respect to the factorising integral transform M. First of all, we describe the procedure of lowering the order of (basic) hypergeometric functions which will lead us to these new sets of interesting polynomials in one variable. We will use (becoming already) standard notations of [8] for the (basic) hypergeometric series and other formulas of the q-analysis.

In 1927 Fox [6] found an interesting relation between hypergeometric functions:

$$
{}_pF_q\left[\begin{matrix} a_1 + m_1, a_2, \ldots, a_p \\ b_1, \ldots, b_q \end{matrix}; y\right] = \sum_{j=0}^{\infty} \frac{(-y)^j(-m_1)_j}{j!} \frac{(a_2)_j \cdots (a_p)_j}{(b_1)_j \cdots (b_q)_j} \tag{2.1}
$$

$$
\times\ {}_pF_q\left[\begin{matrix} a_1 + j, \ldots, a_p + j \\ b_1 + j, \ldots, b_q + j \end{matrix}; y\right].
$$

When $a_1 = b_1$ (2.1) gives the expansion of ${}_pF_q\left[\begin{matrix} a \\ b \end{matrix}; y\right]$ in terms of functions of the type ${}_{p-1}F_{q-1}\left[\begin{matrix} a \\ b \end{matrix}; y\right]$. When m_1 is a positive integer, the series on the right of (2.1) terminates, and we have (for the case of $a_1 = b_1$ and $q = p-1$) the following relation

$$
{}_pF_{p-1}\left[\begin{matrix} b_1 + m_1, a_2, \ldots, a_p \\ b_1, b_2 \ldots, b_{p-1} \end{matrix}; y\right] \tag{2.2}
$$

$$
= \sum_{j=0}^{m_1} \frac{(-y)^j(-m_1)_j}{j!} \frac{(a_2)_j \cdots (a_p)_j}{(b_1)_j \cdots (b_{p-1})_j}\ {}_{p-1}F_{p-2}\left[\begin{matrix} a_2 + j, \ldots, a_p + j \\ b_2 + j, \ldots, b_{p-1} + j \end{matrix}; y\right].
$$

Supposing that $a_2 = b_2 + m_2$, $a_3 = b_3 + m_3, \ldots, a_{p-1} = b_{p-1} + m_{p-1}$ with some non-negative integers m_k, $k = 2, \ldots, p-1$, we then iterate the relation (2.2) further on and, finally, using the binomial theorem,

$$
{}_1F_0\left[\begin{matrix} a \\ - \end{matrix}; y\right] = (1-y)^{-a},
$$

conclude that the function

$$
f_{m_1, \ldots, m_{p-1}}
$$
$$
\equiv (1-y)^{a_p + \sum_{j=1}^{p-1} m_j}\ {}_pF_{p-1}\left[\begin{matrix} b_1 + m_1, b_2 + m_2, \ldots, b_{p-1} + m_{p-1}, a_p \\ b_1, b_2 \ldots, b_{p-1} \end{matrix}; y\right]
$$

is a polynomial in y of the cumulative degree $\sum_{j=1}^{n-1} m_j$.

As for a q-analog of the reduction formula (2.2) we refer to (1.9.4) in [8] (see also [7]). It was proved for the first time in [4] in a more general case (although in different notations than ones in [8] which are adopted here). Following [8], consider the q-analog of the Vandermonde formula in the form

$$
{}_2\phi_1\left[\begin{matrix} q^{-n}, q^{-m} \\ b_{p-1} \end{matrix}; q, q\right] = \frac{(b_{p-1}q^m; q)_n}{(b_{p-1}; q)_n} q^{-mn} \tag{2.3}
$$

where m and n are non-negative integers such that $m \geq n$, then we can reduce the order of the basic hypergeometric function through the following equalities ($|y| < 1$):

$$
{}_p\phi_{p-1}\left[\begin{matrix} a_1,\ldots,a_{p-1},b_{p-1}q^m \\ b_1,\ldots,b_{p-2},b_{p-1} \end{matrix}; q,y\right] \tag{2.4}
$$

$$
= \sum_{n=0}^{\infty} \frac{(a_1,\ldots,a_{p-1};q)_n}{(q,b_1,\ldots,b_{p-2};q)_n}\, y^n \sum_{k=0}^{n} \frac{(q^{-n},q^{-m};q)_k}{(q,b_{p-1};q)_k}\, q^{mn+k}
$$

$$
= \sum_{n=0}^{\infty}\sum_{k=0}^{m} \frac{(a_1,\ldots,a_{p-1};q)_n(q^{-m};q)_k}{(b_1,\ldots,b_{p-2};q)_n(q;q)_{n-k}(q,b_{p-1};q)_k}\, y^n\,(-1)^k q^{mn+k-nk+\binom{k}{2}}
$$

$$
= \sum_{k=0}^{m} \frac{(q^{-m},a_1,\ldots,a_{p-1};q)_k}{(q,b_1,\ldots,b_{p-1};q)_k}\,(-yq^m)^k q^{-\binom{k}{2}}
$$

$$
\times\; {}_{p-1}\phi_{p-2}\left[\begin{matrix} a_1 q^k,\ldots,a_{p-1}q^k \\ b_1 q^k,\ldots,b_{p-2}q^k \end{matrix}; q,yq^{m-k}\right].
$$

Then again, iterating it and using the q-binomial theorem,

$$
{}_1\phi_0\left[\begin{matrix} a \\ - \end{matrix}; q,y\right] = \frac{(ay;q)_\infty}{(y;q)_\infty}, \qquad |y| < 1, \qquad |q| < 1,
$$

we conclude that, for non-negative integers m_k, $k = 1,\ldots,p-1$, the function

$$
\varphi_{m_1,\ldots,m_{p-1}} \equiv \frac{(y;q)_\infty}{(ya_p q^{m_1+\ldots+m_{p-1}};q)_\infty}\; {}_p\phi_{p-1}\left[\begin{matrix} b_1 q^{m_1},\ldots,b_{p-1}q^{m_{p-1}}, a_p \\ b_1,\ldots,b_{p-1} \end{matrix}; q,y\right]
$$

is a polynomial in y of the cumulative degree $\sum_{j=1}^{n-1} m_j$.

Now we can connect integers m_j to a partition $\lambda = \{\lambda_1,\ldots,\lambda_n\}$, namely:

$$
m_j = \lambda_{j+1} - \lambda_j \quad (\geq 0), \qquad j = 1,\ldots,n-1.
$$

Put also ($g \in \mathbb{R}$)

$$
b_j = \lambda_1 - \lambda_{j+1} + 1 - jg, \quad a_j = b_j + m_j, \quad j = 1,\ldots,n-1, \quad a_n = \lambda_1 - \lambda_n + 1 - ng.
$$

Let us define the following function

$$
f_\lambda(y) := y^{\lambda_1}(1-y)^{1-ng}\, {}_nF_{n-1}\left[\begin{matrix} b_1 + m_1,\ldots,b_{n-1} + m_{n-1}, a_n \\ b_1,\ldots,b_{n-1} \end{matrix}; y\right].
$$

Obviously, this function is a polynomial in y of the form

$$
\sum_{k=\lambda_1}^{\lambda_n} \chi_k\, y^k. \tag{2.5}
$$

The polynomials $f_\lambda(y)$ were introduced in [13] and they are factorised polynomials of one variable for the multivariable Jack polynomials. In [22, 13] the corresponding sine-kernel (of Gegenbauer type) factorising the A_2 Jack polynomials was described in detail. In the sequel we concentrate on the q-analogs of these polynomials, the $\varphi_\lambda(y)$, which were introduced in [14].

Put again $m_j = \lambda_{j+1} - \lambda_j$, $j = 1, \ldots, n-1$, but

$$b_j = q^{\lambda_1 - \lambda_{j+1} + 1 - jg}, \quad a_j = b_j q^{m_j}, \quad j = 1, \ldots, n-1, \quad a_n = q^{\lambda_1 - \lambda_n + 1 - ng}.$$

We will also use the parameter t which is connected to g in the following way

$$t := q^g.$$

Then the polynomials φ_λ are defined as follows:

$$\varphi_\lambda(y) := y^{\lambda_1} (y; q)_{1-ng} \, {}_n\phi_{n-1}\left[\begin{matrix} b_1 q^{m_1}, \ldots, b_{n-1} q^{m_{n-1}}, a_n \\ b_1, \ldots, b_{n-1} \end{matrix} ; q, y \right], \qquad (2.6)$$

and, like $f_\lambda(y)$, expand in y as (2.5).

In [14] we have found several useful representations for these polynomials in addition to their definition (2.6). Let us list here some of them. Introduce notations:

$$\lambda_{ij} := \lambda_i - \lambda_j, \qquad |\lambda| = \sum_{j=1}^{n} \lambda_j.$$

First of all, the coefficients χ_k in (2.5) have the following explicit representation:

$$\chi_k = \left(q^{-1}t^n\right)^{\lambda_1 - k} \frac{(q^{-1}t^n; q)_{k-\lambda_1}}{(q; q)_{k-\lambda_1}} \, {}_{n+1}\phi_n\left[\begin{matrix} q^{\lambda_1 - k}, a_1, \ldots, a_n \\ q^{\lambda_1 - k + 2}t^{-n}, b_1, \ldots, b_{n-1} \end{matrix} ; q, q \right]. \tag{2.7}$$

It is easy to give simpler expressions for some of χ_k such as

$$\chi_{\lambda_1} = 1, \qquad \chi_{\lambda_n} = t^{n\lambda_1 - |\lambda|} \prod_{j=1}^{n-1} \frac{(t^j; q)_{\lambda_j - \lambda_1} (t^j; q)_{\lambda_n - \lambda_{n-j}}}{(t^j; q)_{\lambda_{j+1} - \lambda_1} (t^j; q)_{\lambda_n - \lambda_{n-j+1}}}. \tag{2.8}$$

Let us give few first polynomials in the case $n = 3$

$$\varphi_{000} = 1,$$

$$\varphi_{001} = 1 + \frac{1}{t(1+t)} \, y,$$

$$\varphi_{011} = 1 + \frac{1+t}{t^2} \, y,$$

$$\varphi_{002} = 1 + \frac{(1+q)(1-t)}{t(1-qt^2)} \, y + \frac{1-qt}{t^2(1-qt^2)(1+t)} \, y^2,$$

$$\varphi_{012} = 1 + \frac{1 + qt + t - qt^2 - t^2 - qt^3}{t^2(1-qt^2)} \, y + \frac{1}{t^3} \, y^2,$$

$$\varphi_{022} = 1 + \frac{(1+q)(1-t^2)}{t^2(1-qt)}\, y + \frac{(1+t)(1-qt^2)}{t^4(1-qt)}\, y^2\,,$$

$$\varphi_{003} = 1 + \frac{(1+q+q^2)(1-t)}{t(1-q^2t^2)}\, y + \frac{(1+q+q^2)(1-t)}{t^2(1+qt)(1-qt^2)}\, y^2$$

$$+ \frac{1-q^2t}{t^3(1+t)(1+qt)(1-qt^2)}\, y^3\,.$$

There is also a simple formula for $\varphi_\lambda(t^n)$

$$\varphi_\lambda(t^n) = t^{n\lambda_1}(t^n; q)_{\lambda_{n1}} \prod_{j=1}^{n-1} \frac{(t^j; q)_{\lambda_j - \lambda_1}}{(t^j; q)_{\lambda_{j+1} - \lambda_1}}\,. \tag{2.9}$$

There is a nice representation of the polynomials φ_λ in terms of the ϕ_D q-Lauricella function [1, 19, 5]

$$\varphi_\lambda(y) = y^{\lambda_1}\, \frac{(qt^{-n}q^{\lambda_{1n}}y; q)_{\lambda_{n1}}}{\prod_{j=1}^{n-1}(q^{\lambda_1 - \lambda_{n-j+1}+1}t^{j-n}; q)_{\lambda_{n-j+1} - \lambda_{n-j}}}$$

$$\times \phi_D \begin{bmatrix} y; b'_1, \ldots, b'_{n-1} \\ qt^{-n}q^{\lambda_{1n}}y \end{bmatrix} ; q; a'_1, \ldots, a'_{n-1} \end{bmatrix},$$

with

$$a'_j = qt^{j-n}q^{\lambda_1 - \lambda_{n-j}}, \qquad b'_j = q^{\lambda_{n-j} - \lambda_{n-j+1}}, \qquad j = 1, \ldots, n-1\,. \tag{2.10}$$

The ϕ_D q-Lauricella function of $n-1$ variables z_i is a multivariable generalisation of the basic hypergeometric series and is defined by the formula

$$\phi_D \begin{bmatrix} a'; b'_1, \ldots, b'_{n-1} \\ c \end{bmatrix} ; q; z_1, \ldots, z_{n-1} \end{bmatrix}$$

$$:= \sum_{k_1,\ldots,k_{n-1}=0}^{\infty} \frac{(a'; q)_{k_1+\ldots+k_{n-1}}}{(c; q)_{k_1+\ldots+k_{n-1}}} \prod_{j=1}^{n-1} \frac{(b'_j; q)_{k_j} z_j^{k_j}}{(q; q)_{k_j}}\,,$$

using which we get the most explicit representation of our polynomials φ_λ

$$\varphi_\lambda(y) = y^{\lambda_1} \left(\prod_{j=1}^{n-1} (q^{\lambda_1 - \lambda_{n-j+1}+1}t^{j-n}; q)_{\lambda_{n-j+1} - \lambda_{n-j}} \right)^{-1}$$

$$\times \sum_{k_1=0}^{\lambda_n - \lambda_{n-1}} \cdots \sum_{k_{n-1}=0}^{\lambda_2 - \lambda_1} (qt^{-n}q^{\lambda_{1n}+k_1+\ldots+k_{n-1}}y; q)_{\lambda_{n1}-k_1-\ldots-k_{n-1}} (y; q)_{k_1+\ldots+k_{n-1}}$$

$$\times \prod_{j=1}^{n-1} \frac{(q^{\lambda_{n-j} - \lambda_{n-j+1}}; q)_{k_j}(qt^{j-n}q^{\lambda_1 - \lambda_{n-j}})^{k_j}}{(q; q)_{k_j}}\,. \tag{2.11}$$

Finally, these polynomials satisfy the following q-difference equation

$$\sum_{k=0}^{n} (-1)^k \, t^{-(n-1)k/2} \, (1 - q^k t^{-k} y) \, (y; q)_k \, (q^{k+1} t^{-n} y; q)_{n-k} \, h_{n-k;\lambda} \, \varphi_\lambda(q^k y) = 0$$

$$(2.12)$$

where $h_{k;\lambda}$ are given by (1.5) and we assume $h_{0;\lambda} \equiv 1$.

Let us give few remarks about this new class of basic hypergeometric polynomials in one variable.

Remark 1. First of all, we stress that our way of extracting polynomials from hypergeometric series is quite different from the usual one. Indeed, all classical orthogonal q-polynomials of one variable are obtained by just terminating the corresponding hypergeometric series. For instance, the generic Askey-Wilson orthogonal q-polynomials which appear on the $_4\phi_3$ level are defined as follows

$$p_n(x; a, b, c, d | q) = const \cdot {}_4\phi_3 \left[\begin{matrix} q^{-n}, abcdq^{n-1}, ae^{i\theta}, ae^{-i\theta} \\ ab, ac, ad \end{matrix} \, ; q, q \right], \qquad x = \cos\theta.$$

In contrast, to extract the polynomials φ_λ at the level $_n\phi_{n-1}$ we use the procedure of order reduction (2.4) of the basic hypergeometric functions with the specific choice of upper and lower parameters, namely: when the ratio of an upper parameter and one of the lower parameters is equal to q^{m_i} where m_i are non-negative integers. The procedure of order reduction is thus an important second possibility (in addition to simple termination) in order to get polynomials out of hypergeometric series.

Remark 2. The importance of polynomials φ_λ stems from the fact that they are the *factorised polynomials* of one variable for the multivariable Macdonald polynomials. Notice that the polynomials φ_λ, as well as Macdonald polynomials P_λ, are labelled by the dominant weights. The multivariable polynomials Φ_λ combined from the one-variable polynomials φ_λ:

$$\Phi_\lambda(y_1, \ldots, y_n) := y_n^{|\lambda|} \prod_{k=1}^{n-1} \varphi_\lambda(y_k),$$

satisfy the following multiparameter spectral problem (cf. (2.12))

$$\Phi_\lambda(y_1, \ldots, y_{n-1}, q y_n) = h_{n;\lambda} \, \Phi_\lambda(y_1, \ldots, y_n), \qquad (2.13)$$

$$\sum_{k=0}^{n} (-1)^k \, t^{-(n-1)k/2} \, (1 - q^k t^{-k} y_j) \, (y_j; q)_k \, (q^{k+1} t^{-n} y_j; q)_{n-k} \, h_{n-k;\lambda}$$

$$\times \Phi_\lambda(y_1, \ldots, q^k y_j, \ldots, y_n) = 0, \qquad j = 1, \ldots, n-1,$$

with *the same set of spectral parameters* $(h_{1;\lambda}, \ldots, h_{n;\lambda})$ (1.5) as in the spectral problem (1.1) for the Macdonald polynomials P_λ. Hence, one can introduce the commuting operators $H_i^{(y)}$, $i = 1, \ldots, n$ defined by their eigenfunctions $\Phi_\lambda(y)$ and eigenvalues $h_{k;\lambda}$. Since the spectrum $(h_{1;\lambda}, \ldots, h_{n;\lambda})$ coincides with that (1.5) of the Macdonald polynomials, the two sets of commuting operators are isomorphic and there has to exist an intertwining linear operator M

$$M \, H_i^{(x)} = H_i^{(y)} \, M \qquad (H_i^{(x)} \equiv H_i) \,.$$

Actually, to any choice of the normalisation coefficients c_λ in

$$M : \; P_\lambda(x) \mapsto c_\lambda \, \Phi_\lambda(y)$$

there corresponds some intertwiner M. The problem is to select a factorising operator M having an explicit description in terms of its integral kernel or its matrix in some basis in $\mathbb{K}[x]^W$. In [14, 15] we have found such an operator as well as its inversion in the first two non-trivial cases, when $n = 2$ and $n = 3$. It appears that the factorising operator M can be expressed in those cases through the Askey-Wilson operator.

3. The cases $n = 2$ and $n = 3$

Let us first describe the integral operator M_ξ performing the separation of variables in the A_1 Macdonald polynomials (we skip the trivial case of the purely coordinate SoV $x_{1,2} \to x_\pm \equiv (x_1 x_2^{\pm 1})^{1/2}$). Our main technical tool is the famous Askey-Wilson integral identity [3, 8]

$$\frac{1}{2\pi i} \int_{\Gamma_{abcd}} \frac{dx}{x} \frac{(x^2, x^{-2}; q)_\infty}{(ax, ax^{-1}, bx, bx^{-1}, cx, cx^{-1}, dx, dx^{-1}; q)_\infty}$$
$$= \frac{2(abcd; q)_\infty}{(q, ab, ac, ad, bc, bd, cd; q)_\infty} \,. \tag{3.1}$$

The cycle Γ_{abcd} depends on complex parameters a, b, c, d and is defined as follows. Let $C_{z,r}$ be the counter-clockwise oriented circle with the center z and radius r. If $\max(|a|, |b|, |c|, |d|, |q|) < 1$ then $\Gamma_{abcd} = C_{0,1}$. The identity (3.1) can be continued analytically for the values of parameters a, b, c, d outside the unit circle provided the cycle Γ_{abcd} is deformed appropriately. In general case

$$\Gamma_{abcd} = C_{0,1} + \sum_{z=a,b,c,d} \sum_{\substack{k \geq 0 \\ |z|q^k > 1}} \left(C_{zq^k, \varepsilon} - C_{z^{-1}q^{-k}, \varepsilon} \right),$$

ε being small enough for $C_{z^{\pm 1}q^{\pm k}, \varepsilon}$ to encircle only one pole of the denominator.

Put

$$a = yq^{\alpha/2}, \qquad b = y^{-1}q^{\alpha/2}, \qquad c = rq^{\beta/2}, \qquad d = r^{-1}q^{\beta/2}.$$

We will use the notation $\Gamma^{ry}_{\alpha\beta}$ for the contour obtained from Γ_{abcd} by these substitutuions. Introduce also the following useful notation

$$\mathcal{L}_q(\nu; x, y) := (\nu xy, \nu xy^{-1}, \nu x^{-1}y, \nu x^{-1}y^{-1}; q)_\infty. \qquad (3.2)$$

Now let

$$\alpha = \beta = g, \quad y = y_-, \quad x = x_-, \quad r = t^{-1}y_+,$$
$$x_\pm \equiv (x_1 x_2^{\pm 1})^{1/2}, \quad y_\pm \equiv (y_1 y_2^{\pm 1})^{1/2}.$$

Introduce the kernel $\mathcal{M}(y_+, y_-|x_-)$:

$$\mathcal{M}(y_+, y_-|x_-) = \frac{(1-q)(q;q)_\infty^2 (x_-^2, x_-^{-2}; q)_\infty \mathcal{L}_q(t; y_-, t^{-1}y_+)}{2B_q(g,g) \mathcal{L}_q(t^{1/2}; y_-, x_-) \mathcal{L}_q(t^{1/2}; x_-, t^{-1}y_+)}. \qquad (3.3)$$

Assuming ξ to be an arbitrary complex parameter, we introduce the integral operator M_ξ acting on functions $f(x_1, x_2)$ by the formula

$$(M_\xi f)(y_1, y_2) \equiv \frac{1}{2\pi i} \int_{\Gamma^{t^{-1}y_+, y_-}_{g,g}} \frac{dz}{z} \mathcal{M}(y_+, y_-|z) f(t^{-1/2}\xi y_+ z, t^{-1/2}\xi y_+ z^{-1}). \qquad (3.4)$$

Theorem 1 ([15]) *The operator M_ξ (3.4),(3.3) performs the factorisation of (or, in other words, the SoV for) the A_1 Macdonald polynomials:*

$$M_\xi : P_\lambda(x_1, x_2) \to c_{\lambda,\xi} \varphi_\lambda(y_1) \varphi_\lambda(y_2), \qquad (3.5)$$

where $\varphi_\lambda(y)$ is the factorised (or separated) polynomial and the normalisation coefficient $c_{\lambda,\xi}$ is equal to

$$c_{\lambda,\xi} = t^{-2\lambda_1 + \lambda_2}\xi^{|\lambda|} \frac{(t; q)_{\lambda_{21}}}{(t^2; q)_{\lambda_{21}}}. \qquad (3.6)$$

Note that the relation (3.5) is equivalent [15] to the product formula for the continuous q-ultraspherical polynomials [20, 8].

The kernel of the inverse operator M_ξ^{-1} has the form

$$\widetilde{\mathcal{M}}(x_+, x_-|y_-) = \frac{(1-q)(q;q)_\infty^2 (y_-^2, y_-^{-2}; q)_\infty \mathcal{L}_q(t^{1/2}; x_-, t^{-1/2}\xi^{-1}x_+)}{2B_q(-g, 2g) \mathcal{L}_q(t^{-1/2}; y_-, x_-) \mathcal{L}_q(t; y_-, t^{-1/2}\xi^{-1}x_+)},$$

with the following substitutions for the contour Γ:

$$\alpha = -g, \qquad \beta = 2g, \qquad x = y_-, \qquad y = x_-, \qquad r = t^{-1/2}\xi^{-1}.$$

Theorem 2 ([15]) *The inversion of the operator M_ξ (3.4) is given by the formula*

$$(M_\xi^{-1} f)(x_1, x_2) = \frac{1}{2\pi i} \int_{\Gamma^{t^{-1/2}\xi^{-1}x_+,x_-}_{-g,2g}} \frac{dz}{z}$$

$$\times \widetilde{\mathcal{M}}(x_+, x_- | z) f(t^{1/2}\xi^{-1}x_+ z, t^{1/2}\xi^{-1}x_+ z^{-1}). \quad (3.7)$$

The operator M_ξ^{-1} provides an integral representation for the A_1 Macdonald polynomials in terms of the factorised polynomials $\varphi_{\lambda_1,\lambda_2}(y)$

$$M_\xi^{-1} : c_{\lambda,\xi} \, \varphi_\lambda(y_1) \, \varphi_\lambda(y_2) \to P_\lambda(x_1, x_2). \quad (3.8)$$

In contrast to the formula (3.5) which paraphrases an already known result, the formula (3.8) leads to a new integral relation for the q-ultraspherical polynomials. Note that for positive integer g the operator M_ξ^{-1} becomes a q-difference operator of the order g (cf. [14] and [15]).

Now we describe the factorising operator M and its inversion in the case of A_2 Macdonald polynomials. Introduce the following operator M acting on functions $f(x_1, x_2, x_3)$ by the formula

$$(Mf)(y_1, y_2, y_3) = \frac{1}{2\pi i} \int_{\Gamma^{t^{-3/2}y_+,y_-}_{g,2g}} \frac{dx_-}{x_-} \mathcal{M}((y_1 y_2)^{1/2}, (y_1/y_2)^{1/2} | x_-)$$

$$\times f(t^{-3/2} y_3 (y_1 y_2)^{1/2} x_-, t^{-3/2} y_3 (y_1 y_2)^{1/2} x_-^{-1}, y_3)$$

with the kernel

$$\mathcal{M}(y_+, y_- | x_-) = \frac{(1-q)(q;q)_\infty^2 (x_-^2, x_-^{-2}; q)_\infty \, \mathcal{L}_q\left(t^{3/2}; y_-, y_+ t^{-3/2}\right)}{2 B_q(g, 2g) \, \mathcal{L}_q(t^{1/2}; y_-, x_-) \, \mathcal{L}_q(t; x_-, y_+ t^{-3/2})}$$

and the following substitutions: $\alpha = g$, $\beta = 2g$, $r = t^{-3/2} y_+$, $y = y_-$, $x = x_- \equiv (x_1 x_2^{-1})^{1/2}$, $x_+ \equiv (\frac{x_1 x_2}{x_3^2})^{1/2}$, $y_\pm \equiv (y_1 y_2^{\pm 1})^{1/2}$.

Theorem 3 ([14]) *The operator M transforms any A_2 Macdonald polynomial P_λ into the product*

$$M : P_\lambda(x_1, x_2, x_3; q, t) \to c_\lambda \, y_3^{\lambda_1 + \lambda_2 + \lambda_3} \, \varphi_\lambda(y_1) \, \varphi_\lambda(y_2)$$

of factorised polynomials $\varphi_{\lambda_1 \lambda_2 \lambda_3}(y)$ of one variable

$$\varphi_\lambda(y) = y^{\lambda_1} \, (y; q)_{1-3g} \; {}_3\phi_2 \left[\begin{array}{c} t^{-3}q^{1-\lambda_{31}}, t^{-2}q^{1-\lambda_{21}}, t^{-1}q \\ t^{-2}q^{1-\lambda_{31}}, t^{-1}q^{1-\lambda_{21}} \end{array} ; q, y \right].$$

The normalisation coefficient c_λ equals

$$c_\lambda = t^{\lambda_2 - 4\lambda_1} \frac{(t^2; q)_{\lambda_{31}} (t^2; q)_{\lambda_{32}} (t; q)_{\lambda_{21}}}{(t^3; q)_{\lambda_{31}} (t; q)_{\lambda_{32}} (t^2; q)_{\lambda_{21}}}.$$

Theorem 4 ([14]) *The inverse operator M^{-1} is the integral operator*

$$(M^{-1}f)(x_1, x_2, x_3) = \frac{1}{2\pi i} \int_{\Gamma^{x_+, x_-}_{-g, 3g}} \frac{dy_-}{y_-} \, \widetilde{\mathcal{M}}\left(\frac{(x_1 x_2)^{1/2}}{x_3}, \left(\frac{x_1}{x_2}\right)^{1/2} \Big| y_- \right)$$

$$\times f\left(\frac{t^{3/2}(x_1 x_2)^{1/2} y_-}{x_3}, \frac{t^{3/2}(x_1 x_2)^{1/2}}{x_3 y_-}, x_3\right)$$

with the kernel

$$\widetilde{\mathcal{M}}(x_+, x_- \mid y_-) = \frac{(1-q)(q;q)_\infty^2 (y_-^2, y_-^{-2}; q)_\infty \, \mathcal{L}_q(t; x_-, x_+)}{2 B_q(-g, 3g) \, \mathcal{L}_q(t^{-1/2}; y_-, x_-) \, \mathcal{L}_q(t^{3/2}; y_-, x_+)}. \tag{3.9}$$

It provides a new integral representation for the A_2 Macdonald polynomials in terms of the factorised polynomials $\varphi_{\lambda_1, \lambda_2, \lambda_3}(y)$ of one variable

$$M^{-1}: \; c_\lambda \, y_3^{\lambda_1 + \lambda_2 + \lambda_3} \, \varphi_\lambda(y_1) \, \varphi_\lambda(y_2) \mapsto P_\lambda(x_1, x_2, x_3; q, t). \tag{3.10}$$

For positive integer g the operator M^{-1} turns into the q-difference operator of order g:

$$M^{-1}: \; f(y_1, y_2, y_3) \mapsto \sum_{k=1}^{g} \xi_k\left(\frac{(x_1 x_2)^{1/2}}{x_3}, \left(\frac{x_1}{x_2}\right)^{1/2}\right) f\left(q^{g+k}\frac{x_1}{x_3}, q^{2g-k}\frac{x_2}{x_3}, x_3\right)$$

$$\tag{3.11}$$

where $\xi_k(r, y)$ is given by

$$\xi_k(r, y) = (-1)^k q^{-k(k-1)/2} \begin{bmatrix} g \\ k \end{bmatrix}_q y^{-2k}(1 - q^{g-2k} y^{-2})$$

$$\times \frac{(try, tr^{-1}y; q)_k (try^{-1}, tr^{-1}y^{-1}; q)_{g-k}}{(t^2; q)_g (q^{-k}y^{-2}; q)_{g+1}}.$$

Remark 3. As was found in [14, 15] the operators M_ξ and M are closely related to a slightly more general integral operator $M_{\alpha\beta}$ which, in turn, is closely related to the fractional q-integration operator I^α.

Remark 4. The apparent similarity of formulas for the operator M in cases $n = 2$ and $n = 3$ is not a coincidence. Its explanation relies on the reduction $gl(2) \subset gl(3)$, see [15].

Acknowledgments

Authors thank M. E. H. Ismail for interesting discussions on the subject of special functions and for bringing the reference [6] to their attention. VBK wish to acknowledge the support of EPSRC.

References

[1] G. Andrews, *Summations and transformations for basic Appell series*, J. London Math. Soc. (2) **4** (1972), 618–622.

[2] R. Askey and M. E. H. Ismail, *A generalisation of ultraspherical polynomials*, in Studies in Pure Mathematics, P. Erdős, ed., Birkhäuser, 55–78, 1983.

[3] R. Askey and J. A. Wilson, Some basic hypergeometric orthogonal polynomials that generalize Jacobi polynomials, Memoirs of the Amer. Math. Soc. **319**, Providence RI, 1985.

[4] M. Chakrabarty, *Formulae expressing generalised hypergeometric functions in terms of those of lower order*, Nederl. Akad. Wetensch. Proc. **A77** (1974), 199–202.

[5] R. Floreanini, L. Lapointe and L. Vinet, *A quantum algebra approach to basic multivariable special functions*, J. Phys. A: Math.Gen. **27** (1994), 6781–6797.

[6] C. Fox, *The expression of hypergeometric series in terms of similar series*, Proc. London Math. Soc. (2) **26** (1927), 201–210.

[7] G. Gasper, *Orthogonality of certain functions with respect to complex valued weights*, Cand. J. Math. **33** (1981), 1261–1270.

[8] G. Gasper and M. Rahman. Basic hypergeometric series, Cambridge University Press, 1990.

[9] R. Koekoek and R. F. Swarttouw, *The Askey-scheme of hypergeometric orthogonal polynomials and its q-analogue*, Report 94-05, Delft University of Technology, 1994.

[10] T. H. Koornwinder, *Askey-Wilson polynomials for root systems of type BC*, in: Hypergeometric functions on domains of positivity, Jack polynomials, and applications (D. St. P. Richards, ed.), Contemp. Math. **138**, AMS, Providence, R. I., (1992), 189–204.

[11] A. N. Kirillov and M. Noumi, *Affine Hecke algebras and raising operators for Macdonald polynomials*, preprint, q-alg/9605004.

[12] A. N. Kirillov and M. Noumi, *q-Difference raising operators for Macdonald polynomials and the integrality of transition coefficients*, preprint, q-alg/9605005.

[13] V. B. Kuznetsov and E. K. Sklyanin, *Separation of variables in A_2 type Jack polynomials*, RIMS Kokyuroku **919** (1995), 27–34.

[14] V. B. Kuznetsov and E. K. Sklyanin, *Separation of variables for the A_2 Ruijsenaars model and a new integral representation for the A_2 Macdonald polynomials*, J. Phys. A: Math. Gen. **29** (1996), 2779–2804.

[15] V. B. Kuznetsov and E. K. Sklyanin, *Separation of variables and integral relations for special functions*, October 1996, submitted to The Ramanujan Journal.

[16] L. Lapointe and L. Vinet, *Creation operators for the Macdonald and Jack polynomials*, preprint, `q-alg/9607024`.

[17] I. G. Macdonald, Symmetric functions and Hall polynomials, Oxford University Press, second edition, 1995.

[18] W. D. Niven, *On ellipsoidal harmonics*, Phil. Trans. **CLXXXII** (1891), 231.

[19] M. Noumi, *Quantum grassmannians and q-hypergeometric series*, CWI Quarterly **5** (1992), 293–307.

[20] M. Rahman and A. Verma, *Product and addition formulas for the continuous q-ultraspherical polynomials*, SIAM J. Math. Anal. 17 (1986), 1461–1474.

[21] S. N. M. Ruijsenaars, *Complete integrability of relativistic Calogero-Moser systems and elliptic function identities*, Commun. Math. Phys. **110** (1987), 191–213.

[22] E. K. Sklyanin, *Separation of variables. New trends*, Progr. Theor. Phys. Suppl. **118** (1995), 35–60.

[23] E. T. Whittaker and G. N. Watson, A course in modern analysis, Cambridge University Press, 4th ed., 1988.

Chapter 10

Quantum Aspects and Yang-Baxter Equations

Schrödinger equation on quantum homogeneous spaces

F. Bonechi, R. Giachetti, E. Sorace and M. Tarlini
Dipartimento di Fisica,
Università di Firenze and INFN–Firenze

1 Introduction

The homogeneous spaces of the Lie groups of the classical kinematics symmetries are between the most relevant topics in theorethical and mathematical physics. Indeed the equations generated by the action of the algebra on such spaces give rise to the fundamental equations of the classical physics and wave mechanics. Therefore it is a scientific meaningful program to study the analogous structures in the quantum groups framework.

The first step is the building of the quantum counterpart of the noncompact Lie groups. This task is now satisfactorily accomplished, at present we know explicitly the quantum versions - recovered by means of various methods - of all the relevant inhomogeneous Lie groups: from Heisenberg [1] and Galilei [2,3] to Euclides [4,5,6,7] and Poincaré [8], (1+1)-dimensional and (3+1)-dimensional.

The second step is connected to the definition of the quantum homogeneous spaces, once a quantum group is given. Indeed, owing to the non commutativity of the group parameters, the quantum manifolds do not exist. However we can try to deal with manifolds in the quantum world by translating definitorial relations from the spaces to the functions on them and by using the duality between the "functions on the group" and the envelopping algebra. Therefore a crucial point is the injection of the algebra of the quantum functions on the homogeneous space into the algebra of the quantum functions on the group. The methodical approach is to express the classical construction of the homogeneous spaces in the language of Hopf algebras, that is in a way independent from the commutativity of the functions on the group. One can then extend these notions to quantum case thus getting the non commutative algebra of the functions on the quantum homogeneous spaces. The study of the action of the enveloping algebra of the quantum group on these "functions" gives rise to a quantum harmonic analysis and it makes possible to write and discuss the analogous of the invariant equations. Our group has analized the cases of the two possible quantum two dimensional Euclidean groups $E_q(2)$ and $E_l(2)$ [9,10]. We have then written and solved the associated invariant equations, the analogous of the stationary free Schrödinger equation. The solutions are very interesting as they are written in terms of known deformed special functions which are therefore shown to be well rooted in the quantum symmetries.

Let us now give you a short survey of the definitions and concepts used in the construction of quantum homogeneous spaces [11,12]: we begin by rephrasing the case of Lie groups. Let G be a Lie group and $m : G \times G \to G$ its composition law. Let M be a (left) G-space, with action $a : G \times M \to M$. The comultiplication $\Delta : \mathcal{F}(G) \to \mathcal{F}(G) \otimes \mathcal{F}(G)$ and the pullback of the action $\delta : \mathcal{F}(M) \to \mathcal{F}(G) \otimes \mathcal{F}(M)$ or coaction are then defined on the space of functions on G and M respectively. The associativity of the action $a \circ (id \times a) = a \circ (m \times id)$, implies the coassociativity for the coaction, $(id \otimes \delta) \circ \delta = (\Delta \otimes id) \circ \delta$. A manifold M is a homogeneous G-space whenever the action a is transitive, i.e. whenever the map $a \circ (id \times j_p) : G \to M$ is surjective for some $p \in M$ (j_p is the injection of p in M).

Similarly, we shall say that a coaction δ is *transitive* if there exists a character $\tilde{\varepsilon}$ of $\mathcal{F}(M)$ for which the corresponding $\Psi = (id \otimes \tilde{\varepsilon}) \circ \delta : \mathcal{F}(M) \to \mathcal{F}(G)$ is injective $(\tilde{\varepsilon} = j_p{}^*)$.

The properties of homogeneous spaces can be resumed by saying that the image $\Psi(\mathcal{F}(M))$ is a *-subalgebra of $\mathcal{F}(G)$ and a left coideal of $\mathcal{F}(G)$, since $\Delta(\Psi(f)) \in \mathcal{F}(G) \otimes \Psi(\mathcal{F}(M))$ for any $f \in \mathcal{F}(M)$.

We present now an *infinitesimal* treatment of the classical situation that can be followed also at the quantum level [12,13]. Again this formulation does not depend upon the commutativity of $\mathcal{F}(G)$ and it will be explicitly used to produce homogeneous spaces in the quantum deformed cases.

If $f \in \mathcal{F}(G)$ satisfies $f(xy) = f(x)$ for $x \in G$, $y \in H$, then it satisfies also $Y \cdot f \equiv D_t f(x\,e^{tY})|_{t=0} = 0$ for any $Y \in \mathrm{Lie}\,H$. Now we see that we can write the action of any $X \in \mathrm{Lie}\,G$ as $D_t f(x\,e^{tX})|_{t=0} = D_t \Delta f(x, e^{tX})|_{t=0} = \sum_{(f)} f_{(1)}(x)\, D_t f_{(2)}(e^{tX})|_{t=0}$, namely

$$X \cdot f = \sum_{(f)} f_{(1)} \langle X, f_{(2)} \rangle, \tag{1.1}$$

where the map $(X, f) \mapsto \langle X, f \rangle = D_t f(e^{tX})|_{t=0}$ can be extended to a canonical and nondegenerate duality pairing of Hopf algebras $\mathcal{U}(\mathrm{Lie}\,G) \times \mathcal{F}(G) \to \mathbf{C}$. Conversely, once we are given a nondegenerate duality pairing of Hopf algebras $\mathcal{H}_1 \times \mathcal{H}_2 \to \mathbf{C}$, it is a simple matter of computation to verify that (1.1) defines an action of $\mathcal{H}_1$ on $\mathcal{H}_2$, independently of the commutativity of these algebras. It is therefore natural to call an element $f \in \mathcal{H}_2$ *infinitesimally invariant* with respect to an element $X \in \mathcal{H}_1$ if $X \cdot f = \varepsilon(X) f$.

Let us now return to the $*$-subalgebra and left coideal $\mathcal{M} = \Psi(\mathcal{F}(M)) \subseteq \mathcal{F}(G)$ that we have previously defined and consider the subset $K_\mathcal{M} \subseteq \mathcal{U}(\mathrm{Lie}\, G)$ of those elements for which $\mathcal{M}$ is infinitesimally invariant. Letting $\tau = * \circ S$, it follows from the definitions that $K_\mathcal{M}$ is a τ-invariant two-sided coideal and a left ideal in $\mathcal{U}(\mathrm{Lie}\, G)$. The converse of this statement is relevant for the applications and will be used in the following. Observe that if K is a τ-invariant two-sided coideal,

$$\mathcal{M}_K = \{ f \in \mathcal{F}(G) \,|\, K \cdot f = 0 \}$$

is a $*$-subalgebra and left coideal, and hence it defines a homogeneous space for the group G. The non trivial problem in using this definition in quantum groups is to find K and then to determine the space of solutions of $K \cdot f = 0$.

2 The homogeneous spaces of $E_q(2)$

The Hopf algebra $\mathcal{F}_q(E(2))$ [6] is generated by v, $\bar{v}$, n, $\bar{n}$, with the relations

$$
\begin{aligned}
& vn = q^2 nv\,, && v\bar{n} = q^2 \bar{n} v\,, && n\bar{n} = q^2 \bar{n} n\,, \\
& \bar{n}\bar{v} = q^2 \bar{v}\bar{n}\,, && n\bar{v} = q^2 \bar{v} n\,, && v\bar{v} = \bar{v}v = 1\,,
\end{aligned}
$$

the coalgebra operations

$$
\begin{aligned}
& \Delta v = v \otimes v\,, && \Delta \bar{v} = \bar{v} \otimes \bar{v}\,, \\
& \Delta n = n \otimes 1 + v \otimes n\,, && \Delta \bar{n} = \bar{n} \otimes 1 + \bar{v} \otimes \bar{n}, \\
& \varepsilon(v) = \varepsilon(\bar{v}) = 1 && \varepsilon(n) = \varepsilon(\bar{n}) = 0
\end{aligned}
$$

and the antipode map

$$ S(v) = \bar{v}\,, \qquad S(\bar{v}) = v\,, \qquad S(n) = -\bar{v}n\,, \qquad S(\bar{n}) = -v\bar{n}\,. $$

Assuming from now on a real q, a compatible involution is given by $v^* = \bar{v}$, $n^* = \bar{n}$.

The *quantized enveloping algebra* $\mathcal{U}_q(E(2)) \equiv E_q(2)$ is generated by the unity and the three elements $P_\pm$, J, such that

$$ [P_+, P_-] = 0\,, \qquad [J, P_\pm] = \pm P_\pm $$

and

$$
\begin{aligned}
& \Delta J = J \otimes 1 + 1 \otimes J\,, && \Delta P_\pm = q^{-J} \otimes P_\pm + P_\pm \otimes q^J \\
& S(J) = -J\,, && S(P_\pm) = -q^{\pm 1} P_\pm,
\end{aligned}
$$

with vanishing counit and involution: $J^* = J$, $P_\pm^* = P_\mp$. Since J is primitive in $E_q(2)$, $q^{\pm J}$ are group-like moreover $P_\pm$ are twisted-primitive with respect to q^{-J}.

The explicit relations of the duality between $E_q(2)$ and $\mathcal{F}_q(E(2))$, are given by

$$\langle J, v^r n^s \bar{n}^t \rangle = -r \, \delta_{s,0} \delta_{t,0} \, ,$$
$$\langle P_-, v^r n^s \bar{n}^t \rangle = -q^{r-1} \delta_{s,1} \delta_{t,0} \, ,$$
$$\langle P_+, v^r n^s \bar{n}^t \rangle = q^r \, \delta_{s,0} \delta_{t,1}.$$

We shall now consider the following two left actions of $\mathcal{U}_q(()G)$ on $\mathcal{F}(G)$

$$\ell(X)f \;=\; (id \otimes X) \circ \Delta f = \sum_{(f)} f_{(1)} \, \langle X, f_{(2)} \rangle \, ,$$
$$\lambda(X)f \;=\; (S(X) \otimes id) \circ \Delta f = \sum_{(f)} \langle S(X), f_{(1)} \rangle \, f_{(2)}.$$

In the classical case and for a group-like element $x \in G$, ℓ is the multiplication to the right of the argument of f by x and λ the multiplication to the left of the argument by x^{-1}.

The definitions of the q-combinatorial quantities that will be used in the following are :

$$[\alpha]_q = (q^\alpha - q^{-\alpha})/(q - q^{-1}) \, ,$$
$$[s]_q! = [s]_q \, [s-1]_q \cdot [1]_q,$$
$$(\alpha; q)_s = \prod_{j=1}^{s} (1 - q^{j-1}\alpha),$$

with $s \in \mathbf{N}$.

It is then not difficult, by using the duality relations, to calculate explicitly the actions of ℓ and λ for the case of $E_q(2)$:

$$\ell(q^{\pm J}) \, v^r n^s \bar{n}^t = q^{\mp r} \, v^r n^s \bar{n}^t \, ,$$
$$\ell(P_-) \, v^r n^s \bar{n}^t = -[s]_q \, q^{r-s} \, v^{r+1} n^{s-1} \bar{n}^t,$$
$$\ell(P_+) \, v^r n^s \bar{n}^t = [t]_q \, q^{r+2s+t-1} \, v^{r-1} n^s \bar{n}^{t-1}$$

and

$$\lambda(q^{\pm J}) \, v^r n^s \bar{n}^t = q^{\pm(r+s-t)} \, v^r n^s \bar{n}^t$$
$$\lambda(P_-) \, v^r n^s \bar{n}^t = [s]_q \, q^{r+t-2} \, v^r n^{s-1} \bar{n}^t,$$
$$\lambda(P_+) \, v^r n^s \bar{n}^t = -[t]_q \, q^{r+s+1} \, v^r n^s \bar{n}^{t-1}.$$

We are thus able to study the homogeneous spaces of $\mathcal{F}_q(E(2))$ by using the infinitesimal invariance.

Let us define $X_\mu = -\bar{\mu} [J]_q + P_+ + q \, (\bar{\mu}/\mu) \, P_-$, where μ can assume the values $-\rho$ and $i\rho$ for positive real ρ. In [9] we have proved that the linear span of the elements X_μ and $X_\infty = [J]_q$ constitute a τ-invariant two-sided coideal of $E_q(2)$, twisted-primitive with respect to q^{-J}.

It is then possible to get the algebras of the functions on the quantum homogeneous spaces of $\mathcal{F}_q(E(2))$ by solving $\ell(X_\mu)f = 0$ and $\ell(X_\infty)f = 0$.

Define z and $\bar{z}$ as follows:

$$z = v + \mu n, \qquad \bar{z} = \bar{v} + \bar{\mu}\bar{n}, \qquad (|\mu| < \infty),$$
$$z = n, \qquad \bar{z} = \bar{n}, \qquad (|\mu| = \infty).$$

Then $(z, \bar{z})$ satisfy the relations

$$\begin{aligned}(qH):& \qquad z\bar{z} = q^2\,\bar{z}z + (1 - q^2), \\ (qP):& \qquad z\bar{z} = q^2\,\bar{z}z\end{aligned}$$

for $|\mu| < \infty$ and $|\mu| = \infty$ respectively. Moreover they are connected by the involution $*$ and generate the $*$-invariant subalgebra and left coideal

$$B = \{f \in \mathcal{F}_q(E(2)) \mid \ell(X_\mu)f = 0\}$$

of $\mathcal{F}_q(E(2))$.

They thus define quantum homogeneous spaces respectively called quantum hyperboloid (qH) and quantum plane (qP).

The explicit forms of the coactions read

$$\begin{aligned}(qH):& \qquad \delta z = v \otimes z + \mu n \otimes 1, \qquad \delta\bar{z} = \bar{v} \otimes \bar{z} + \bar{\mu}\bar{n} \otimes 1, \\ (qP):& \qquad \delta z = v \otimes z + n \otimes 1, \qquad \delta\bar{z} = \bar{v} \otimes \bar{z} + \bar{n} \otimes 1.\end{aligned}$$

3   The q-Schrödinger equation

A canonical action of $E_q(2)$ on the homogeneous spaces (qP) and (qH) is given by the left action λ defined in the last section. We can then study the action of the Casimir of $E_q(2)$ and the corresponding eigenvalue problem. This can be considered the natural q-analog of the free Schrödinger equation for the functions on the plane; its eigenfunctions will be analyzed by diagonalizing the Casimir in two bases which are the q-counterparts of the plane waves and of the angular momentum bases. The equation we want to study is

$$\lambda(P_+P_-)\,\psi = E\psi. \tag{3.1}$$

From the explicit formulas of the action λ on $\mathcal{F}_q(E(2))$ we have for (qH)

$$\begin{aligned}\lambda(q^{\pm J})\,\bar{z}^j z^m &= q^{\mp(j-m)}\,\bar{z}^j z^m, \\ \lambda(P_-)\,\bar{z}^j z^m &= \mu\,[m]_q\,q^{-j-2}\,\bar{z}^j z^{m-1}, \\ \lambda(P_+)\,\bar{z}^j z^m &= -\bar{\mu}\,[j]_q\,q^{-m+1}\,\bar{z}^{j-1} z^m,\end{aligned} \tag{3.2}$$

the analogous relations for (qP) are obtained with $\mu = 1$.

The λ-action of the Casimir $P_+ P_-$ on a formal series ψ is

$$
\begin{aligned}
\lambda(P_+ P_-)\,\psi &= \lambda(P_+ P_-) \sum_{j,m} c_{j,m}\, \bar{z}^j z^m \\
&= \sum_{j,m} c_{j,m}\, (-\mu\bar{\mu})\, q^{-j-m}\, [j]_q [m]_q\, \bar{z}^{j-1} z^{m-1},
\end{aligned}
$$

so that, when substituted in (3.1) with the position $\mathcal{E} = -E/(\mu\bar{\mu})$, one gets:

$$
q^{-j-m-2}\, [j+1]_q\, [m+1]_q\, c_{j+1,m+1} = \mathcal{E}\, c_{j,m}. \tag{3.3}
$$

(*i*) *The plane wave states.* First of all we search for solutions factorizable in the variables z and $\bar{z}$, thus we introduce coefficients of the form

$$
c_{j,m} = \frac{k^j \tilde{k}^m}{[j]_q!\, [m]_q!}\, q^{-\vartheta(j,m)}, \tag{3.4}
$$

where k and $\tilde{k}$ are to be determined while $q^{\vartheta(j,m)}$ factorizes in its arguments.

By substituting (3.4) in equation (3.3) we get

$$
q^{\vartheta(j,m)-\vartheta(j+1,m+1)}\, k\tilde{k} = q^{j+m+2}\,\mathcal{E}. \tag{3.5}
$$

Now we choose to diagonalize, in addition to the Casimir, $b_- = -q^{-J}P_-$ and $b_+ = P_+ q^{J}$. The new eigenvalue equation is $\lambda(b_-)\psi = \beta\psi$. For the coefficients of the expansion ψ in powers of z and $\bar{z}$ we find the following relation

$$
-\mu\, q^{-m-2}\, [m+1]_q\, c_{j,m+1} = \beta\, c_{j,m}. \tag{3.6}
$$

Using (3.4), we deduce the system

$$
\begin{aligned}
\vartheta(j,m) - \vartheta(j+1,m+1) - j - m &= const_1 \\
\vartheta(j,m) - \vartheta(j,m+1) - m &= const_2,
\end{aligned}
$$

where, up to an inessential rescaling, the two constants can be fixed to zero. The solution satisfied the factorization requirement previously made, yielding

$$
\vartheta(j,m) = -\frac{1}{2} j(j-1) - \frac{1}{2} m(m-1)
$$

and finally, from (3.5)

$$
\mathcal{E} = k\tilde{k}/q^2.
$$

From $\mathcal{E} < 0$ we have $k\tilde{k} < 0$. Thus in the plane waves expansion the eigenvalue equation (3.1) is satisfied by the states

$$
\psi_{k\tilde{k}} = E_{q^2}[(1-q^2)\,\tilde{k}\bar{z}]\, E_{q^2}[(1-q^2)\,kz],
$$

where $E = -\mu\bar{\mu}k\tilde{k}/q^2$ and $E_q(x) = {}_0\phi_0(-;-;q,-x)$ denotes a q-exponential [13].

The elements $\psi_{k\tilde{k}}$ are also eigenstates of $\lambda(b_-)$ with eigenvalue $-\mu\tilde{k}/q^2$ and of $\lambda(b_+)$ with eigenvalue $-\bar{\mu}k$. (*ii*) *The angular momentum states.* In

this case we solve (3.1) with the additional diagonalization of q^J. A solution ψ of the eigenvalue equation for q^J has the coefficients $c_{j,m}$ that satisfy the condition $m - j = \pm r = const$, $r > 0$, and it has eigenvalue $q^{\pm r}$. So with $c_{j,m} = \delta_{m-j,\pm r}\, d_j$, from (3.3) we get

$$q^{-2(j+1)-r}\, [j+1]_q\, [j+r+1]_q\, d_{j+1} = \mathcal{E}\, d_j$$

solved by

$$d_j = \frac{[r]_q!}{[j]_q!\, [j+r]_q!}\, q^{j(j+1)}\, (\mathcal{E}q^r)^j\,.$$

We summarize by saying that the eigenvalue equation (3.1) is satisfied by the states

$$\psi_r = \frac{q^{2r}\,\mathcal{E}^{r/2}}{[r]_q!}\, J_r^{(q)}\, z^r \quad \text{and} \quad \psi_{-r} = \frac{q^{2r}\,\mathcal{E}^{r/2}}{[r]_q!}\, \bar{z}^r\, J_r^{(q)}$$

where

$$J_r^{(q)} = \sum_{j=0}^{\infty} \frac{[r]_q!}{[j]_q!\, [j+r]_q!}\, q^{j(j+1)}\, (\mathcal{E}q^r)^j\, \bar{z}^j z^j\,. \tag{3.7}$$

Moreover $\psi_{\pm r}$ are also eigenfunctions of q^J with eigenvalues $q^{\pm r}$. The series (3.7) reduces to the Bessel function $J_r(\bar{z}z)$ for $q \to 1$.

The expression $J_r^{(q)}$ can be written in terms of the single variable $\bar{z}z$. In the case of (qP) $\bar{z}^j z^j = q^{-j(j-1)}(\bar{z}z)^j$, so that

$$\begin{aligned} J_r^{(q)} &= \sum_{j=0}^{\infty} (-)^j \frac{q^{j(j-1)}}{(q^2;q^2)_j\, (q^{2(r+1)};q^2)_j}\, [q^{2r}(1-q^2)^2\, E]^j\, (q^2\bar{z}z)^j \\ &= {}_1\phi_1(0;q^{2(r+1)};q^2,q^{2(r+1)}(1-q^2)^2\, E\, \bar{z}z), \end{aligned}$$

this is the Hahn-Exton q-Bessel function [13,14]. For the (qH) case we have the following relation:

$$\bar{z}^j z^j = (1 - \bar{z}z;q^{-2})_j$$

obtained solving the recurrence equation

$$\bar{z}z\, P_{n-1}(q^{-2}\bar{z}z + (1-q^{-2})) = P_n(\bar{z}z)\,,$$

that is deduced by $P_n(\bar{z}z) = \bar{z}^n z^n$ and $\bar{z}z(\bar{z}z)^i = \bar{z}(z\bar{z})^i z$. Substituting it in (3.7) we get a deformed q-hypergeometric function in which the expression $\bar{z}z$ appears in a parameter rather than in the argument.

Finally we notice that the states ψ_r give rise to the unitary representation of $E_q(2)$

$$\lambda(J)\,\psi_r = r\,\psi_r\,, \qquad \lambda(P_+)\,\psi_r = \bar{R}\,\psi_{r+1}\,, \qquad \lambda(P_-)\,\psi_r = R\,\psi_{r-1}\,, \qquad (3.8)$$

where $R = \mu \mathcal{E}^{1/2}$.

4 Invariant equation on Quantum homogeneous spaces of $E_\ell(2)$

We now apply the same metod to recover a quantum homogeneous space of $E_\ell(2)$ and the invariant equation thereon [10]. A different real form of this group – reproducing a 1+1 deformation of the Poincaré group – has been applied to phonons and it has been proved to describe the kinematical symmetry for these quasi-particles on the one dimensional lattice [15].

We recall the Hopf algebra structure of the algebra of function $\mathcal{F}_\ell(E(2))$, of the quantum enveloping algebra and the duality pairing [6,7]. The algebra is generated by $e^{-i\theta}$, a_1, a_2, with relations

$$[e^{-i\theta}, a_1] = \frac{z}{2}\,(1 - e^{-i\theta})^2\,,$$

$$[e^{-i\theta}, a_2] = i\frac{z}{2}\,(e^{-2i\theta} - 1)\,,$$

$$[a_1, a_2] = i\,z\,a_1\,,$$

coalgebra operations

$$\Delta(e^{-i\theta}) = e^{-i\theta} \otimes e^{-i\theta}\,,$$
$$\Delta(a_1) = \cos(\theta) \otimes a_1 - \sin(\theta) \otimes a_2 + a_1 \otimes 1\,,$$
$$\Delta(a_2) = \sin(\theta) \otimes a_1 + \cos(\theta) \otimes a_2 + a_2 \otimes 1\,,$$

and antipode

$$S(a_1) = -\cos(\theta)\,a_1 - \sin(\theta)\,a_2\,,$$
$$S(e^{-i\theta}) = e^{i\theta}\,,$$
$$S(a_2) = \sin(\theta)\,a_1 - \cos(\theta)\,a_2\,,$$

where z is now a dimensional deformation parameter, it has the dimension of a_1, a_2.

With z real, a compatible involution is given by $a_1^* = a_1$, $a_2^* = a_2$, $\theta^* = \theta$.

The quantized enveloping algebra $\mathcal{U}_\ell(E(2))$ is generated by the unity and the three elements P_1, P_2, J satisfying

$$[J, P_1] = (i/z)\sinh(zP_2)\,, \qquad [J, P_2] = -i\,P_1\,, \qquad [P_1, P_2] = 0\,,$$

and such that

$$
\begin{aligned}
\Delta P_1 &= e^{-z\,P_2/2} \otimes P_1 + P_1 \otimes e^{z\,P_2/2}, \\
\Delta P_2 &= P_2 \otimes 1 + 1 \otimes P_2, \\
\Delta J &= e^{-z\,P_2/2} \otimes J + J \otimes e^{z\,P_2/2}, \\
S(P_2) &= -P_2, \\
S(P_1) &= -P_1, \\
S(J) &= -J - i\,z\,P_1/2,
\end{aligned}
$$

with vanishing counit and involution

$$
J^* = J, \qquad P_1^* = P_1, \qquad P_2^* = P_2.
$$

We finally write the duality pairing between $\mathcal{U}_\ell(E(2))$ and $\mathcal{F}_\ell(E(2))$

$$
\begin{aligned}
\langle \nu_1, \theta^r a_1^s a_2^t \rangle &= \delta_{r,0}\delta_{s,1}\delta_{t,0}, \\
\langle \nu_2, \theta^r a_1^s a_2^t \rangle &= \delta_{r,0}\delta_{s,0}\delta_{t,1},
\end{aligned}
$$

$$
\langle \tau, \theta^r a_1^s a_2^t \rangle = \delta_{r,1}\delta_{s,0}\delta_{t,0},
$$

where

$$
\tau = -i\,e^{-z\,P_2/2} \left(J - i(z/4)\,P_1 \right),
$$

and

$$
\nu_1 = -i\,e^{-z\,P_2/2}\,P_1, \qquad \nu_2 = -i\,P_2.
$$

Observing that $\Delta\tau = e^{-iz\,\nu_2} \otimes \tau + \tau \otimes 1$ and using the condition $\langle u^*, a \rangle = \overline{\langle u, (S(a))^* \rangle}$, with $u \in \mathcal{U}_\ell(E(2))$ and $a \in \mathcal{F}_\ell(E(2))$, we have that $\nu_1^* = -\nu_1$, $\nu_2^* = -\nu_2$, $\tau^* = -\tau - i\,z\,\nu_1$.

The action of ℓ and of λ on $\mathcal{F}_\ell(E(2))$ can be directly calculated and again the infinitesimal method described in the previous sections can be used to obtain the quantum homogeneous spaces of $E_\ell(2)$ [2]. As a first step we find that the linear span of $X = J - i(z/4)\,P_1$ constitutes a $(* \circ S)$-invariant two-sided coideal of $\mathcal{U}_\ell(E(2))$, twisted primitive with respect to $e^{-z\,P_2/2}$.

Defining $x = a_1 - i\,a_2$, $\bar{x} = a_1 + i\,a_2$, then $x^* = \bar{x}$ and

$$
[x, \bar{x}] = -z\,(x + \bar{x}).
$$

It can be shown that x and $\bar{x}$ generate the invariant subalgebra and left coideal

$$
B_X = \{ f \in \mathcal{F}_\ell(E(2)) | \ell(X)\,f = 0 \}
$$

so that they define a quantum homogeneous space of $E_\ell(2)$.

The Casimir of $\mathcal{U}_\ell(E(2))$ is

$$
\mathcal{C} = 4H^+H^- = (4/z^2)\,\sinh^2((z/2)\,P_2) + P_1^2,
$$

where the elements

$$H^+ = \frac{1}{2z}(e^{zP_2} - 1) - \frac{1}{2}i\,e^{zP_2/2}\,P_1\,,$$

$$H^- = \frac{1}{2z}(1 - e^{-zP_2}) + \frac{1}{2}i\,e^{-zP_2/2}\,P_1$$

are the deformations of the holomorphic and antiholomorphic operators $P_2/2 \mp iP_1/2$.

The coproducts of $H^\pm$ are

$$\Delta H^+ = 1 \otimes H^+ + H^+ \otimes e^{zP_2}\,, \qquad \Delta H^- = e^{-zP_2} \otimes H^- + H^- \otimes 1\,.$$

Thus the z-deformed free Schrödinger equation is:

$$4\,\lambda(H^+H^-)\,\psi = E\,\psi\,. \tag{4.1}$$

In this case too the Casimir operator in (4.1) can be diagonalized in the plane wave and angular momentum bases. It is convenient to define for the ℓ-quantum plane the following variables: $\chi = x/z\,, \quad \bar{\chi} = -\bar{x}/z$. (i) The

plane wave states.

The plane wave basis is defined diagonalizing H^+ and H^-, whose actions, obtained by direct computations, are:

$$\lambda(H^+)\,(\bar{\chi})_n = 0\,,$$

$$\lambda(H^+)\,(\chi)_n = -\frac{n}{z}\,(\chi)_{n-1}\,,$$

$$\lambda(H^+)\,(1-\chi)_n = -\frac{n}{z}\,(1-\chi)_n/(1-\chi)\,,$$

and

$$\lambda(H^-)\,(\chi)_n = 0\,,$$

$$\lambda(H^-)\,(\bar{\chi})_n = \frac{n}{z}\,(\bar{\chi})_n/(\bar{\chi})\,,$$

$$\lambda(H^-)\,(1-\bar{\chi})_n = \frac{n}{z}\,(1-\bar{\chi})_{n-1}\,.$$

where we used the Pochammer symbol $(a)_n = \prod_{k=0}^{n-1}(a+k)$.

We search for eigenstates in the form: $\psi_{h^+h^-} = \sum_{m,n} h_{mn}\,(\chi)_n\,(1-\bar{\chi})_m\,,$ where the coefficients h_{mn} derive from the equation

$$\lambda(H^\pm)\,\psi_{h^+h^-} = h^\pm\,\psi_{h^+h^-}\,,$$

and have the following expression: $h_{mn} = \frac{1}{m!n!}(-zh^+)^m (zh^-)^n$. The eigenvalues equation of the Casimir on these states are:

$$\lambda(\mathcal{C})\,\psi_{h^+h^-} = 4\,\lambda(H^+H^-)\,\psi_{h^+h^-} = 4h^+h^-\,\psi_{h^+h^-}\,.$$

and the $\psi_{h^+h^-}$ can be expressed in the form of a hypergeometric series.
Indeed it turns out that

$$\begin{aligned}
\psi_{h^+h^-} &= \sum_{m=0}^{\infty} \frac{(-zh^+)^m}{m!}(\chi)_m \sum_{n=0}^{\infty} \frac{(zh^-)^n}{n!}(1-\bar\chi)_n \\
&= (1+zh^+)^{-\chi}(1-zh^-)^{\bar\chi-1}
\end{aligned}$$

According to the general definition of the classical hypergeometric series in terms of Pochammer symbols it can be written as

$$\psi_{h^+h^-} = {}_1F_0\left[\begin{matrix}\chi\\-\end{matrix}\;;\;-z\,h^+\right]\,{}_1F_0\left[\begin{matrix}1-\bar\chi\\-\end{matrix}\;;\;z\,h^-\right]\,.$$

The non commuting parameters appear as parameters of the hypergeometric function; in the classical limit $z \to 0$ we recover the usual plane waves $e^{i(h^+x-h^-\bar x)}$.

(*ii*) *The angular momentum states.*

The duality and the involution relations indicate that the angular momentum states are obtained by diagonalizing the operator

$$\mathcal{J} = e^{-zP_2/2}(J - i(z/4)\,P_1)\,.$$

Observing that $\mathcal{J}^* = \mathcal{J}$, let us therefore discuss the equation (4.1) together with

$$\lambda(\mathcal{J})\,\psi = r\,\psi\,.$$

By induction one can prove that the polynomials $(\chi)_n$ and $(\bar\chi)_n$ are eigenstates of $\lambda(\mathcal{J})$ with eigenvalues $-n$ and n respectively; moreover the polynomial $\rho_n = (\bar\chi)_n\,(1-\chi)_n = (\chi)_n\,(1-\bar\chi)_n$ is invariant under the action of $\lambda(\mathcal{J})$, *i.e.* $\lambda(\mathcal{J})\,\rho_n = 0$. Moreover it can be written as $\rho_n = \rho(\rho+2)(\rho+6)\cdots(\rho+n(n-1))$, where $\rho = \bar\chi(1-\chi)$. The states that diagonalize $\lambda(\mathcal{J})$ and $\lambda(H^+H^-)$ are then

$$\psi_{-r} = \sum_{\ell=0}^{\infty} c^{\ell}_{-r}\,\rho_\ell\,(\chi)_r\,, \qquad \psi_r = \sum_{\ell=0}^{\infty} c^{\ell}_r\,\rho_\ell\,(\bar\chi)_r\,,$$

where

$$c^{\ell}_{-r} = (-kz/\bar\epsilon)^r \frac{(kz^2)^\ell}{\ell!(\ell+r)!} \qquad c^{\ell}_r = (-kz/\epsilon)^r \frac{(kz^2)^\ell}{\ell!(\ell+r)!}$$

and k is related to the complex variable ϵ by

$$|\epsilon|^2 = \epsilon\bar{\epsilon} = k(1 + z^2 k).$$

Even in the angular momentum basis the states $\psi_{\pm r}$ are classical hypergeometric series with noncommutative coefficients. Indeed, by the explicit ρ_ℓ we get:

$$\psi_{-r} = \sum_{\ell=0}^{\infty}(-kz/\bar{\epsilon})^r \frac{(kz^2)^\ell}{\ell!(\ell+r)!}(\bar{\chi})_\ell\,(1-\chi)_\ell\,(\chi)_r$$

$$= {}_2F_1\left[\begin{matrix}\bar{\chi} & 1-\chi \\ & r+1\end{matrix}\; ; kz^2\right](-kz/\bar{\epsilon})^r\,\frac{(\chi)_r}{r!}\,,$$

and

$$\psi_r = {}_2F_1\left[\begin{matrix}\bar{\chi} & 1-\chi \\ & r+1\end{matrix}\; ; kz^2\right](-kz/\epsilon)^r\,\frac{(\bar{\chi})_r}{r!}\,.$$

Actually the situation is very different from the q-Euclidean case described in the previous section: here the *quantum* deformation appears from the noncommutative variables χ and $\bar{\chi}$ that enter in two of the parametres of ${}_2F_1$. The other coefficient and the hypergeometric variable are instead numbers.

The classical limit $z \to 0$ of the hypergeometric ${}_2F_1$ yield again the usual Bessel functions in the commutative variable $\bar{x}x$, this result is due to a specific confluence phenomenon caused by the non commutativity of the arguments.

5   The special functions as matrix elements of T

Let us now recall that, considering a representation $\mathcal{R}$ of $\mathcal{U}_q(\mathrm{Lie}\,G)$ and the canonical element [16] or *universal matrix* T [17,18,19] in $\mathcal{F}_q(G)\otimes\mathcal{U}_q(\mathrm{Lie})$, the elements $t_{rs}^{\mathcal{R}} = ((1 \otimes \mathcal{R})T)_{rs} \in \mathcal{F}_q(G)$ satisfy the usual definition $\langle\, X, t_{rs}^{\mathcal{R}}\rangle = \mathcal{R}(X)_{rs}$ for every $X \in \mathcal{U}_q(\mathrm{Lie}\,G)$: we now exploit this construction for the case of $E_q(2)$.

We take

$$J, \quad b_- = -q^{-J}P_-, \quad b_+ = P_+q^J$$

as the generators of $\mathcal{U}_q(E(2))$ and π, $\pi_{\pm}$ as their duals. Introducing the q-exponential

$$e_q(x) = \sum_{j=0}^{\infty} x^j/(q;q)_j = {}_1\phi_0(0; -; q, x)$$

we find the universal T-matrix [18]

$$T = e_{q^2}[(1 - q^2)\,\pi_- \otimes b_-]\ e^{\pi\otimes J}\ e_{q^{-2}}[(1 - q^{-2})\,\pi_+ \otimes b_+]\,,$$

where π, $\pi_\pm$ close an algebra, that we call $\tilde{\mathcal{F}}_q(E(2))$, containing $\mathcal{F}_q(E(2))$: at the classical level they are the canonical coordinates of the second kind. The algebra $\mathcal{F}_q(E(2))$ is obtained by letting

$$v = e^{-\pi}, \qquad n = \pi_-, \qquad \bar{n} = e^{\pi}\pi_+.$$

The matrix elements obtained from the universal T-matrix using the representation (3.8) are given in terms of Hahn-Exton q-Bessel functions and coincide with the result of [14] provided the identifications $v = \alpha^2$, $\bar{v} = \delta^2$, $n = -q^{-1/2}\beta\alpha$, $\bar{n} = q^{1/2}\delta\gamma$, $r = -i$ and $s = -j$. Our results agree with those of [20], so that their q-exponentials, although unrelated to canonical construction, prove to be efficient in getting the matrix elements of the representations.

We conclude this talk by a general definition of spherical elements based on the use of the T-matrix. Given a τ invariant two sided coideal $K_\mathcal{M}$ in $\mathcal{U}_q(\text{Lie}\,G))$, consider a unitary representation $\mathcal{R}$ of $\mathcal{U}_q(\text{Lie}\,G)$ and suppose there exists an element ξ spanning a one dimensional kernel of $K_\mathcal{M}$ in the representation space of $\mathcal{R}$. Denoting by $(\ ,\ )$ the scalar product of $\mathcal{R}$, we define the *zonal spherical function* $t^{\mathcal{R}}_{\text{zon}}$ of the representation $\mathcal{R}$ with respect to $K_\mathcal{M}$ as follows:

$$t^{\mathcal{R}}_{\text{zon}} = (\xi, (1 \otimes \mathcal{R})\,T\,\xi).$$

We then call *associated spherical functions* the elements

$$t^{\mathcal{R}}_k = (\xi_k, (1 \otimes \mathcal{R})\,T\,\xi),$$

where $\{\xi_k\}$ is a basis of the representation $\mathcal{R}$.

With these definitions we have proved in [9] that for any $Y \in K_\mathcal{M}$

$$\lambda(Y)\,t^{\mathcal{R}}_{\text{zon}} = 0, \qquad\qquad \ell(Y)\,t^{\mathcal{R}}_{\text{zon}} = 0,$$

and

$$\ell(Y)\,t^{\mathcal{R}}_k = 0, \text{ for any } Y \in K_\mathcal{M}.$$

We use the results of [14] to compute the kernel of X_μ in the representation (3.8) with $R = i\sqrt{E}\,(\mu/|\mu|)$.: it is given by

$$\xi = \sum_r q^{r(r-1)/2}\,i^r\,J^{(2)}_r(\sigma; q^2)\psi_r,$$

where $J^{(2)}_r$ are the Jackson q-Bessel functions and $\sigma = 2q\,(q - q^{-1})\,\sqrt{E}/|\mu|$.

The zonal spherical function is then

$$t^{\lambda}_{\text{zon}} = \sum_{r,s} q^{r(r-1)/2}\,q^{s(s-1)/2}\,i^{s-r}\,J^{(2)}_r(\sigma; q^2)\,J^{(2)}_s(\sigma; q^2)\,t^{\lambda}_{rs},$$

and the associated spherical functions are

$$t_k^\lambda = \sum_s q^{s(s-1)/2}\, i^s\, J_s^{(2)}(\sigma; q^2)\, t_{ks}^\lambda \,.$$

At the end of this report in which we have shown a physically fruitful approach to the noncommutative geometry of quantum groups we want to stress that all the results here presented, with the concrete determination of deformed special functions solving the eigenvalue problem of the invariant element, come out quite naturally by following step by step the canonical procedure which gives the definition of the quantum homogeneous spaces, the action of the enveloping algebra and of the quantum invariant elements.

References

[1] Celeghini E., Giachetti R., Sorace E. and Tarlini M., *J. Math. Phys.*, **32**, 1155 (1991); Bonechi F., Giachetti R., Sorace E. and Tarlini M., *Commun. Math. Phys*, **169** (1995) 627

[2] Celeghini E., Giachetti R., Sorace E. and Tarlini M., *"Contractions of quantum groups"*, in Lecture Notes in Mathematics n. 1510, 221, (Kulish P.P. ed., Springer–Verlag, 1992.

[3] Bonechi F., Celeghini E., Giachetti R., Sorace E. and Tarlini M., *Phys. Rev. B*, **32** (1992) 5727 and *J.of Phys.A*, **25**, (1992) L939.

[4] Vaksman L.L. and Korogodski L.I., *Sov. Math. Dokl.*, **39** (1989) 173.

[5] Celeghini E., Giachetti R., Sorace E. and Tarlini M., *J. Math. Phys.*, **31**, 2548 (1990).

[6] Woronowicz S.L., *Lett. Math. Phys.*, **23** (1991) 251; *Commun. Math. Phys.*, **144** (1992) 417; *Commun. Math. Phys.*, **149** (1992) 637.

[7] Ballesteros A., Celeghini E., Giachetti R., Sorace E. and Tarlini M. *J. Phys. A: Math. Gen.*, **26** (1993) 7495.

[8] Celeghini E., Giachetti R., Sorace E. and Tarlini M., *J. Math. Phys.*, **32**, 1159 (1991), Lukierski J., Ruegg H., Nowicki A., Tolstoy V.N., *Phys. Lett.B*, **264** (1991) 331.

[9] Bonechi F.,Ciccoli N., Giachetti R., Sorace E. and Tarlini M., *Commun. Math. Phys*, **175** (1996) 161.

[10] Bonechi F., del Olmo M.A., Giachetti R., Sorace E. and Tarlini M., *J.of Phys.A*, **29** (1996) 7973.

[11] Dijkhuizen M.S., Koornwinder T.H., *Geom. Dedicata*, **52**, (1994), 291.

[12] Dijkhuizen M.S., *"On compact quantum groups and quantum homogeneous spaces"*, Thesis (Amsterdam University, 1994).

[13] Koornwinder T.H., *"Quantum groups and q-special functions"* in *Representations of Lie groups and quantum groups*, Baldoni V. and Picardello M. A. (eds), Pitman Research Notes in Mathematical Series 311, (Longman Scientific & Technical, 1994), pp. 46–128

[14] Koelink H.T., *"On quantum groups and q-special functions"*, Thesis (Leiden University, 1991) and *Duke Math. J.*, **76** (1994) 483

[15] Bonechi F., Celeghini E., Giachetti R., Sorace E. and Tarlini M., *Phys. Rev. Lett.*, **68** (1992) 3718.

[16] N.Yu Reshetikhin and M.A. Semenov-Tian-Shansky, *Journal Geom. Phys.*, **5** (1988) 533.

[17] Fronsdal C. and Galindo A., *Lett. Math. Phys.*, **27** (1993) 59.

[18] Bonechi F., Celeghini E., Giachetti R., Pereña C.M., Sorace E. and Tarlini M., *J. Phys. A: Math. Gen.*, **27** (1994) 1307.

[19] Sorace E. *CRM Proceedings and Lecture Notes*, **9** (1996) 353. Talk given at the Workshop on Symmetries and Integrability of Difference, May 22-29, 1994, Québec, Canada.

[20] Floreanini, R., Vinet, L.: *Lett. Math. Phys.*, **27** (1993) 179.

Local Yang-Baxter relations associated with Hirota's discrete equation

R.M. Kashaev*[†]

Laboratoire de Physique Théorique ENSLAPP[‡]

ENSLyon, 46 Allée d'Italie,

69007 Lyon, FRANCE

Abstract

The discrete Hirota equation is associated with infinite dimensional solutions of the local Yang-Baxter equation both in classical and quantum cases. The corresponding solution to the tetrahedron equation reproduces the infinite dimensional solution found recently by Sergeev et al.

1   Introduction

The local Yang-Baxter equation (LYBE) has been introduced by Maillet and Nijhoff [1, 2] as a generalization of the discrete zero curvature (Lax) relation to three dimensions. Recently Korepanov [3] has constructed a family of (free-fermion) solutions for LYBE and has demonstrated by algebro-geometric methods integrability of the corresponding dynamical systems. The simplest solution in this family is related with the star-triangle relation (STR) in the Ising model. In [4] an infinite dimensional solution to LYBE has been constructed. It is related to the STR in electrical networks, and the corresponding dynamical system appeared to be nothing else than Miwa's discrete equation of the BKP hierarchy [5].

---

*On leave of absence from St. Petersburg Branch of the Steklov Mathematical Institute, Fontanka 27, St. Petersburg 191011, RUSSIA

†The work is supported by the Programme TEMPRA-Europe de l'Est from the Région Rhône-Alpes.

‡URA 14-36 du CNRS, associée à l'E.N.S. de Lyon, au LAPP d'Annecy et à l'Universitè de Savoie

In this article we construct an infinite dimensional solution for LYBE, which directly leads to the Hirota's discrete equation of the generalized Toda system [6]. The quantized Hirota equation [8] can be obtained from the straightforward q-deformation of the obtained solution. From the latter one can derive also a solution to the tetrahedron equation found by Sergeev et al [7].

The author is grateful to L.D. Faddeev, J.M. Maillet, A.Yu. Volkov for useful discussions.

2 LYBE and Hirota's discrete equation

First recall what we mean by LYBE [1, 2]. Let $\mathcal{A}$ be an associative algebra, and $\mathfrak{M}$, a lattice in $\mathbb{R}^3$, generated by three linearly independent vectors $e_i \in \mathbb{R}^3$, $i = 1, 2, 3$:

$$\mathfrak{M} = \{ n = \sum_{i=1}^{3} n_i e_i | n_i \in \mathbb{Z} \}. \tag{2.1}$$

Consider three functions

$$L, M, N \colon \mathfrak{M} \to \mathcal{A}. \tag{2.2}$$

We say that L, M, N satisfy LYBE, if the following identity holds for any $n \in \mathfrak{M}$:

$$N(n)M(n + e_2)L(n) = L(n + e_1)M(n)N(n + e_3). \tag{2.3}$$

This equation reduces to the discrete zero curvature equation, if we put $N(n) = 1$. A solution to (2.3) we will be dealing with in this paper can be described as follows.

2.1 The classical case

Consider a Lie algebra $\mathfrak{h}$ over $\mathbb{C}$, generated by two elements A, B, satisfying the Heisenberg (Serre) commutation relations:

$$[A, [A, B]] = [[A, B], B] = 0, \tag{2.4}$$

which mean that the element $[A, B]$ lies in the center of $\mathfrak{h}$. Elements of the universal enveloping algebra $U(\mathfrak{h})$, which are exponentials of elements from $\mathfrak{h}$, form the Heisenberg group $\mathcal{H} = \exp(\mathfrak{h}) \subset \mathcal{U}(\mathfrak{h})$ with the multiplication law:

$$\exp(X)\exp(Y) = \exp(X + Y + \frac{1}{2}[X, Y]). \tag{2.5}$$

The following identity in $\mathcal{H}$ can be easily proved by using (2.5):

$$\exp(x_3 B)\exp(y_2 A)\exp(x_1 B) = \exp(y_1 A)\exp(x_2 B)\exp(y_3 A), \tag{2.6}$$

where complex parameters $x_i, y_i \in \mathbb{C}$, are constrained by the relations:

$$x_1 + x_3 = x_2, \tag{2.7}$$

$$x_3 y_2 = x_2 y_3, \quad y_2 x_1 = y_1 x_2. \tag{2.8}$$

To interpret (2.6) as a LYBE let the algebra $\mathcal{A}$ be such that $U(\mathfrak{h})^{\otimes 3} \subset \mathcal{A}$. Take some mappings

$$\mu_i \colon \mathfrak{M} \to \mathbb{C}, \quad i = 1, 2, 3, \tag{2.9}$$

and define functions (2.2) as follows:

$$L(n) = \exp(\mu_1(n) C_{23}) P_{23}, \quad M(n) = \exp(\mu_2(n) C_{13}) P_{13},$$

$$N(n) = \exp(\mu_3(n) C_{12}) P_{12}, \tag{2.10}$$

where

$$C_{ij} = A_i + B_j, \quad i, j = 1, 2, 3;$$

$\{A_i, B_i\}$ generate Lie algebras $\mathfrak{h}_i$, the three mutually commuting copies of $\mathfrak{h}$, while $P_{ij} = P_{ji}$ are (additional) elements in $\mathcal{A}$ defined by

$$P_{ij}^2 = 1, \quad P_{ij} X_i = X_j P_{ij}, \quad [P_{ij}, X_k] = 0, \quad \{i, j, k\} = \{1, 2, 3\}, \tag{2.11}$$

for any $X_i \in \mathfrak{h}_i$. Substitute (2.10) into (2.3). Using (2.11) we can cancel all P-factors from the both sides of the equation. The resulting identity coincides with (2.6) provided

$$A = C_{23}, \quad B = C_{12}, \tag{2.12}$$

and

$$x_i = \mu_i(n), \quad y_i = \mu_i(n + e_i), \quad i = 1, 2, 3. \tag{2.13}$$

Thus, we obtain a system of three discrete equations (2.7),(2.8) on three functions (2.9). Equations (2.8), however, can be solved by the substitutions

$$\mu_i(n) = (-)^i \alpha_i \frac{\tau(n) \tau(n + e_j + e_k)}{\tau(n + e_j) \tau(n + e_k)}, \quad \{i, j, k\} = \{1, 2, 3\}, \tag{2.14}$$

with some $\tau \colon \mathfrak{M} \to \mathbb{C}$, and α_i being complex parameters. Equation (2.7) then reduces to Hirota's discrete equation of the generalized Toda system [6]:

$$\sum_{i=1}^{3} \alpha_i \tau(n + e_i) \tau(n + e_0 - e_i) = 0, \quad e_0 = e_1 + e_2 + e_3. \tag{2.15}$$

2.2 The quantum case

Let a complex parameter q be such that $|q| < 1$. We shall find it convenient to use the notation $q' \equiv 1 - q$. Define a function

$$\psi(x) = 1/(-x;q)_\infty = 1/\prod_{n=0}^{\infty}(1 + xq^n). \tag{2.16}$$

It can be considered both as a q-exponential and q-dilogarithm because of the following two operator functional equations it solves:

$$\psi(v)\psi(u) = \psi(v + u), \tag{2.17}$$

for operators u and v satisfying the Weyl exchange relation

$$uv = qvu, \tag{2.18}$$

and

$$\psi(a)\psi(b) = \psi(b)\psi([b,a]/q')\psi(a), \quad q' = 1 - q, \tag{2.19}$$

for a and b satisfying the q-deformed Serre relations:

$$ab^2 + qb^2a = (1+q)bab, \quad a^2b + qba^2 = (1+q)aba, \tag{2.20}$$

see, for example, [9] and references therein.

Denote by $U_q(\mathfrak{h})$ the algebra generated by two elements a, b subject to the q-deformed Serre relations (2.20), and by P_q, the algebra generated by t_1, t_2, t_3 with defining relations of the form:

$$t_1t_2 = qt_2t_1, \quad t_2t_3 = t_3t_2, \quad t_3t_1 = qt_1t_3. \tag{2.21}$$

Then the following identity in $P_q \otimes U_q(\mathfrak{h})$ naturally generalizes (2.6):

$$\psi(t_1b)\psi(t_2a)\psi(t_3b) = \psi(s_1a)\psi(s_2b)\psi(s_3a), \tag{2.22}$$

where $s_1, s_2, s_3 \in P_q$ are given by rational expressions:

$$s_1 = t_2t_3(t_1 + t_3)^{-1}, \quad s_2 = t_1 + t_3, \quad s_3 = (t_1 + t_3)^{-1}t_1t_2. \tag{2.23}$$

In a slightly different form formula (2.22) first appeared in [10] (see also [11]). Here we prove it by straightforward application of formulas (2.17) and (2.19).

For this note that formulas (2.23) imply the following exchange relations for elements s_1, s_2, s_3:

$$s_1s_2 = s_2s_1, \quad s_2s_3 = qs_3s_2, \quad s_3s_1 = qs_1s_3. \tag{2.24}$$

Thus, we can apply (2.19) to the last two factors of the LHS, and to the first two factors of the RHS of (2.22):

$$\psi(t_1 b)\psi(t_3 b)\psi(t_2 t_3[b,a]/q')\psi(t_2 a) = \psi(s_2 b)\psi(s_1 s_2[b,a]/q')\psi(s_1 a)\psi(s_3 a).$$
$$(2.25)$$

Apply now (2.17) to the first two factors of the LHS, and to the last two factors of the RHS:

$$\psi((t_1 + t_3)b)\psi(t_2 t_3[b,a]/q')\psi(t_2 a) = \psi(s_2 b)\psi(s_1 s_2[b,a]/q')\psi((s_1 + s_3)a).$$
$$(2.26)$$

This relation becomes identity provided the following equations hold

$$s_2 = t_1 + t_3, \quad s_1 s_2 = t_2 t_3, \quad s_1 + s_3 = t_2, \tag{2.27}$$

which are equivalent to (2.23).

To get the LYBE from (2.22), we have to make similar substitutions as we did in the classical case, the main difference being that quantities $\mu_i(n)$ are non-commuting elements of an algebra A_o (algebra of observables):

$$\mu_i \colon \mathfrak{M} \to A_o. \tag{2.28}$$

Define an auxiliary algebra A_a, which is $U_q(\mathfrak{h})^{\otimes 3}$, extended by permutation elements P_{ij} between different factors in the tensor product (see (2.11), with $X_i \in U_q(\mathfrak{h})_i$, $i = 1, 2, 3$). Then, the elements in $A_o \otimes A_a$, defined by

$$L(n) = \psi(\mu_1(n)a_2 b_3)P_{23}, \quad M(n) = \psi(\mu_2(n)a_1 b_3)P_{13},$$

$$N(n) = \psi(\mu_3(n)a_1 b_2)P_{12}, \tag{2.29}$$

where $\{a_i, b_i\}$ generate $U_q(\mathfrak{h})_i$, the ith factor in A_a, satisfy the LYBE (2.3) provided the operator valued functions $\mu_i(n)$ satisfy the following exchange relations:

$$\mu_1(n)\mu_2(n+e_2) = \mu_2(n+e_2)\mu_1(n), \quad \mu_3(n)\mu_2(n+e_2) = q\mu_2(n+e_2)\mu_3(n),$$

$$\mu_1(n)\mu_3(n) = q\mu_3(n)\mu_1(n), \tag{2.30}$$

and the quantum (Heisenberg) equations of motion:

$$\mu_3(n)\mu_2(n+e_2) = \mu_2(n)\mu_3(n+e_3), \quad \mu_2(n+e_2)\mu_1(n) = \mu_1(n+e_1)\mu_2(n),$$

$$\mu_2(n) = \mu_1(n) + \mu_3(n). \tag{2.31}$$

There exists an algebraic structure in A_o, consistent with (2.30) and (2.31), which can be described as follows. Let us choose a time variable:

$$t = n_1 + n_3 - n_2, \tag{2.32}$$

and introduce new, more explicit notations for the dynamical variables:

$$\mu_i(t, n_1, n_3) = \begin{cases} \mu_i(n), & i = 1, 3; \\ \mu_2(n + e_2), & i = 2. \end{cases} \tag{2.33}$$

Define integer valued functions $G_{ij}(k, l)$, $i, j = 1, 2, 3$; $k, l \in \mathbb{Z}$:

$$G_{ij}(k, l) = -G_{ji}(-k, -l),$$

$$G_{ii}(k, l) = \delta(k + l)\epsilon(l), \quad i = 1, 2, 3,$$

$$G_{12}(k, l) = \theta(k)\delta(k + l) - \theta(l)\delta(k + l - 1),$$

$$G_{23}(k, l) = \theta(k + 1)\delta(k + l + 1) - \theta(l + 1)\delta(k + l),$$

$$G_{13}(k, l) = \delta(k + l)(\epsilon(l) + \delta(l)), \tag{2.34}$$

where

$$\epsilon(k) = \begin{cases} 0, & k = 0; \\ k/|k|, & k \neq 0; \end{cases} \qquad \delta(k) = \begin{cases} 1, & k = 0; \\ 0, & k \neq 0; \end{cases} \tag{2.35}$$

$$2\theta(k) = \epsilon(k) - \delta(k) + 1. \tag{2.36}$$

Then, the following equal-time exchange relations:

$$\mu_i(t, k, l)\mu_j(t, k', l') = q^{G_{ij}(k-k', l-l')}\mu_j(t, k', l')\mu_i(t, k, l) \tag{2.37}$$

are consistent with (2.30) and (2.31). Thus, we have recovered the classical and quantum Hirota's dynamical systems from the LYBE. The quantum version of this system, expressed in terms of other variables, first has been constructed in [8] in the context of the quantum inverse scattering method.

3 Relation to the tetrahedron equation

Here we rewrite identity (2.22) in a refined form, which will reveal a connection with a particular solution of the tetrahedron equation [12].

First, define a q-deformed oscillator algebra O_q, generated by a, b, c with the commutation relations:

$$[a, b] = q^c, \quad [c, a] = -a, \quad [c, b] = b. \tag{3.1}$$

There is an evident homomorphism of algebras $U_q(\mathfrak{h}) \to O_q$, given by identification of the generators a and b:

$$U_q(\mathfrak{h}) \supset \{a, b\} \mapsto \{a, b\} \subset O_q. \tag{3.2}$$

Next, let V be the universal enveloping algebra of a nonabelian two-dimensional Lie algebra:

$$[h, u] = u. \tag{3.3}$$

Take now for the auxiliary algebra A_a the algebra $O_q^{\otimes 3}$, again extended by permutation operators P_{ij}, acting between different O_q factors (see (2.11), with $X_i \in (Q_q)_i)$. Let $\{u_i, h_i\} \subset V^{\otimes 3}$, $i = 1, 2, 3$, generate three mutually commuting copies of the algebra V. So for each i, u_i and h_i satisfy the commutation relation (3.3), while the generators with different indices commute. Let $\{\tilde{u}_i, \tilde{h}_i\} \in V^{\otimes 3}$, $i = 1, 2, 3$, be another set of generators with the same commutation relations, as for the untilded ones. Note, that both, tilded and untilded generators, belong to one and the same algebra $V^{\otimes 3}$.

Define the elements

$$L_{ij}(u, h) = \psi(ua_i b_j)q^{h c_j} P_{ij} \in V^{\otimes 3} \otimes A_a, \quad i, j = 1, 2, 3, \qquad (3.4)$$

where $\{u, h\}$ stand for any pair from the six ones defined above. Impose now the following equation:

$$L_{12}(u_1, h_1)L_{13}(u_2, h_2)L_{23}(u_3, h_3) = L_{23}(\tilde{u}_3, \tilde{h}_3)L_{13}(\tilde{u}_2, \tilde{h}_2)L_{12}(\tilde{u}_1, \tilde{h}_1), \quad (3.5)$$

which, after cancellation of the permutation operators P_{ij}, reduces to equation (2.22), provided u and h variables satisfy the equations

$$h_1 + h_3 = \tilde{h}_2, \quad h_2 = \tilde{h}_1 + \tilde{h}_3,$$

$$u_1 u_2 q^{h_3} = \tilde{u}_1 \tilde{u}_2 q^{\tilde{h}_3}, \quad u_2 u_3 = \tilde{u}_2 \tilde{u}_3, \quad u_1 + u_3 q^{h_1} = \tilde{u}_2, \qquad (3.6)$$

the corresponding identifications being

$$t_1 = u_1, \quad t_2 = u_2 q^{-h_1}, \quad t_3 = u_3 q^{h_1},$$

$$s_1 = \tilde{u}_3, \quad s_2 = \tilde{u}_2, \quad s_3 = \tilde{u}_1 q^{\tilde{h}_3 - \tilde{h}_2}, \qquad (3.7)$$

and

$$a = a_2 b_3, \quad b = a_1 b_2. \qquad (3.8)$$

The structure of the identity (3.5) suggests that the intertwining operator R, relating tilded and untilded generators in $V^{\otimes 3}$,

$$\tilde{X} = R_{123}^{-1} X R_{123}, \qquad (3.9)$$

where X stands for any of the generators from $\{u_i, h_i\}$, $i = 1, 2, 3$, should satisfy the tetrahedron equation. The general solution of the linear equation (3.9) has the form

$$R_{123} = \varphi(u_1 u_3^{-1} q^{-h_1}) P_{23} q^{h_1(h_2 - h_3)} u_3^{h_1} u_2^{-h_1} \psi(u_1 u_3^{-1} q^{-h_1}), \qquad (3.10)$$

with $\varphi(x)$ being arbitrary function of its argument. For two choices, either $\varphi(x) = 1$ or $\varphi(x) = 1/\psi(x)$, R_{123} satisfies the tetrahedron equation:

$$R_{123} R_{145} R_{246} R_{356} = R_{356} R_{246} R_{145} R_{123}. \qquad (3.11)$$

These solutions first were found in [7].

References

[1] J.M. Maillet, F.W. Nijhoff, *Integrability for multidimensional lattice models*, Phys. Lett. B224 (1989) 389-396

[2] J.M. Maillet, *Integrable systems and gauge theories*, Nucl. Phys. (Proc. Suppl.) 18B (1990) 212-241

[3] I.G. Korepanov, *Algebraic integrable dynamical systems, 2+1-dimensional models in wholly discrete space-time, and inhomogeneous models in 2-dimensional statistical physics*, doctoral dissertation, solv-int/9506003

[4] R.M. Kashaev, *On Discrete 3-Dimensional Equations Associated with the Local Yang-Baxter Relation*, ENSLAPP-L-569/95, solv-int/9512005, to be published in Lett. Math. Phys.

[5] T. Miwa, *On Hirota's Difference Equations*, Proc. Japan Acad., 58, Ser. A (1982) 9-12

[6] R. Hirota, *Discrete analogue of a generalized Toda equation*, J. Phys. Soc. Jpn., Vol. 50, No. 11, (1981) 3785-3791

[7] S.M. Sergeev, V.V. Bazhanov, H.E. Boos, V.V. Mangazeev, Yu.G. Stroganov, *Quantum dilogarithm and tetrahedron equation*, IHEP 95-129

[8] R.M. Kashaev, N.Yu. Reshetikhin, *Affine Toda field theory as a 3-dimensional integrable system*, preprint ENSLAPP-L-548/95, to be published in Lett. Math. Phys.

[9] T. Koornwinder, *Special functions and q-commuting variables*, Institut Mittag-Leffler, Djursholm, Sweden, 1995/96. no.1, q-alg/9608008

[10] A. Morozov, L. Vinet, *Free-Field Representation of Group Element for Simple Quantum Group*, ITEP-M3/94 and CRM-2202, hep-th/9409093

[11] A. Berenstein, *Group-like elements in quantum groups, and Feigin's conjecture*, BRE-96-04, q-alg/9605016

[12] A.B. Zamolodchikov, *Tetrahedron equations and the relativistic S-matrix of straight-strings in 2 + 1 dimensions*, Commun. Math. Phys. 79 (1981) 489-505

Some Algebraic Solutions of Discrete Equations from Anticommuting Variables

C. Viallet

Centre National de la Recherche Scientifique
Laboratoire de Physique Théorique et des Hautes Energies
Université de Paris VI, Tour 16, 1er étage, boîte 126.
4 Place Jussieu/ F–75252 PARIS Cedex 05

Abstract

We describe specific solutions of some discrete-invariance equations. The equations are related the symmetries of the Yang-Baxter equations. The solutions are inspired from the so-called free-fermions models of lattice statistical mechanics, and make use of anticommuting variables.

1   Introduction

Among the discrete equations subject to current study, many lead to the following problem: find the invariants of rational and rationally invertible transformations (see the other contributions to these proceedings and for instance [1, 2, 3, 4, 5]). As an example, all of the discrete Painlevé equations presented in [6] are invariance equations under birational transformations. The analysis of such equations often reduces to the analysis of the iterates of a mapping, and sometimes to the study of arborescent iterations.

It was noted in [4, 7] that the "integrability" of a mapping is related to its *low* complexity, measured for example by the rate of growth of the degree of the successive iterates.

We describe here a somewhat extreme case of low complexity: finite orbits. For rational transformations, finite orbits automatically provide algebraic invariant varieties. They are unfortunately not so easy to find: calculating the iterates of the transformations, just to write down the finiteness conditions for the orbits, is more than often beyond reach. We will nevertheless produce solutions of the finiteness conditions. One novel feature of these solutions is that they are obtained through the use of anticommuting variables.

We first describe the transformations we have in mind. We then recall which discrete equations are to be solved. We then give the construction of finite orbits [8], inspired from the so-called free-fermion conditions of lattice statistical mechanics [9, 10, 11, 12, 13, 14, 15, 16], and analyze them. We conclude with some comments on the use of these solutions.

2 Some discrete groups

We start with the following ingredients, which come from lattice statistical mechanics, and more precisely the so-called vertex models:

- a vertex drawn in d-dimensional space and of even coordination 2ν (see figure below)

- a number q of "colors" which is the number of states each bond can take

- an assignment of numbers, arranged in a matrix R, to each vertex configuration. These numbers are complex or real in general. They would be the Boltzmann weights of the configuration in statistical mechanics

- a group of rational transformations built out of the matrix inversion I (defined up to an overall multiplicative factor), and simple permutations of the entries of R. The latter are induced by the reflections of the vertex (see [17, 18, 19])

The vertex looks like:

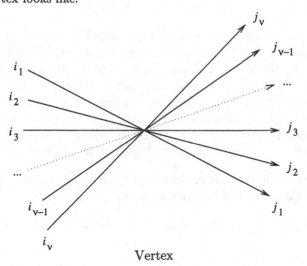

Vertex

The above picture was intentionally drawn without reference to the dimension d. To the above configuration we associate the component

$$R_{j_1 j_2 \ldots j_\nu}^{i_1 i_2 \ldots i_\nu} \tag{1}$$

Warning: We will always work in projective space, and consider the entries of R as defined up to a constant (the dimension of the projective space is $q^{4\nu} - 1$). The generator I is then to be thought of as the inverse up to an overall constant. In other words it may be written in a polynomial way. This has some deep implications, and permits to talk about the degree of the realizations of the group elements, as in [7, 4, 20]

Each bond reaching the vertex bears a color taking values $[1, 2, \ldots, q]$. The size $(q^{2\nu} \times q^{2\nu})$ of the matrix R depends on ν and on the number q of colors, but not on the dimension d: for example the vertex of a planar triangular lattice and the vertex of a three-dimensional cubic lattice have R-matrices of the same form (and same size for identical values of q).

For each geometry of the vertex (different dimensions d) we construct an infinite group Γ realized by birational transformations. The group is generated by $\nu + 1$ involutions:

$$\text{The matrix inverse} \quad I : R \longrightarrow R^{-1}$$
$$\nu \ \ \text{reflections} \quad t_k, \quad k = 1 \ldots \nu$$

The first generator (I) is non-linear. The partial transposition t_k is the reflection of the vertex which reverses the orientation of the arrow $i_k j_k$, and permutes the other arrows according to the geometry of the vertex. The ν reflections t_k are projective linear transformations (see section 3). The realization of a generic group elements γ of Γ is essentially non-linear, because one of the generators of Γ (namely I) is non-linear.

The basic problem may then be summarized as follows: we have an infinite discrete group Γ of birational transformations of some projective space P_n generated by $\nu + 1$ involutions I_k, $k = 0 \ldots \nu$ (setting $I_0 = I$ and $I_k = t_k \quad k = 1 \ldots \nu$). Find all kinds of Γ-automorphic (Γ-covariant) objects, like polynomials (i.e invariant algebraic varieties of codimension 1), invariant rational functions (i.e. foliation of the space by invariant algebraic varieties of codimension 1), invariant functions (transcendants), ideals of polynomials (i.e. invariant varieties of codimension higher than 1), and so on.

There can be algebraic invariants, leading to foliation of the space of parameters. Suppose we look for such an invariant Δ, i.e.

$$\Delta = P/Q, \qquad \Delta(\gamma\, x) - \Delta(x) = 0 \qquad \forall x \in P_n, \quad \forall \gamma \in \Gamma. \tag{2}$$

To solve this discrete-invariance equation, we have to solve the discrete-covariance equation

$$V(I_k(x)) = a(I_k, x) \cdot V(x) \quad \forall k, \quad \forall x \in P_n \tag{3}$$

We may proceed in two steps: first find the possible factors $a(I_k, x)$ and then solve the discrete-covariance condition.

The factor $a(I_k, x)$ verifies a cocycle condition

$$a(\gamma_1\gamma_2, x) = a(\gamma_1, \gamma_2 x)a(\gamma_2, x) \tag{4}$$

Indeed

$$
\begin{aligned}
V(\gamma_1\gamma_2 x) &= a(\gamma_1\gamma_2, x)V(x) \tag{5}\\
&= a(\gamma_1, \gamma_2 x)V(\gamma_2 x) \tag{6}\\
&= a(\gamma_1, \gamma_2 x)a(\gamma_2, x)V(x) \tag{7}
\end{aligned}
$$

and one may simplify by V.

Taking $\gamma_1 = \gamma_2 = I_k$ and using $I_k^2 \simeq \phi_k$, we get

$$a(I_k, I_k(x))\, a(I_k, x) = a(I_k^2, x) = \phi_k^d(x) \tag{8}$$

We see that $a(I_k, x)$ divides some power of ϕ_k^d. This solves the cocycle condition, and the problem has become linear (the covariance condition).

There can also be (possibly isolated) invariant manifolds of codimension higher than 1, with equation

$$\Sigma = \{\Pi_1(x) = 0, \Pi_2(x) = 0, \ldots, \Pi_k(x) = 0\} \tag{9}$$

We have the multicomponent equation

$$\Pi(\gamma(x)) = A(\gamma, x)\Pi(x) \quad \forall \gamma \in \Gamma, \forall x \in CP_n \tag{10}$$

Then the cocycle condition reads:

$$A(\gamma_1\gamma_2 x)\Pi(x) = A(\gamma_1, \gamma_2 x)A(\gamma_2, x)\Pi(x) \tag{11}$$

Take $\gamma_1 = \gamma_2 = I_k$ as above

$$A(I_k, I_k x)A(I_k, x)\Pi(x) = diag(\phi_k^{d_1}, \ldots, \phi_k^{d_2})\,\Pi(x) \tag{12}$$

We cannot divide by Π, and we have no general solution of this much harder problem.

Our point is to produce particular solutions of this problem. We could just write the conditions saying that the orbit is finite, but we do not know a priori which ones to choose. Moreover it quickly gets impossible to write anything explicit, and the intersection of all conditions may reduce to points. We want something more effective, leading to varieties of dimension larger than 1.

We present here a construction using anticommuting variables, taken from the so-called free-fermion models of statistical mechanics. It will produce specific solutions where the orbits are finite, as an effect of a degeneration of the inverse, which becomes a linear transformation.

3   More about the reflections

We should insist on the effect of the value of the dimension on the definition of the partial transpositions t_k. To illustrate this, let us fix $\nu = 3$ and examine $d = 2$ and $d = 3$. The following picture represents graphically the effect of t_1 for $d = 2$ and $d = 3$ respectively.

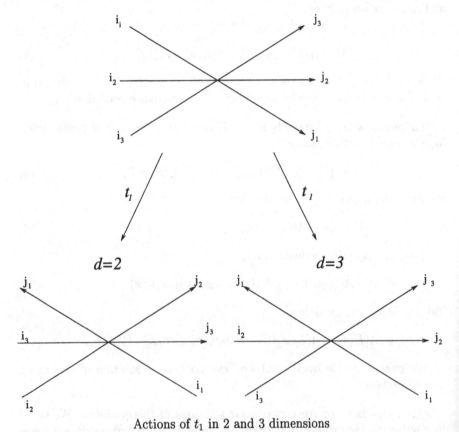

Actions of t_1 in 2 and 3 dimensions

At this point it is useful to recall that we may construct the group Γ for all values of ν and d, *even if there is no regular lattice with these vertices*. We free ourselves from this constraint on the vertices, and thus go beyond applications to statistical mechanics on the lattice. The examples we will use to illustrate our claims however come from existing lattices.

4   The linearization phenomenon

Here is a construction with generalizes the one of [13, 14], for $q = 2$. The two states of the bonds are denoted "+" and "-".

One may then proceed as follows:

- number the bonds of the vertex (from 1 to the coordination number 2ν).

- introduce 2ν anticommuting objects $\{\alpha_1, \alpha_2, ... \alpha_{2\nu}\}$, in a definite (normal) order.

- to each configuration is associated a matrix element. Look at the bonds in the state "+", and take the monomial obtained by multiplying the corresponding α's.

- combine all this into an element ρ of the Grassmann algebra on $\{\alpha_1, \alpha_2, ..., \alpha_{2\nu}\}$.

For the sake of clarity, let us describe the association for one explicit configuration

$$
\begin{array}{c|c}
4 & v = - \\
\hline
\begin{array}{c} i = + \\ 1 \end{array} & \begin{array}{c} u = + \\ 3 \end{array} \\
\hline
2 & j = -
\end{array}
$$

one associates $R_{+-}^{+-} \, \alpha_1 \, \alpha_3$, and so on to produce all term of ρ.

Suppose R is of the form

$$
R = \begin{pmatrix}
a_1 & a_2 & b_1 & b_2 \\
a_3 & a_4 & b_3 & b_4 \\
c_1 & c_2 & d_1 & d_2 \\
c_3 & c_4 & d_3 & d_4
\end{pmatrix}
\tag{13}
$$

If we write the element of the Grassmann algebra associated to R with the normal order $\alpha_1\alpha_2\alpha_4\alpha_3$, we get some

$$\rho = \rho_0 + \rho_1 + \rho_2 + \rho_3 + \rho_4. \tag{14}$$

We may require that

$$\rho = a_1 \cdot \exp(\gamma), \quad \text{with} \quad \gamma = \gamma_1 + \gamma_2, \tag{15}$$

i.e:

$$\rho = a_1 \left[1 + \gamma_1 + \gamma_2 + \frac{1}{2}\left(\gamma_2^2 + 2\gamma_1\gamma_2\right)\right], \tag{16}$$

or equivalently

$$\rho_0\,\rho_3 = \rho_1\,\rho_2, \qquad \rho_0\,\rho_4 = \frac{1}{2}\,\rho_2^2. \tag{17}$$

Notice that we obtain quadratic relations between the entries.

From equ. (17) we see that the quantities ρ_0, ρ_1, ρ_2 are free, and ρ_3, ρ_4 may then be evaluated unambiguously. The matrices R verifying equ. (17) constitute an 11-parameter family, containing non-bidiagonal matrices.

Claim: If (17) is verified, then the inverse linearizes on R. It is proportional to

$$l(R) \simeq \begin{pmatrix} d_1 & -d_3 & -b_1 & b_3 \\ -d_2 & d_4 & b_2 & -b_4 \\ -c_1 & c_3 & a_1 & -a_3 \\ c_2 & -c_4 & -a_2 & a_4 \end{pmatrix} \tag{18}$$

and matrix inversion reduces to a mixture of permutations of entries with changes of signs.

Remark 1. When we say "linearization", we of course refer to a linear map acting on the unconstrained model with fully generic entries, and implicitly use the natural embedding of the specific model in the more general one.

Remark 2. Any linearization condition of the matrix inverse leads to quadratic relations between the entries of the matrix, since it reads

$$R \cdot l(R) \simeq 1 \tag{19}$$

Remark 3. We may construct similar generalized families for larger sizes of matrices. We have done it for size 8×8, and there exists a linear transformation l such that relation (19) holds.

5   Finite orbits

We have not, up to now, examined how the inversion and the other generators of the groups Γ marry, and this is crucial for the stability of the linearization conditions we are writing.

Let us first recall what these other generators are. Contrary to the inversion, they are sensitive to the geometry of the vertex, and in particular to its dimension. For example, the transformations appearing for the triangular lattice and the cubic lattice differ from each other, although the apparent form of R is the same.

For the square lattice, we have two linear transformations, the partial transpositions $t_1 = t_l$ and $t_2 = t_r$ [18]:

$$(t_l R)^{ij}_{uv} = R^{uj}_{iv}, \qquad (t_r R)^{ij}_{uv} = R^{iv}_{uj} \qquad i,j,u,v = 1..q \qquad (20)$$

The product $t_l t_r$ is the matrix transposition (l stand for 'left' and r stands for 'right' in the standard tensor product structure of R).

For the triangular lattice, we have three linear transformations $t_1 = \tau_l$, $t_2 = \tau_m$, $t_3 = \tau_r$ [8]:

$$(\tau_l R)^{ijk}_{uvw} = R^{ukj}_{iwv}, \qquad (\tau_m R)^{ijk}_{uvw} = R^{wvu}_{kji}, \qquad (\tau_r R)^{ijk}_{uvw} = R^{jiw}_{vuk}. \qquad (21)$$

Finally, for the cubic lattice, we have three linear transformations $t_1 = t_l$, $t_2 = t_m$, $t_3 = t_r$ [18]:

$$(t_l R)^{ijk}_{uvw} = R^{ujk}_{ivw}, \qquad (t_m R)^{ijk}_{uvw} = R^{ivk}_{ujw}, \qquad (t_r R)^{ijk}_{uvw} = R^{ijw}_{uvk}, \qquad (22)$$

and the product $t_l\, t_m\, t_r$ of the three partial transpositions is the matrix transposition.

What is not guaranteed, when considering all the quadratic relations described above, is that they will be stable by the groups Γ. Notice that, if they are, the realization of Γ automatically gets finite.

What is particularly interesting is the question of the stability of the conditions obtained in [11, 12] for the triangular lattice, and used in [16] for the cubic lattice. These conditions are stable by $\Gamma_{triangular}$ and this provides us with an invariant algebraic variety of large dimension (fifteen) and large codimension (sixteen). The same conditions *are not stable by* Γ_{cubic}!

All our further explorations lead to the following conclusion: the linearization conditions are stable by the reflections if and only if the vertex is considered as planar.

This is of importance in view of the study of the symmetries of integrability. Indeed the known solution of tetrahedron equations for vertex models enjoy a degeneration of the symmetry group Γ similar to the one we describe here. However the 'fermionic' interpretation of the phenomenon does not work for $d = 3$, and it is still an open question to understand why.

6 Conclusion

More could be said about the solution of the tetrahedron equation we analyzed in [8], but this would not fit here.

One point could be made about the stability of the linearization conditions. We may trade the reflections t_k for superreflexions defined by permutations of the Grassmann variables. By construction, they will leave the linearization conditions stable. The question then is to understand what is the geometrical meaning of these superreflections.

The construction we presented applies only to $q = 2$. For larger values of q, it may be necessary to use not Grassmannian variables, but rather Z_q graded algebras.

Beyond the applications to statistical mechanics, we may also emphasize that the construction we described is the only one we know at present producing varieties of large dimension and large codimension obeying the discrete-invariance condition presented above.

References

[1] G.R.W. Quispel, J.A.G. Roberts, and C.J. Thompson, *Integrable Mappings and Soliton Equations*. Phys. Lett. A **126** (1988), p. 419.

[2] A. Ramani, B. Grammaticos, and J. Hietarinta, *Discrete versions of the Painlevé equations*. Phys. Rev. Lett. **67** (1991), pp. 1829–1832.

[3] G.R.W. Quispel and F. W. Nijhoff. Phys. Lett. A **161** (1992), p. 419.

[4] G. Falqui and C.-M. Viallet, *Singularity, complexity, and quasi-integrability of rational mappings*. Comm. Math. Phys. **154** (1993), pp. 111–125.

[5] C.-M. Viallet, *On some rational Coxeter Groups*. CRM Proceedings and Lecture Notes **9** (1996), pp. 377–388. Centre de Recherches Mathématiques.

[6] B. Grammaticos, F.W. Nijhoff, and A. Ramani. Discrete painlevé equations. In R. Conte, editor, *The Painlevé property, one century later*, (1996). Cargèse proceedings, to appear in CRM proceedings and lecture notes.

[7] A.P. Veselov, *Growth and Integrability in the Dynamics of Mappings*. Comm. Math. Phys. **145** (1992), pp. 181–193.

[8] J.-M. Maillard and C.-M. Viallet, *A comment on free-fermion conditions for lattice models in two and more dimensions*. Phys. Lett. **B**(381) (1996), pp. 269–276. available from e-print -database hep-th/9603162.

[9] T.D. Schultz, D.C. Mattis, and E.H. Lieb, *Two dimensional Ising model as a soluble problem of many fermions*. Rev. Mod. Phys. **36** (1964), pp. 856–871.

[10] C. Fan and F.Y. Wu, *General lattice statistical model of phase transition*. Phys. Rev. **B 2** (1970), pp. 723–733.

[11] C. A. Hurst, *New Approach to the Ising Problem*. J. Math. Phys **7**(2) (1966), pp. 305–310.

[12] J.E. Sacco and F. Y. Wu, *32-vertex model on the triangular lattice*. J. Phys. **A 8**(11) (1975), pp. 1780–1787.

[13] S. Samuel, *The use of anticommuting variable integrals in statistical mechanics. I: The computation of partition functions, II: The computation of correlation functions, III: Unsolved models*. J. Math. Phys. **21** (1980), pp. 2806–2814,2815–2819,2820–2833.

[14] S. Samuel, *The pseudo-free 128 vertex model*. J. Phys. A **14** (1981), pp. 191–209.

[15] S. Samuel, *The correlation functions in the 32-vertex model*. J. Phys. A **14** (1981), pp. 211–218.

[16] V.V. Bazhanov and Y.G. Stroganov, *Free fermions on a three-dimensional lattice and tetrahedron equations*. Nucl. Phys. **B**(230) (1984), pp. 435–454.

[17] M.P. Bellon, J-M. Maillard, and C-M. Viallet, *Infinite Discrete Symmetry Group for the Yang-Baxter Equations: Spin models*. Physics Letters **A 157** (1991), pp. 343–353.

[18] M.P. Bellon, J-M. Maillard, and C-M. Viallet, *Infinite Discrete Symmetry Group for the Yang-Baxter Equations: Vertex Models*. Phys. Lett. B **260** (1991), pp. 87–100.

[19] M.P. Bellon, J-M. Maillard, and C-M. Viallet, *Rational Mappings, Arborescent Iterations, and the Symmetries of Integrability*. Physical Review Letters **67** (1991), pp. 1373–1376.

[20] S. Boukraa, J-M. Maillard, and G. Rollet, *Integrable mappings and polynomial growth*. Physica A **209** (1994), pp. 162–222.

Q-Combinatorics and Quantum Integrability

A.Yu. Volkov[*]
Lab de Physique Theorique ENSLAPP
ENSLyon, 46 Alle d'Italie
Lyon 69007, France

Abstract

The idea that a Dynkin diagram can provide one of the 'spatial' variables for an integrable difference-difference system is no news. I propose a 'model' where the only variable is of this sort.

It has been observed by S. Fomin and An. Kirillov [FK] that if in some algebra there were a bunch of elements $R_n(\lambda)$

(i) obeying Artin-Yang-Baxter's commutation relations

$$R_{n+1}(\lambda - \mu)R_n(\lambda)R_{n+1}(\mu) = R_n(\mu)R_{n+1}(\lambda)R_n(\lambda - \mu)$$

$$R_m R_n = R_n R_m \quad \text{if} \quad |m - n| \neq 1$$

(ii) and depending on the spectral parameter in an exponential way

$$R_n(\lambda) = R_n(\lambda - \mu)R_n(\mu)$$

then ordered products of those R-matrices would all commute with each other:

$$\left(R_N(\lambda)\ldots R_2(\lambda)R_1(\lambda)\right)\left(R_N(\mu)\ldots R_2(\mu)R_1(\mu)\right)$$
$$= \left(R_N(\mu)\ldots R_2(\mu)R_1(\mu)\right)\left(R_N(\lambda)\ldots R_2(\lambda)R_1(\lambda)\right).$$

[*]On leave of absence from Steklov Mathematical Institute, St. Petersburg

Unfortunately, this was not to become an algebraic skeleton of quantum integrability. Conditions (i) and (ii) proved too restrictive and all the examples that emerged were of a nondeformable classical sort. Fortunately, it is reparable. Let us go back to basics and make R-matrices depend on two spectral parameters so that (i) and (ii) rather read

$$R_{n+1}(\lambda, \mu)R_n(\lambda, \nu)R_{n+1}(\mu, \nu) = R_n(\mu, \nu)R_{n+1}(\lambda, \nu)R_n(\lambda, \mu)$$

$$R_m R_n = R_n R_m \quad \text{if} \quad |m - n| \neq 1 \tag{i}$$

$$R_n(\lambda, \nu) = R_n(\lambda, \mu)R_n(\mu, \nu). \tag{ii}$$

Proposition remains: ordered products

$$Q(\lambda, \mu) \equiv R_N(\lambda, \mu) \ldots R_2(\lambda, \mu)R_1(\lambda, \mu)$$

commute whenever their second 'arguments' coincide:

$$Q(\lambda, \nu)Q(\mu, \nu) = Q(\mu, \nu)Q(\lambda, \nu).$$

Proof:

$$
\begin{aligned}
Q&(\lambda, \nu)Q(\mu, \nu) \\
&= R_N(\lambda, \nu)\,(R_{N-1}(\lambda, \nu)R_N(\mu, \nu)) \ldots (R_1(\lambda, \nu)R_2(\mu, \nu))\,R_1(\mu, \nu) \\
&= R_N(\mu, \nu)R_N(\lambda, \mu)\,(R_{N-1}(\lambda, \nu)R_N(\mu, \nu)) \ldots (R_1(\lambda, \nu)R_2(\mu, \nu))\,R_1(\mu, \nu) \\
&= R_N(\mu, \nu)\,(R_{N-1}(\mu, \nu)R_N(\lambda, \nu)) \ldots (R_1(\mu, \nu)R_2(\lambda, \nu))\,R_1(\lambda, \mu)R_1(\mu, \nu) \\
&= R_N(\mu, \nu)\,(R_{N-1}(\mu, \nu)R_N(\lambda, \nu)) \ldots (R_1(\mu, \nu)R_2(\lambda, \nu))\,R_1(\lambda, \nu) \\
&= Q(\mu, \nu)Q(\lambda, \nu) \quad \square
\end{aligned}
$$

It is plain to see that this would turn right back into Fomin-Kirillov's case if I added the usual (iii) $R(\lambda, \mu) = R(\lambda - \mu)$. Naturally, I do not.

Proposition: in the algebra whose only two generators are bound by Serre-style commutation relations

$$x_1 x_1 x_2 + x_2 x_1 x_1 q = x_1 x_2 x_1 (1 + q)$$
$$x_1 x_2 x_2 + x_2 x_2 x_1 q = x_2 x_1 x_2 (1 + q)$$

criteria (i) and (ii) are met by the elements

$$R_n(\lambda, \mu) = (x_n)_\lambda^\mu \equiv \prod_{j=\mu}^{\lambda-1} (1 - x_n q^j).$$

(ii) comes free of charge, *proof* of (i) (only the first line applies) starts with **lemma** establishing something like Campbell-Hausdorff multiplication rules:

$$(x_1)_\lambda^\mu (x_2)_\lambda^\mu = \prod_{j=\mu}^{\lambda-1} \left(1 + cq^{2j} - (x_1 + x_2 + kq^\lambda)q^j \right)$$

$$(x_2)_\lambda^\mu (x_1)_\lambda^\mu = \prod_{j=\mu}^{\lambda-1} \left(1 + cq^{2j} - (x_1 + x_2 + kq^\mu)q^j \right)$$

with

$$k = \frac{x_1 x_2 - x_2 x_1}{1 - q} \qquad c = \frac{x_1 x_2 - x_2 x_1 q}{1 - q},$$

the element c actually being central. Proof is by induction for there is nothing but polynomials in here. Let me omit it. So,

$$
\begin{aligned}
(x_2)_\lambda^\mu (x_1)_\lambda^\nu (x_2)_\mu^\nu &= (x_2)_\lambda^\mu (x_1)_\lambda^\mu (x_1)_\mu^\nu (x_2)_\mu^\nu \\
&= \prod_{j=\nu}^{\lambda-1} \left(1 + cq^{2j} - (x_1 + x_2 + kq^\mu)q^j \right) \\
&= (x_1)_\mu^\nu (x_2)_\mu^\nu (x_2)_\lambda^\mu (x_1)_\lambda^\mu = (x_1)_\mu^\nu (x_2)_\lambda^\nu (x_1)_\lambda^\mu \qquad \square
\end{aligned}
$$

The two propositions combine into the message of this note: the algebra whose r generators commute like this

$$x_n x_n x_{n+1} + x_{n+1} x_n x_n q = x_n x_{n+1} x_n (1 + q)$$
$$x_n x_{n+1} x_{n+1} + x_{n+1} x_{n+1} x_n q = x_{n+1} x_n x_{n+1} (1 + q)$$
$$x_m x_n = x_n x_m \qquad \text{if} \qquad |m - n| \neq 1$$

contains a good supply

$$Q(\lambda) = (x_r)_\lambda \ldots (x_2)_\lambda (x_1)_\lambda \qquad (\cdot)_\lambda \equiv (\cdot)_\lambda^0$$

of mutually commuting elements

$$Q(\lambda)Q(\mu) = Q(\mu)Q(\lambda).$$

In conclusion, some remarks. The definition of $(\cdot)_\lambda^\mu$ required integer λ, μ bound by $\lambda \geq \mu$. It would be more practical (if less stylish) to do without those limitations. Formal power series in x's could help. An obvious identity

$$(x)_\lambda^\mu = \frac{(xq^\mu)_\infty}{(xq^\lambda)_\infty} \qquad (x)_\infty \equiv \prod_{j=0}^{\infty} (1 - xq^j)$$

would then double as a definition of $(\cdot)_\lambda^\mu$ for non-integer λ, μ. As a matter of fact, AYB relations

$$(x_{n+1})_\lambda^\mu (x_n)_\lambda^\nu (x_{n+1})_\mu^\nu = (x_n)_\mu^\nu (x_{n+1})_\lambda^\nu (x_n)_\lambda^\mu$$

survive the extrapolation. Anyway, it is perhaps more important to guess where the commutation relations governing the x's belong. Quantum A_r algebra provides

$$e_n e_n e_{n+1} + e_{n+1} e_n e_n = e_n e_{n+1} e_n (q^{\frac{1}{2}} + q^{-\frac{1}{2}})$$

$$e_n e_{n+1} e_{n+1} + e_{n+1} e_{n+1} e_n = e_{n+1} e_n e_{n+1} (q^{\frac{1}{2}} + q^{-\frac{1}{2}})$$

and with a little help of suitable 'quantum coordinates'

$$\chi_m e_n = e_n \chi_m \qquad \chi_n \chi_{n+1} = q^{\frac{1}{2}} \chi_{n+1} \chi_n \qquad n = 1, 2, \ldots, r - 1$$

'vectors' $\chi_n e_n$ just fit the commutation relations prescribed for the x's. $Q(\infty)$ becomes a piece of 'quantized' and 'bosonized' Gauss decomposition

$$\begin{aligned}
\mathbf{g} = \ & (\chi_r e_r)_\infty \cdots (\chi_2 e_2)_\infty (\chi_1 e_1)_\infty \\
& \times (\chi_{2r-1} e_r)_\infty \cdots (\chi_{r+1} e_2)_\infty \\
& \ \vdots \\
& \times (\chi_{r(r+1)/2} e_r)_\infty \\
& \times \mathbf{g}_d \mathbf{g}_l
\end{aligned}$$

a là Morozov-Vinet [MV]. Let me decline further comments on this issue. Instead, let me mention that the same 'R-matrix' $(\cdot)_\lambda^\mu$ used along the guidelines of [V] provides the quantization of a very major nonlinear difference-difference system [H]

$$1 + \psi_{m+1,n} \psi_{m,n+1} + \psi_{m+1,n+1} \psi_{m,n} = -\Lambda \psi_{m+1,n} \psi_{m,n+1} \psi_{m+1,n+1} \psi_{m,n}.$$

The case $\Lambda = 1$ is actually linear, $\Lambda = 0$ (corresponding, by the way, to $\lambda = \infty$) approximates the Liouville equation while everything else is the sine-Gordon equation. I think we've got a few more pieces of the big puzzle called Quantum Solitons.

Acknowledgements

I would like to thank L. Faddeev, An. Kirillov, R. Kashaev, J.-M. Maillet, A. Morozov and V. Tarasov for stimulating discussions.

References

[FK] S. Fomin and An. Kirillov, Discrete Mathematics 153 (1996).

[MV] A. Morozov and L. Vinet, hep-th/9409093.

[V] A.Yu. Volkov, hep-th/9509024.

[H] R. Hirota, J. Phys. Soc. Japan **43** (1977).

Printed in the United States
By Bookmasters